I0894554

Intelligent Internet of Everything for Automated and Sustainable Farming

Biplob Ray
Central Queensland University, Australia

Jahan Hassan
Central Queensland University, Australia

Hailong Huang
Hong Kong Polytechnic University, Hong Kong

Nahina Islam
Central Queensland University, Australia

Zaglul Shahadat
Rajshahi University of Engineering and Technology, Bangladesh

IGI Global
Scientific Publishing
Publishing Tomorrow's Research Today

Vice President of Editorial	Melissa Wagner
Director of Acquisitions	Mikaela Felty
Director of Book Development	Jocelynn Hessler
Production Manager	Mike Brehm
Cover Design	Phillip Shickler

Published in the United States of America by
IGI Global Scientific Publishing
701 East Chocolate Avenue
Hershey, PA, 17033, USA
Tel: 717-533-8845
Fax: 717-533-7115
Website: https://www.igi-global.com E-mail: cust@igi-global.com

Copyright © 2025 by IGI Global Scientific Publishing. All rights reserved. No part of this publication may be reproduced, stored or distributed in any form or by any means, electronic or mechanical, including photocopying, without written permission from the publisher.

Product or company names used in this set are for identification purposes only. Inclusion of the names of the products or companies does not indicate a claim of ownership by IGI Global Scientific Publishing of the trademark or registered trademark.

Library of Congress Cataloging-in-Publication Data

Names: Ray, Biplob, 1979- editor. | Hassan, Jahan, 1973- editor. | Huang, Hailong, 1988- editor. | Islam, Nahina, 1986- editor. | Shahadat, Mhia, 1977- editor.
Title: Intelligent internet of everything for automated and sustainable farming / edited by Biplob Ray, Jahan Hassan, Hailong Huang, Nahina Islam, Mhia Shahadat.
Description: Hershey, PA : IGI Global Scientific Publishing, 2025. | Includes bibliographical references and index. | Summary: "This book aims to provide innovative theoretical, application, and policy research findings in the farming industry. One of the key objectives of this edited book is to feature multidisciplinary research demonstrating the innovative use of emerging technology to solve farming challenges that cannot be addressed otherwise"-- Provided by publisher.
Identifiers: LCCN 2025000071 (print) | LCCN 2025000072 (ebook) | ISBN 9798337300207 (hardcover) | ISBN 9798337300214 (paperback) | ISBN 9798337300221 (ebook)
Subjects: LCSH: Agricultural innovations. | Precision farming. | Internet of things. | Sustainable agriculture.
Classification: LCC S494.5.I5 I5772 2025 (print) | LCC S494.5.I5 (ebook) | DDC 630.2/0854678--dc23/eng/20250218
LC record available at https://lccn.loc.gov/2025000071
LC ebook record available at https://lccn.loc.gov/2025000072

British Cataloguing in Publication Data
A Cataloguing in Publication record for this book is available from the British Library.

All work contributed to this book is new, previously-unpublished material.
The views expressed in this book are those of the authors, but not necessarily of the publisher.
This book contains information sourced from authentic and highly regarded references, with reasonable efforts made to ensure the reliability of the data and information presented. The authors, editors, and publisher believe the information in this book to be accurate and true as of the date of publication. Every effort has been made to trace and credit the copyright holders of all materials included. However, the authors, editors, and publisher cannot assume responsibility for the validity of all materials or the consequences of their use. Should any copyright material be found unacknowledged, please inform the publisher so that corrections may be made in future reprints.

Table of Contents

Detailed Table of Contents

This chapter examines the impact of environmental degradation and DTs on cereal yield across countries. Per capita CO2 emissions are considered as a proxy variable for environmental degradation, while digital infrastructure related indicators are included as proxy variables for DTs in empirical investigation. Foreign direct investment, population growth, application of fertilizer, per capita GDP and inflation are considered as control variables in the empirical models. Log-linear regression equation is applied to examine the impact of mentioned variables on cereal yield as compiling a country-wise panel data during 2000 - 2020. CO2 emissions show a negative impact on cereal yield. It found a negative relationship of FDI, and population growth with cereal yield. Cereal yield is expected to be increased as application of DTs in the agricultural sector increases. Finally, several policy implications are provided to reduce environmental degradation and increase agricultural sustainability.

Weed is a challenge for production and quality of crops in modern agriculture. The effective weed removing solutions are crucial for food security. To address this problem an IoT based automatic moveable weeding hoe for targeted weed removing is depicted in this chapter. Basically it is a smart agriculture firming tool which is incorporated with cutting-edge technologies such as path trajectory, automatic navigation and obstacle detection. The integrated GPS-based IoT guidance in this tool causes precise identification of the weeds. The device receives weed location's coordinates from a drone-based image processing unit. The tool is shaped like a four wheel drive vehicle driven by four individual motors. Two types of blades are used for cutting and rooting of weeds in which solar power is considered for sustainable energy supply. The device is also incorporated with a fertilizer spraying system to support integrated smart farming.

Modern agriculture faces challenges requiring innovative solutions to balance productivity with sustainability. Optimizing irrigation is critical, as traditional systems lead to water waste and inefficiencies, necessitating smart technologies. Key advancements include Dual Electromagnetic (DUAL-EM) scanning for soil conductivity mapping, optimizing sensor placement, reducing costs, and ensuring data quality. Noise reduction techniques like Isolation Forest and LOF enhance data reliability, achieving 81.02% accuracy and 0.328 MAE. Incremental learning models such as LightGBM and XGBoost dynamically refine irrigation schedules, optimizing water application under changing conditions. IoT integration with soil sensors, weather stations, and automated valves ensures precise irrigation. These strategies conserve water, reduce costs, and align with environmental goals, transforming farming practices and advancing sustainable agriculture.

This study explores the transformative potential of integrating Artificial Intelligence (AI) with precision agriculture to address key challenges in farming, such as disease detection, resource optimization, and yield improvement. By leveraging the Real-Time Detection Transformer (RT-DETR) framework, the study combines high-resolution imaging, hyperspectral scanners, and real-time data processing to enable efficient and accurate detection of tomato leaf diseases. A robotic system, equipped with autonomous navigation and non-invasive diagnostic tools, was developed to

advancements, such as MangoYOLO, improve fruit detection, size estimation, and yield forecasting, addressing challenges like environmental variability. Case studies in mango production highlight the transformative impact of these innovations on orchard automation and market efficiency. Portable NIR devices with GPS offer rapid, cost-effective quality assessments, while chemometric modeling and multi-source data integration enhance predictive accuracy.

This chapter aims to create a system that uses AI and IoT technologies to monitor and understand chicken behaviors. It will use computer vision and machine learning to recognize common actions like pecking, scratching, resting, and feeding. The system will also track environmental factors such as temperature, humidity, and light to see how they affect behavior. Farmers will get real-time updates and insights through an easy-to-use dashboard or mobile app. Additionally, the system will keep improving over time by using advanced deep learning to make behavior detection more accurate and dependable. The system will use advanced deep learning models to detect chicken behaviors in real time. To ensure accurate results, it will rely on datasets with labeled examples of different chicken actions for precise training. It will also identify unusual behaviors, such as reduced feeding, and send timely alerts to help farmers take preventive steps. Based on the behavior analysis, the system will offer actionable suggestions to address potential issues.

Quantum computing is revolutionizing computation and communication, with its integration into drones transforming agrotech and other industries. The Internet of Quantum Drones (IoQDs) enables ultra-secure communication, rapid data processing, and precise navigation through quantum computing, quantum communication, and quantum sensors. This integration tackles challenges in intelligent farming, such as real-time decision-making, ecological complexity, and efficient data management. IoQDs support applications in precision agriculture, disaster management, environmental monitoring, and defense using quantum-based image processing and machine learning. This chapter explores advancements, case studies, and IoQD integration with 6G and satellites, ensuring low-latency global connectivity. Challenges—scalability, cost, hardware downsizing, and regulations are analyzed alongside potential solutions. A case study on IoQD-based weed detection demonstrates its agricultural potential. Future directions highlight quantum technologies' transformative role in drone networks across industries.

Quantum image processing has recently received much attention due to the opportunity to place image data in quantum mechanics systems. The literature survey has shown that the complexity of preparing image data raises the question of how and in what efficient way a medium- and high-size image is represented inside the quantum domain, which is still a big question. Quantum properties such as entanglement, superposition, and parallelism make the quantum system faster and more robust. Qubit is the central basic unit that maps the image data in the quantum processor, which propagates from left to right. The number of qubits and their connection determine the efficiency of the circuit architecture. This chapter aims to investigate the performance of the transformed coefficient-based quantum image representation and compression approaches in terms of the required number of gates and PSNR values. Moreover, the challenges and prospects of quantum image compression and processing will be discussed as a future research direction.

Agriculture is a fundamental global primary industry, evolving with advances in ICT and automation. The rise in 'smart farming' integrates cutting-edge technologies

Preface

In recent years, agriculture has emerged as a critical domain for digital transformation, driven by the growing urgency to produce more food sustainably while responding to challenges such as climate change, population growth, and resource scarcity. The convergence of advanced technologies—including the Internet of Everything (IoE), artificial intelligence, quantum computing, and virtual reality—has opened unprecedented opportunities for reimagining farming systems. This edited volume, *Intelligent Internet of Everything for Automated and Sustainable Farming*, brings together innovative research and real-world applications at the intersection of agriculture, automation, and intelligent systems to offer a comprehensive view of the future of smart farming.

The chapters in this book collectively reflect a diverse spectrum of global perspectives, practical innovations, and theoretical insights. From AI-driven behavioral monitoring in poultry farms to the application of quantum communication in drone-assisted agriculture, each contribution demonstrates how intelligent, connected systems are reshaping the agricultural landscape. These technologies are not only improving crop yields and livestock welfare but are also supporting climate resilience, economic viability, and ethical practices in food production. The authors come from various academic and research institutions across Australia, Bangladesh, China, India, and Sri Lanka, offering readers a truly international outlook.

We, the editors, curated this volume with a strong belief in interdisciplinary collaboration. Our aim was to present a resource that not only supports academic research but also resonates with practitioners, policy-makers, and technology developers. By integrating case studies, empirical investigations, and technological breakthroughs, this book serves as a vital contribution to the evolving discourse on intelligent agriculture. It is our hope that this collection will spark further research, encourage innovation, and foster collaboration among stakeholders committed to building a more sustainable and food-secure future.

CHAPTER OVERVIEW

Chapter 1: Empirical Insights into Environmental Degradation, Digital Technologies, and Agricultural Productivity

This chapter presents a comprehensive empirical study on the interplay between environmental degradation, digital technology adoption, and cereal yield across countries. Using CO_2 emissions as a proxy for environmental harm and various indicators of digital infrastructure, the authors apply a log-linear regression model to panel data from 2000 to 2020. Findings indicate that while CO_2 emissions, FDI, and population growth negatively impact cereal yields, the integration of digital technologies shows a positive correlation. The study concludes with policy recommendations to curb environmental degradation and foster sustainable agricultural practices.

Chapter 2: IoT-Powered Autonomous Weeding Hoe for Precision Agriculture

Focusing on automated weed management, this chapter introduces a solar-powered IoT-based weeding hoe designed for targeted, efficient weed removal. The autonomous tool integrates GPS navigation, obstacle detection, and drone-fed weed localization for precise operation. With its smart mobility and dual-blade mechanism for cutting and rooting weeds, the system also features a fertilizer spraying unit, reinforcing its utility in integrated smart farming.

Chapter 3: Improving Smart Irrigation with Sensor Optimization and Incremental Learning

This chapter addresses inefficiencies in irrigation practices by proposing an IoT-enabled system enhanced through optimized sensor placement, noise reduction techniques, and adaptive machine learning models. Leveraging soil conductivity mapping and algorithms like LightGBM and XGBoost, the system dynamically adjusts irrigation schedules. Demonstrating notable improvements in accuracy and water use efficiency, the approach supports sustainable and cost-effective agricultural water management.

Chapter 4: AI-Driven Tomato Leaf Disease Detection for Smarter Crop Management

This chapter explores the integration of AI and robotics in detecting tomato leaf diseases. Employing models like YOLOv8 and RT-DETR on benchmark datasets, the study demonstrates high accuracy in disease identification through real-time, non-invasive diagnostic tools. The robotic system not only classifies diseases but also aids in decision-making, offering a practical application of AI to enhance yield quality and agricultural sustainability.

Chapter 5: Deep Learning for Tea Leaf Quality Assessment in Smart Agriculture

Focusing on tea production, this chapter investigates the use of deep learning to classify and evaluate tea leaf quality based on age. The research offers insights into the correlation between leaf maturity and quality indicators such as flavor and texture. By automating quality assessments, the study paves the way for scalable, accurate, and efficient quality control solutions in large-scale tea agriculture.

Chapter 6: Monitoring Fruit Load and Quality in Precision Agriculture

This chapter details the use of machine vision and AI technologies to assess fruit yield and quality throughout the growth cycle. It highlights innovations like MangoYOLO, hyperspectral imaging, and portable NIR devices that enable non-invasive, real-time monitoring. The integration of decision support systems enhances predictive capabilities, optimizing orchard management and postharvest operations for better market outcomes.

Chapter 7: Ethical AI-Driven Poultry Monitoring and Management

This chapter introduces an AI and IoT-based system for behavioral monitoring in poultry farming. Using deep learning models and environmental sensors, the system detects routine and abnormal behaviors, issuing alerts to support timely interventions. A user-friendly dashboard provides real-time insights, aiming to improve animal welfare, productivity, and ethical management in poultry operations.

Chapter 8: Quantum Drones in Precision Farming: A Leap Toward Ultra-Secure, Real-Time Agriculture

This chapter explores the transformative role of quantum communication and processing in agriculture, particularly through the concept of the Internet of Quantum Drones (IoQDs). Merging quantum technologies with 6G connectivity and satellite integration, IoQDs enhance precision, security, and data handling in agrotech. A case study on weed detection showcases the practical potential of quantum-powered drones, while the chapter also addresses scalability, cost, and regulatory challenges.

Chapter 9: Quantum Image Representation for Efficient Agricultural Imaging

Delving into quantum computing applications, this chapter examines new methods for representing and compressing agricultural images in quantum circuits. It analyzes performance metrics such as PSNR and gate requirements in transformed coefficient-based quantum image processing. The discussion also highlights the potential of entanglement and parallelism to improve the efficiency of high-resolution image analysis in future agricultural technologies.

Chapter 10: Cybersecurity Threats in Smart Farming: A Regional Australian Perspective

As agriculture embraces digital transformation, this chapter addresses the critical issue of cybersecurity in smart farming systems, particularly in regional Australia. Highlighting vulnerabilities in IoT, drone, and blockchain-based agricultural systems, the authors call for stronger awareness, best practices, and cybersecurity integration to safeguard sensitive farm data and maintain trust in emerging technologies.

Chapter 11: Virtual Reality as a Bridge Between Farmers and Global Crop Buyers

This chapter explores the innovative use of virtual reality to connect farmers with international crop buyers. By showcasing produce in immersive, real-time environments and integrating IoT and blockchain technologies, the proposed VR platform enhances transparency, traceability, and buyer trust. The prototype demonstrates the potential to revolutionize global crop trading by offering a scalable, inclusive solution for small-scale farmers.

As the chapters unfold, readers will encounter technologies that blend sustainability with automation—from sensor-optimized irrigation systems to robotic disease detection tools. These chapters illustrate the profound impact of smart technologies in refining traditional practices and enabling data-informed decision-making. For instance, autonomous weeding tools and precision fruit monitoring systems not only enhance productivity but also minimize environmental footprints. In many cases, solar power integration and eco-conscious design highlight how digital innovation can align seamlessly with green objectives.

This book also brings attention to the ethical and cybersecurity challenges inherent in digitizing agriculture. With increasing reliance on interconnected systems and data infrastructures, smart farming must navigate new vulnerabilities, particularly in rural and regional contexts. The chapters dedicated to cybersecurity and ethical management underscore the need for holistic approaches that balance technological adoption with privacy, equity, and trust. Likewise, the inclusion of quantum and virtual reality technologies signals a forward-looking perspective, inviting readers to consider what lies ahead as agriculture converges with the next frontier of computing.

In closing, we extend our deepest gratitude to the contributing authors for their outstanding research and commitment, and to the institutions that supported their work. We also thank the reviewers and production team whose efforts were essential in bringing this volume to life. It is our sincere hope that *Intelligent Internet of Everything for Automated and Sustainable Farming* will inspire further exploration, inform impactful policies, and accelerate the global transition to a smarter, more sustainable agricultural future. May this work serve as a catalyst for innovation, collaboration, and lasting change in one of humanity's most vital sectors.

Chapter 1 Empirical Investigation for Assessing the Impact of Environmental Degradation and Digital Technologies on the Agricultural Sector

Ajay Kumar Singh
https://orcid.org/0000-0003-0429-0925
Department of Humanities and Social Sciences, Graphic Era University, Dehradun, India

S. Srinivasan
https://orcid.org/0009-0002-0179-9849
Department of Humanities and Social Sciences, Graphic Era University, Dehradun, India

ABSTRACT

This chapter examines the impact of environmental degradation and DTs on cereal yield across countries. Per capita CO2 emissions are considered as a proxy variable for environmental degradation, while digital infrastructure related indicators are included as proxy variables for DTs in empirical investigation. Foreign direct investment, population growth, application of fertilizer, per capita GDP and inflation are considered as control variables in the empirical models. Log-linear regression equation is applied to examine the impact of mentioned variables on cereal yield as compiling a country-wise panel data during 2000 - 2020. CO2 emissions

DOI: 10.4018/979-8-3373-0020-7.ch001

Copyright © 2025, IGI Global Scientific Publishing. Copying or distributing in print or electronic forms without written permission of IGI Global Scientific Publishing is prohibited.

show a negative impact on cereal yield. It found a negative relationship of FDI, and population growth with cereal yield. Cereal yield is expected to be increased as application of DTs in the agricultural sector increases. Finally, several policy implications are provided to reduce environmental degradation and increase agricultural sustainability.

1. INTRODUCTION

Environmental development and digital technologies (DTs) are essential to meet the input requirements of the agricultural sector in multidimensional aspects (Chandio et al., 2024). For example, water, air, soil fertility and rainfall are natural inputs for effective growth of the agricultural sector (Kumar & Singh, 2023). DTs are helping farmers to communicate with different stakeholders like agricultural development agencies, agri-industries, suppliers and government. Farmers are being more aware of the government policies using DTs. For example, social media is useful for farmers to increase their understanding to be a beneficiary of government policies. Social media is providing multiple information like appropriate price of products, marketing, seed quality, sowing time of a crop, etc. to farmers (Ashraf & Singh, 2022). Furthermore, there are many DTs like mobile Apps, artificial intelligence and robotics, sensors that are providing desired information to the farmers on regular time. Accordingly, farming communities are taking favourable decisions to maintain cultivation of crops in different weather seasons.

The agricultural sector has a greater dependence on environment and ecosystem services, and DTs (Biró & Csete, 2023). Digitalization is solving different challenges in the agri-food-industries and farming sector (Annosi et al., 2020). For example, digitalization determines all dimensions of sustainability, and increases resource efficiency and efficient agricultural development (Biró & Csete, 2023). Application of ICT in the agricultural sector is helping to increase its growth (Chandio et al., 2024). Digitalization is helping to reduce CO_2 emissions from this sector (Ali et al., 2023; Chen & Li, 2024). Therefore, at present dependence of the agricultural sector on DTs and digitalization are consistently increasing due to climate change, natural disasters and growing population (Rai, 2023). As the growing population is accountable for increasing additional demand for food-grain. The agricultural sector faces numerous problems to meet the food demand of the growing population.

Many countries, thus, are expected to resolve multiple other issues like climate change, food insecurity and resource scarcity using DTs (Geng et al., 2024). DTs are playing a positive role in increasing the supply chain system in the agricultural sector (Đuric, 2020). These are helping farming communities to increase their collaboration with various stakeholders like government, agricultural research institutions and agricultural cooperative societies for mitigating climate change impact in the cultivation (Trendov et al., 2019). DTs are conducive to increase yield and production, efficiency of inputs, and reduce costs and use of irrigation, fertilizer and pesticides. Consequently, DTs enhance food security and sustainability of resources (Trendov et al., 2019). Addison et al. (2024) noticed that digitalization could be helpful to increase the communication of people and their skills in Ghana. DTs, therefore, are playing a positive role to reduce the negative impact of climate change in the agricultural sector, and increasing its sustainability (Biró & Csete, 2023; Šermukšnytė-Alešiūnienė & Melnikienė, 2024; Singh et al., 2024). DTs are reducing the barriers of farmers to be engaged in the agricultural value chain (Smidt & Jokonya, 2021). DTs are creating a platform to increase the direct communication of farmers with traders in the market.

However, environmental degradation is increasing due to multiple reasons and it is creating several barriers for agricultural development (Chen & Li, 2024). Subsequently, agricultural production and its allied activities are adversely affected due to environmental degradation (Mekyassi & Kızıldeniz, 2024). Environmental degradation and pollution are creating multiple channels and barriers for poor growth of the agricultural sector (Ali et al., 2023; Chen & Li, 2024). The overwhelming growth of the agricultural sector is accountable for environmental degradation (Singh et al., 2023). Hence, it is a negative and causal association between environmental degradation and the agricultural sector reported by existing scientists. While DTs and digitalization are found positive to increase the all dimensions of sustainability in the farming and other sectors (Biró & Csete, 2023; Singh et al., 2024). Šermukšnytė-Alešiūnienė and Melnikienė (2024) claimed the introduction of digital innovation could be helpful to increase the sustainability of small farms in EU countries.

Environmental development and ecosystem services are crucial to meet the input requirements of the agricultural sector (Singh et al., 2023). For instance, water, land, and air are crucial factors for consistent growth of the agricultural sector. The efficient growth of this sector requires energy which is generated through ecosystem services. The growth of this sector, therefore, is greatly dependent on environmental development and ecosystem services. On the other hand, chemical-based agriculture is helpful to produce surplus production to ensure global food security. Further, chemical-based agriculture is responsible for increasing multiple diseases like cancer, respiratory, cardiovascular, neurological among the people, and environmental issues. It is caused to increase environmental degradation as chemical-based agriculture contributes CO_2 and GHGs emissions in the atmosphere (Singh et al., 2023).

Environmental degradation also reduces the nutritional contents in the food products. Further, high concentration of CO_2 and GHGs emissions are causing climate change and natural disaster. Therefore, farming communities are facing multiple challenges due to climate change and environmental degradation (Almaraz et al., 2023). Organic agriculture is supported by various international awareness campaigns and training programmes to reduce the negative impact of climate change on the agricultural sector and environment across countries. While low knowledge, subpar inputs, yield and input choice, productivity, and marketplace accessibility are creating barriers to address the issues of climate change in the agricultural sector. For example, India is also promoting organic farming through various programs like Bharatiya Prakritik Krishi Paddhati, and Capital Investment Subsidy Scheme. Moreover, past literature argued that technological advancement would be a significant initiative to reduce climate change impact in the production sectors and to enhance environmental quality (Almaraz et al., 2023). For example, use of green technologies in the production sector would be effective to abate CO_2 emissions. Usage of appropriate technologies would work as adaptation and mitigation approaches in the cultivation (Ashraf & Singh, 2022).

Few studies also recommend that reducing transmission costs of renewable technology would be active to abate CO_2 emissions from the agricultural sector in developing countries. Promotion of green agriculture would increase the sustainability of this sector (Jatav & Kalu, 2023). Moreover, digital and cellular agriculture are crucial to increase agricultural sustainability. Rai (2023) suggested that internet penetration is vital to increase the progress of the agricultural, education and business sectors. Biochar technology is found crucial to reduce GHGs emissions from soil (Shrestha et al., 2023). Torres-Morales et al. (2023) showed that biochar is working as a negative emission technology and mitigating climate change impacts. Most studies clarified that digital technologies (DTs) are worthy to reduce climate change impact in the agricultural sector as increasing environmental sustainability (Mhlanga et al., 2023). Hence, DTs have a significant impact on the agricultural sector. However, previous studies are unable to examine the inclusive impact of environmental degradation and DTs on the agricultural sector. Therefore, the present chapter assesses the association of the agricultural sector with DTs and environmental degradation. It also suggests policy proposals to increase the sustainable transformation of the agricultural sector through application of DTs across countries. It focused on following research questions:

- How would DTs play a crucial role in reducing environmental degradation?
- How is environmental degradation harmful for the agricultural sector and vice-versa?

- How would DTs be effective to reduce the negative impact of climate change in the agricultural sector?
- How application of DTs in the agricultural sector would increase its sustainability, and reduce environmental degradation in future?
- Can DTs work as a mitigation strategy to avoid the negative impact of environmental degradation and climate change in the agricultural sector?

This chapter is written to attain following objectives:

- To assess the role of DTs for reducing environmental degradation as per the existing studies.
- To explain the contribution of DTs in the agricultural sector as per the previous literature.
- To observe the impact of environmental degradation and DTs on the agricultural sector.

2. REVIEW OF LITERATURE

2.1. Environmental Degradation and Agricultural Sector

Environmental degradation creates multiple processes to reduce agricultural production (Mekyassi & Kızıldeniz, 2024). While environmental degradation is continuously increasing due to high concentration of CO_2 and GHGs emissions. While these are increasing due to overwhelming industrialization, urbanization, population growth, intensification and mechanization of the agricultural sector, and infrastructural development (Singh et al., 2019; Raihan, 2023; Jang et al., 2023). Agricultural production and its allied activities are primarily leading to increased environmental degradation due to its large contribution of anthropogenic emissions (Ali et al., 2023; Jang et al., 2023; Raihan, 2023). Food security is adversely affected due to environmental degradation and natural resource depletion (Abbasi et al., 2022; Jatav, 2024). Rice cultivation is caused to increase methane gas emission in the atmosphere (Jang et al., 2023). Boadi et al. (2022) argued that the cocoa crop is adversely affected by climate change and the cultivation of this crop has potential to increase climate change in Ghana. The cultivation of cocoa crops is also causing deforestation in Ghana. Ali et al. (2023) also reported that agricultural development is leading to increased environmental pollution in Africa. Ali et al. (2023) found that fossil fuel energy consumption is contributing to increased CO_2 emissions in Africa. Hence, it is a positive and two ways causal association between agricultural development and environmental degradation. Political systems and public goods are increasing another environmental concern in many agri-based countries.

Over population growth increases enormous pressure on natural resources and the energy sector to meet their needs in terms of food, water, air, shelter, land space for house and road construction (Singh et al., 2019). Energy consumption is extensively increasing due to overwhelming industrialization and population growth in many countries. Urbanization is also increasing labour migration from rural to urban areas (Singh et al., 2023). Pollution, CO_2 and GHGs emissions are also increasing due to the mentioned reasons. Extensive consumption of energy is also accountable for environmental degradation (Raihan, 2023). Past evidence inferred a negative impact of these activities on environmental and natural resources. Over economic growth and development are causing environmental degradation and natural resources depletion (Raihan, 2023). Globalization and global competition are also creating environmental issues in global countries (Singh et al., 2019). High desirability for global countries to achieve high economic growth are increasing over use of resources in production activities (Singh, 2025). Therefore, globalization and foreign trade are causing environmental degradation. Soil quality and fertility, and land productivity are decreasing due to environmental degradation. Further, multiple studies have also reported a negative consequence of soil infertility on agricultural productivity.

2.2. Socio-economic Impacts of Environmental Degradation and DTs on Farmers

Environmental degradation is accountable for climate change and natural disaster (Singh & Jyoti, 2024). Further, climate change is caused to reduce agricultural productivity, food security and nutritional security (Mekyassi & Kızıldeniz, 2024). Agricultural productivity has directly declined due to sudden change in climatic factors and natural disasters. Consequently, environmental degradation has an adverse impact on income, employment, livelihood security and social well-being of farmers. For example, the income of farmers may decline due to climate change and environmental degradation. Employment opportunities in the agricultural sector declines when production is damaged or lost due to climate change. Accordingly, livelihood security and social well-being of people are negatively impacted due to environmental degradation. Application of DTs, therefore, can be efficient to resolve above-mentioned issues of farming communities.

Furthermore, pest and insects are germinating as variability in climatic factors increases (Rossi et al., 2023; Singh & Jyoti, 2024). High temperature and rainfall variability create many diseases in the plants. Consequently, quality of final production declines due to germination of pests and insects in the crops (Singh & Jyoti, 2024). For example, the frequency of multiple diseases is increased especially in vegetable and horticulture crops like potato, cabbage, brinjal, mangoes, guava, etc. as climate change increases. Agricultural production, thus, decreases due to germination of pests and insects. Hence, environmental degradation has a negative impact on food quality and human health. Thereupon, income of farmers declines as climate change increases. Plant's growth, production and yield also decline due to increase in heat wave. Heat waves may increase additional water requirement for irrigation in the farming. Therefore, environmental degradation has a negative impact on water resources and its sustainability (Singh et al., 2024).

The occurrence of pest and insecticides are increased in the agricultural sector as environmental degradation increases. Consequently, food production, food security, income of farmers and economic profits of farming decreases (Abbasi et al., 2022; Singh & Jyoti, 2024). Further, it has an adverse impact on food quality due to increase in food wastage. Food wastage may be harmful for the environment. The summary of related studies is given in Table 1. DTs, therefore, can be effective to identify the pest infected cropped areas. While applications of DTs would be helpful for farmers to increase their understanding to minimize the negative implications of pest and insects in the farming. Use of DTs suggests multiple mechanisms to reduce pests and insects in the cultivation. DTs can be adopted to address current environmental challenges.

Table 1. Summary of review of literature

Author(s)	Research's area	Objectives	Indicators (DVs/IVs)	Main conclusion
Maharjan et al. (2023)	Nepal	Examine the significance of environmental conservation in agriculture	Environmental conservation agriculture (ECA), climatic factors	ECA is favourable for increasing sustainability in the agricultural sector.
Raihan (2023)	Vietnam	Examine the interconnection among the energy use, agricultural sector and CO2 emissions	Economic growth, agricultural value added, energy use and CO_2 emissions	Energy use enhances environmental degradation.
Đuric (2020)	Review	Provide technical overview of DTs in context of agricultural sector	-	DTs are helpful to increase fare food system, increase productivity and food value chain.

continued on following page

Table 1. Continued

Author(s)	Research's area	Objectives	Indicators (DVs/IVs)	Main conclusion
Boadi et al. (2022)	Ghana	Observe the understanding of farmers for applying innovation and technology in cocoa cultivation	-	Farmers are aware towards new technologies in cultivation. Technology and innovation helping to increase cocoa production.
Shrestha et al. (2023)	Review article	Examine the contribution of biochar in the agricultural sector	-	Biochar can be effective to achieve net-zero emissions agriculture.
Torres-Morales et al. (2023)	Colombia	Role of biochar as a practice of negative emission technology in the agricultural sector	Biochar and agricultural sector	Agricultural residual biomass can be re-integrated into soil using biochar technology.
Abiri et al. (2023)	Review	Role of digital agricultural in the cultivation	-	Digital agricultural helps to reduce wastage, and enhance crop production
Rai (2023)	India	Examine the contribution of digitalization in the agricultural sector	-	Income of farmers, food and nutritional security improved as application of digitalization in the agricultural sector.
Vilkov and Tian (2023)	Review of literature	Role of digitalization in emission trading system (ETS)	-	ETS provides an appropriate solution for carbon market.
Šermukšnytė-Alešiūnienė and Melnikienė (2024)	EU countries	Role of digital innovation in the performance of smallholders	Digital innovation and farm income	Digital innovation helps to increase all dimensions of sustainable development in small farm.
Reza et al. (2024)	India	Impact of digital soil mapping approach on soil organic carbon	Digital soil mapping and soil organic carbon	Digital soil mapping approach is effective for climate change mitigation.

continued on following page

Table 1. Continued

Author(s)	Research's area	Objectives	Indicators (DVs/IVs)	Main conclusion
Addison et al. (2024)	Ghana	Examine the agricultural digitalization in livelihood of farmers	DTs, educational attainment, cooperative membership, accessibility of electricity, internet and mobile	Agricultural digitalization is found positive to increase livelihood assets of farmers
Chandio et al. (2024)	SAARC countries	Examine the impact of ICT on grain production	Grain production and ICT	Grain production could be improved significantly as use of ICT in the farming.

2.3. Solution for Reducing Environmental Degradation in the Agricultural Sector

The agricultural sector is a prime driver to increase environmental degradation (Jang et al., 2023; Raihan, 2023). It is a prime contributor to GHGs and CO_2 emissions. Thus, the extensive growth of this sector is significantly accountable for climate change and natural resource depletion. Consequently, environmental degradation is caused to decrease agricultural production, and growth of this sector leads to increased environmental degradation. Hence, this sector should pursue effective strategies and practices to ameliorate environmental development. Purcell et al. (2023) reported that digital twins would be valuable to reduce the challenge of environmental degradation and enhance sustainability. For instance, water management practices may be useful to GHGs emissions from the agricultural sector (Jang et al., 2023). Implementation of smart agroforestry is positive to reduce CO_2 emissions, and increase biodiversity and water conservation (Octavia et al., 2022). Soil carbon sequestration (SCS) is a scientific practice to abate CO_2 emissions (Almaraz et al., 2023). Biochar technology may be useful to increase agricultural growth with net-zero emissions (Shrestha et al., 2023; Torres-Morales et al., 2023). Biochar technology is reducing CO_2 emission (Torres-Morales et al., 2023). DTs, therefore, would reduce waste in the agricultural sector, and help to reduce environmental degradation (Abiri et al., 2023). Application of appropriate technologies would reduce environmental degradation in the agricultural sector (Ashraf & Singh, 2022).

2.4. Technological Development and Agricultural Sector

Earlier studies gave extensive focus on the association of technological development and agricultural sector (Newman et al., 2023; Ashraf & Singh, 2022; Beig et al., 2023). Findings of most studies found a positive implication of technological development on agricultural growth and production (Ashraf & Singh, 2021; Kumar & Singh, 2023). For instance, Du et al. (2019) noted that green technology advancements have a single income-level effect on CO_2. Further, Chandio et al. (2024) suggested that DTs can be effective for increasing agricultural productivity and farm revenue. Application of green technologies in the agricultural sector would be crucial to abate CO_2 and GHGs emissions. Almaraz et al. (2023) argued that new technologies would reduce GHGs emissions. Accordingly, it would be helpful to reduce environmental degradation. Song et al. (2022) applied counterfactual thought-based synthetic control technique to examine the implication on agrarian productivity in China. It argued that agricultural productivity declines due to increase in life-threatening heat. Therefore, the Chinese government suggested that farming communities should pursue high-quality regional green agriculture to avoid the stress of climate change in it.

Extensive agricultural development is caused to increase energy consumption and contribute more CO_2 emissions (Raihan, 2023). Therefore, green energy is the fruit and output of technological development. Green energy would be positive to increase agricultural productivity in a sustainable manner. Consequently, green energy would also increase agricultural and environmental sustainability. Ali et al. (2012) observed the role of renewable and energy in the agricultural sector. Ali et al. (2023) noticed that renewable energy consumption helps to reduce CO_2 emissions and improve environmental quality in Africa. Further, it is concluded that green energy has the potential to increase the revenue of farmers. Use of renewable energy may be positive to increase sustainable agriculture (Chel & Kaushik, 2011). Eyuboglu and Uzar (2020) detected the potential contribution of renewable energy sources in curbing ecological deprivation. Murshed (2021) claimed that LPG can serve as a transition fuel for mitigating CO_2 emissions. Raihan et al. (2022) investigated a significant association between energy consumption, land use expansion for farming, forest loss, and emissions of CO_2 emissions in Malaysia. It noted that a 1% increase in energy consumption is linked to a similar increase of 0.91% in CO_2 emissions.

Li et al. (2022) examined the climate change impact on FDI. It indicates that multinational corporations' investments made in foreign countries exhibit a lower vulnerability to physical climate threats on a global scale. Green finance is playing a prime role to increase the progress of the agricultural sector and reduce environmental degradation. The availability of green finance helps economic agents to implement green projects. Foreign direct investment (FDI) is efficient to increase the technology transfer from one country to another. FDI is a driver to increase job opportunities as technology commercializes. Hence, FDI is vital to increase the overall performance of an economy. FDI, thus, has a crucial impact on the agricultural sector. Dungana and Ghimire (2013); Arora and Wadhwa (2014) explored the implications of FDI in the Indian agricultural sector. It notes that FDI seems positive to increase the overall growth of India. Therefore, large agrarian countries should increase the significant share of FDI in the agricultural sector. Public-private partnerships have the potential to effectively bridge the existing gap in savings and investment in all sectors. While technological development would be crucial to increase the public-private partnership in the agricultural sector. It would also be beneficial for customers and marginalised farmers to increase their economic benefits.

Moreover, FDI helps to promote industrial and infrastructural development. Therefore, FDI may be harmful for environmental sustainability and the agricultural sector. Pazienza (2015) explored the inverse connection of FDI with CO_2 emissions. It claimed that only specific types of FDI could be useful to reduce CO_2 emissions. Most studies also examine the impact of FDI on ecological aspects within the farming and fisheries sectors. The previous findings are highlighted that FDI should be in the domain in sustainability enhancing sectors. Chaudhary (2016) emphasised the imperative of international investments to bridge the void created by insufficient domestic savings in adequately addressing the financial requirements of the agricultural industry. The Indian government has strategically implemented multiple schemes to attract FDI with the objective of bolstering the overall strength of the economy. Warsame (2023) examined the role of FDI and renewable energy consumption for reducing environmental degradation in Somalia.

Newman et al. (2023) highlighted the role of emerging technologies (i.e., digital and cellular agriculture) in the agricultural sector. It noticed that emerging technologies are helping to increase sustainable food systems, conservation of biodiversity and nutritional standards of food. Nanotechnology is significant to increase sustainable agrochemical production (Beig et al., 2023). Mekyassi and Kızıldeniz (2024) observed the impact of climate-smart agriculture (CSA) in agricultural sustainability. It noticed that CSA helps to improve agricultural sustainability by increasing income and production. Smart agriculture is highly crucial for farmers to deal with different challenges with environmental degradation, production and farm production (Abbasi et al., 2022).

2.5. Advanced Technologies, DTs and Environmental Degradation

Advanced technologies and DTs are crucial to reduce environmental degradation (Ali et al., 2023; Vilkov & Tian, 2023). For example, climate-smart agriculture is supportive to reduce environmental degradation in the agricultural sector (Abbasi et al., 2022; Mekyassi & Kızıldeniz, 2024). CSA improves the nutritional contents and overall quality of food (Mekyassi & Kızıldeniz, 2024). Hence, food security is positively associated with CSA (Mekyassi & Kızıldeniz, 2024). CSA would be positive to attain SDGs and mitigate climate change impacts (Mekyassi & Kızıldeniz, 2024). Biró and Csete (2023) argued that digitalization helps to increase sustainable development. Rai (2023) claimed a positive contribution of digitalization to reduce environmental degradation in the cultivation. Digital and cellular agriculture are playing a positive role to reduce land use and environmental footprint (Newman et al., 2023). Ali et al. (2023) examined the role of digitalization for improving environmental quality in Africa. It claims that digitalization helps to increase environmental quality and reduce CO_2 emissions.

Chen and Li (2024) suggested that digital transformation of the agricultural sector would provide the scientific process to reduce CO_2 emissions. Soil carbon sequestration technique may reduce GHGs emission and environmental degradation (Almaraz et al., 2023). Promotion of digital agriculture would reduce wastage from the agricultural sector and reduce environmental degradation (Trendov et al., 2019; Abiri et al., 2023). Digital agriculture would be positive to increase economic benefits as improving agricultural productivity and minimizing cost of production (Trendov et al., 2019; Klerkx et al., 2019). Digital agriculture enhances the agricultural value chain (Dibbern et al., 2024). Digital agriculture would increase the efficiency of agricultural inputs to meet the global food demand and ensure sustainable food production (Abiri et al., 2023). Du et al. (2022) examined the contribution of digital agriculture for promoting food supply chain. It claims that DTs help to increase the food supply chain.

2.6. Advanced DTs and Agricultural Sector

Recent studies are paying significant attention to examine the implications of digital technologies (DTs), digitalization, smart phone, digital platform and digital infrastructure in the agricultural and other sectors (Boadi et al., 2022; Abiri et al., 2023; Biró & Csete, 2023; Rai, 2023; Addison et al., 2024; Singh & Singh, 2025). For instance, artificial intelligence (AI), cloud computing, automation, machine learning, and internet of things (IoT) are increasing the sustainability of the farming sector (Ingram & Maye, 2020; Singh & Kaur, 2021; Abiri et al., 2023;

Gabriel et al., 2023; Dibbern et al., 2024). Use of IoT can be applied to increase productivity, quality and profits of farmers (Abbasi et al., 2022). Cloud computing is internet-based technology that increases the spread of the information using the internet (Mhlanga et al., 2023). AI helps to increase the production process in the agricultural sector. AI can be proactive to increase the operation of various crops in different weather seasons.

Application of AI also supports the farmers to manage the crops to minimize less climate change impact on it. Agri-food sector can increase its progress using AI (Du et al., 2022). Advanced DTs provides the data accessibility for the users (Dibbern et al., 2024). There are other DTs like robotics, sensors and blockchain that are significant for the agricultural sector (Klerkx et al., 2019). Farming methods can be as effective and more productive as using robotics and machine learning (Ibrahim & Truby, 2023). These DTs are playing a key role to increase agri-food sustainability (Ingram & Maye, 2020). Alos, these are helping farmers to select appropriate time for seed sowing, farm preparation and irrigation (Singh & Kaur, 2021). Use of IoT is significant to enhance agricultural productivity (Singh & Kaur, 2021). IoT is helpful to remote control and helps to increase decision capacity of the farmers (Mhlanga et al., 2023).

Rai (2023) proposed that digitalization is helpful to increase productivity and income of farmers, and reduce waste in Indian agriculture. Šermukšnytė-Alešiūnienė and Melnikienė (2024) examined the role of digital innovation in farming activities of smallholders in EU countries. It produces a useful association of digital innovation in the agricultural sector. It noticed that digitalization is found useful to introduce a culture for promoting all components of sustainable development in small farms. DTs could be helpful to increase economic performance, reduce labour cost, and increase investment planning of small farms (Šermukšnytė-Alešiūnienė & Melnikienė, 2024; Singh, 2025). Digital carbon mapping approach is helpful to take necessary action to implement carbon management plan (Reza et al., 2024).

2.7. Application of DTs in the Era of Climate Change in the Agricultural Sector

Digitalization is significant to create multiple ways to increase the sustainability of the agricultural sector in the presence of climate change. Digital agriculture is an important innovation to address the challenge of climate change and to increase sustainability of the food production system (Newman et al., 2023; Ingram & Maye, 2020). Abiri et al. (2023) noticed that digital technologies are crucial to increase sustainable food production systems. Addison et al. (2024) examined the contribution of agricultural digitalization for increasing the livelihood security of smallholders in Ghana. It concluded that livelihood assets of farmers could be improved due to agricultural digitalization. Boadi et al. (2022) investigated the contribution of new technology and smart phones in cocoa farming in Ghana. It suggested that smartphones are helping farmers to avail different training to increase cocoa production and mitigating climate change impact in it.

Digitalization helps to nurture the best ecosystem to mitigate climate change in the farming sector (Šermukšnytė-Alešiūnienė & Melnikienė, 2024). For instance, digital platforms create an appropriate platform of communication among the various stakeholders in the farming sector (Ingram & Maye, 2020). Digital applications are increasing the accessibility and utility of information among the users (Ingram & Maye, 2020). It also reduces the cost of cultivation and enhances the income of the farmers (Šermukšnytė-Alešiūnienė & Melnikienė, 2024). Choudhary and Gupta (2023) examined the climate change impact on agronomic output. This study proposed multiple measures to mitigate the negative consequences of climate change on agronomic output. Gabriel et al. (2023) observed the adoption of DTs and its multiple usages in the agricultural sector in Germany as using field-based data. It indicated that DTs are helping farming communities to reduce workload.

2.8. Few Successful Examples of DTs in Agriculture

Application of DTs are helping to increase farm productivity and enhancing the sustainability of the farming sector. Mhlanga et al. (2023) reported that DTs could be helpful to increase environmentally friendly farming in Africa. There are many successful examples of DTs in the agricultural sector. DTs can be favourable to increase minimum use of irrigation and fertilizer. It is helpful to increase resource management in the farming sector. Agricultural scientists can check the crop health and detect the problems of soil. Appropriate yield can be estimated using artificial intelligence and big data analysis in the farming. Ashraf & Singh (2022) noticed the positive impact of social media and appropriate technologies for increasing agricultural productivity in India. Jyoti et al. (2023) explored the overview of appropriate technologies to reduce the negative impact of climate change and environmental issues in the agricultural sector. Geng et al. (2024) analyze the impact of DTs on economic benefits in the Chinese agricultural sector. It noticed the intensity of DTs could be used to increase economic benefits such as improving yields and quality, and reduce cost of various inputs.

2.9. Drawbacks and Challenges of DTs in the Agricultural Sector

While application of DTs brings new and effective possibilities to increase the production scale of the agricultural sector. Conversely, DTs create multiple challenges and barriers for the farmers due to their low digital literacy, low financial capabilities to afford high cost of digital infrastructure and low education of farmers (Mhlanga et al., 2023; Dibbern et al., 2024). For example, transition of traditional agriculture to digital agriculture is increasing various challenges for the farmers (Dibbern et al., 2024). Dibbern et al. (2024) explored various challenges and barriers in adoption of digital agriculture. It noted that technological infrastructure, technical skills and consistency of technologies play a key role to increase digital agriculture. While farmers are unable to apply agricultural technologies due to limited accessibility of infrastructure and technology (Trendov et al., 2019). Low learning for farmers is also denying the applications of DTs in the agricultural sector. Farmers are unable to apply new technologies due to lack of proper training (Ibrahim & Truby, 2023).

Low digital skills of farmers is another challenge to adopt DTs in the agricultural sector. Farmers do not have appropriate funding to bear the high cost of digital instruments. Therefore, inappropriate financial resources reduce the intention of farmers to adopt DTs (Trendov et al., 2019). Mhlanga et al. (2023) perceived that resource scarcity, inappropriate digital infrastructure and data privacy are the significant concerns in adoption of DTs in the agricultural sector of Africa. Low size of land holding reduces more practices of DTs in the farming (Trendov et al., 2019). Moreover, there is an unavailability of digital infrastructure and internet connectivity, and electricity in rural areas of largely agrarian countries (Mhlanga et al., 2023). Low internet connectivity is a significant barrier in digital agriculture. Inappropriate regulatory framework is increasing the barriers for transforming digital agriculture (Trendov et al., 2019). Regulatory framework and financial insecurity are creating a challenge in digital agriculture. While farmers are unaware about the legal regulation of DTs due to their high digital illiteracy (Ibrahim & Truby, 2023). Moreover, the agricultural sector could not create more agri-startups to increase the income of the farmers and increase their collaboration with different traders in agriculture (Singh, Jyoti, et al., 2025). However, few emerging countries like India are giving appropriate attention to increase Agri-tech startups (Ibrahim & Truby, 2023). Economies of scale is another challenge in adoption of digital technologies in the cultivation (Trendov et al., 2019).

2.10. Role of Government and Institutions for Promoting DTs and Agriculture

The government and institutions play a crucial role for increasing sustainable agriculture. Financial support from the government is essential to increase the process of DTs in the agricultural sector (Geng et al., 2024). The government provides financial support to the agricultural universities and research organizations to discover more technologies that can be useful to increase agricultural production. Geng et al. (2024) suggested that government support is essential to increase the adoption rate of DTs in the agricultural sector. Financial availability would increase the economic capacity of farmers to apply advanced DTs in the cultivation. Banking sector should provide green funds to the farmers to adopt green technologies and DTs to increase sustainable agriculture. Research institutions should organize regular training programmes to the farmers to increase their usability of new technologies in the cultivation. Agricultural development agencies play an active role to provide useful information to the farmers. Social media plays an effective responsibility to disseminate the government policies among the farmers.

3. RESEARCH METHODS AND MATERIALS

3.1. Area of Research

Table 2. List of countries

Countries	Time Period
Canada, Austria, Brazil, China, Colombia, Czechia, Denmark, Finland, France, Germany, Guatemala, Hungary, Indonesia, Iran Islamic Rep., Italy, Japan, Korea Rep., Latvia, Mexico, Netherlands, New Zealand, Nicaragua, Norway, Paraguay, Portugal, Spain, Sweden, Switzerland, United Arab Emirates, United Kingdom, United States, Zambia	2000-2020
Argentina, Australia, Cyprus, Ecuador, Estonia, Georgia, India, Israel, Jamaica, Jordan, Lithuania, Malaysia, Moldova, Mongolia, Poland, Romania, Saudi Arabia, Sri Lanka, Turkiye, Zimbabwe	2001-2020
Azerbaijan, Belarus, Belgium, Bolivia, Burkina Faso, Egypt, Arab Rep., Ireland, Kyrgyz Republic, Luxembourg, Mauritius, Morocco, Qatar, Russian Federation, Senegal, Slovak Republic, Slovenia, South Africa, Suriname, Trinidad and Tobago, Tunisia, Viet Nam	2002-2020
Algeria, Barbados, Belize, Cote d'Ivoire, Croatia, Dominican Republic, El Salvador, Ethiopia, Gabon, Greece, Kazakhstan, Malawi, Uzbekistan	2003-2020
Afghanistan, Bulgaria, Comoros, Gambia The, Ghana, Niger, Rwanda, Uruguay	2004-2020
Albania, Botswana, Cameroon, Fiji, Guyana, Kenya, Lao PDR, Mauritania, Myanmar, Namibia, Nigeria, Pakistan, Tanzania, Timor-Leste, Uganda, Ukraine	2005-2020
Angola, Cambodia, Chile, Iraq, Madagascar, Mali, Mozambique, Nepal, Philippines	2006-2020

continued on following page

Table 2. Continued

Countries	Time Period
Lebanon, Togo	2007-2020
Bangladesh, Bhutan, Burundi, Eswatini, Liberia	2008-2020

This chapter is intended to examine the impact of environmental degradation and digital technologies (DTs) on cereal yield across countries. Hence, the countries are selected as per the availability of statistics of related variables. Accordingly, 126 countries have the data of desired independent and dependent variables. Hence, these countries are comprised for further statistical and empirical analysis in this chapter. Moreover, the statistical data of prescribed variables are not available for an equal time period for these countries. This chapter, therefore, compiled country-wise imbalanced panel data. The time period of respective countries is given in Table 2.

3.2. Source of Data

This chapter attempts to assess the implications of DTs, environmental degradation, and other variables on cereal yield across countries. Therefore, the statistics of related variables are taken from the World Development Indicators (The World Bank), and the FAOSTAT databases.

3.3. Selection of Dependent and Independent Variables

The prime aim of this chapter is to explore the implication of environmental degradation and digital technologies (*DT*s) on cereal yield. Cereal yield is a worthy variable that measures the overall performance of the agricultural sector. Henec, cereal yield is used as output variable. Per capita CO_2 emissions are used to capture the impact of environmental degradation on cereal yield. Many studies are also using per capita CO_2 emissions as a key indicator for environmental degradation in the empirical investigation (Raihan, 2023; Singh et al., 2023). While fixed broadband subscriptions, fixed telephone subscriptions and Individuals using the Internet are used as proxy variables for DTs in regression analysis. Following studies like Gupta et al. (2025); Singh, Srinivasan, et al. (2025) also applied mentioned variables for assessing the impact of DTs on the agricultural and other sectors. Land under cereal production, agriculture, forestry, and fishing, value added (% of GDP), employment in agriculture, and fertilizer consumption are used as crucial inputs of the agricultural sector (Singh et al., 2023; Chandio et al., 2024).

Table 3. Summary of dependent and independent variables

Indicators	Code	Unit
Cereal yield (kg per hectare)	*CY*	Kg./Ha.
Fixed broadband subscriptions (per 100 people)	*FBS*	Number
Fixed telephone subscriptions (per 100 people)	*FTS*	Number
Individuals using the Internet (% of population)	*IUI*	%
CO_2 emissions (metric tons per capita)	CO_2PC	metric tons
Land under cereal production (hectares)	*LUCP*	Ha.
Agriculture, forestry, and fishing, value added (% of GDP)	*AFFVA*	%
Employment in agriculture (% of total employment)	*EA*	%
Fertilizer consumption (kilograms per hectare of arable land)	*FCPH*	Kg./Ha.
GDP per capita (constant 2015 US$)	*GDPPC*	US$
Inflation, GDP deflator (annual %)	*IGDPD*	%
Annual freshwater withdrawals, agriculture (% of total freshwater withdrawal)	*AFWWA*	%
Foreign direct investment, net inflows (% of GDP)	*FDINI*	%
Population growth (annual %)	*PG*	%

Source: Indicators are finalized as per the existing literature and availability of their statistics for selected countries.

Foreign direct investment, net inflows (% of GDP) and GDP per capita are also considered as macro level indicators in the model (Singh et al., 2023). Warsame (2023) examined the contribution of FDI and renewable energy consumption for reducing environmental degradation in Somalia. Population growth is included as a proxy indicator for assessing the impact of demographic change on cereal yield (Singh et al., 2023). Inflation GDP deflator is considered to examine the impact of inflation on cereal yield. Annual freshwater withdrawals are used to examine the impact of water resources on cereal yield. Hence, these indicators are used as control variables in the empirical models. The summary of the variables is presented in Table 3.

3.4. Statistical and Empirical Analysis

Simple statistical analysis is used to explore the collected secondary data. Furthermore, the scientific researchers, environmentalists, economists and social scientists have formulated diverse empirical models to examine the impact of environmental degradation and DTs on farming sectors across countries using secondary and primary data (Dungana & Ghimire, 2013; Arora & Wadhwa, 2014; Song et al., 2022; Raihan et al, 2022; Raihan, 2023; Ali et al., 2023; Newman et al., 2023; Gabriel et al., 2023). The country-wise imbalanced panel data of related indicators are compiled during 2000-2020 in this chapter. While log-linear regression equation is employed to examine the impact of environmental degradation, *DT*s and other control variables on cereal yield. For empirical investigation, this research assumes that cereal yield (*CY*) is a function of the mentioned indicators in Table 3. The mathematical functional association of *CY* with these variables are defined as:

$$(CY) = f\{(FBS), (FTS), (IUI), (CO_2PC), (LUCP), (AFFVA), (EA), (FCPH), (GDPPC), (IGDPD), (AFWWA), (FDINI), (PG)\} \quad (1)$$

The signification of all the variables is provided in Table 3. After adopting, log-linear regression model, the above equation is transformed as:

$$log(CY)_{ct} = \beta_0 + \beta_1 log(FBS)_{ct} + \beta_2 log(FTS)_{ct} + \beta_3 log(IUI)_{ct} + \beta_4 log(CO_2PC)_{ct} + \beta_5 log(LUCP)_{ct} + \beta_6 log(AFFVA)_{ct} + \beta_7 log(EA)_{ct} + \beta_8 log(FCPH)_{ct} + \beta_9 log(GDPPC)_{ct} + \beta_{10} log(IGDPD)_{ct} + \beta_{11} log(AFWWA)_{ct} + \beta_{12} log(FDINI)_{ct} + \beta_{13} log(PG)_{ct} + \mu_{ct} \quad (2)$$

Here, β_0 is constant coefficient that measures the total factor productivity; β_1, β_2, ..., β_{13} are the coefficients of associated explanatory variables; *C* is cross-sectional balanced countries; *t* is time period; μ_{ct} is the residual term in equation (2). Here, β_1, β_2, ..., β_{13} are also estimate the elasticities of cereal yield with respect to corresponding explanatory variables.

3.5. Final Estimation and Use of Statistical Software

The regression coefficients of independent variables with cereal yield are estimated using feasible generalized least square (*FGLS*) regression estimation model. While MS excel, SPSS and STATA statistical software are used to analysis the data and to estimate the parameter of variables.

4. DISCUSSION ON DESCRIPTIVE AND EMPIRICAL RESULTS

4.1. Correlation Coefficients of Cereal Yield with Explanatory Variables

The linear association among the indicators are estimated through correlation coefficients analysis (Table 4). The Karl-Pearson correlation coefficients analysis is used to estimate the association among the variables. The coefficients of *FBS* ($r =$

0.440**), *FTS* (*r* =0.451**) and *IUI* (*r* =0.489**) with *CY* are observed positive and statistically significant at 1% significance level. Hence, DTs are positively associated with cereal yield. The coefficients of *AFFVA* (*r* = -0.393**), *EA* (*r* = - 0.450**), *IGDPD* (*r* = - 0.084**), *AFWWA* (*r* = - 0.229**) and *PG* (*r* = -0.083**) with *CY*. Hence, the estimates demonstrate that cereal yield is negatively correlated with the share of the agricultural sector in GDP, employment in agriculture, inflation, annual freshwater withdrawal and population growth. While the coefficients of CO_2PC (*r* = 0.475**), *FCPH* (*r* = 0.449**) and *GDPPC* (*r* = 0.514**) with *CY*. The estimates suggested that cereal yield is positively associated with per capita CO_2 emissions, fertilizer consumption and per capita GDP.

Table 4. Correlation coefficients among the indictors

Indicators	*CY*	*FBS*	*FTS*	*IUI*	CO_2PC	*LUCP*	*AFFVA*
CY	1	0.440**	0.451**	0.489**	0.475**	0.029	-0.393**
FBS	0.440**	1	0.636**	0.865**	0.411**	-0.001	-0.562**
FTS	0.451**	0.636**	1	0.612**	0.561**	0.008	-0.661**
IUI	0.489**	0.865**	0.612**	1	0.513**	-0.037*	-0.644**
CO_2PC	0.475**	0.411**	0.561**	0.513**	1	0.081**	-0.562**
LUCP	0.029	-0.001	0.008	-0.037*	0.081**	1	0.017
AFFVA	-0.393**	-0.562**	-0.661**	-0.644**	-0.562**	0.017	1
EA	-0.450**	-0.608**	-0.700**	-0.706**	-0.594**	0.051**	0.833**
FCPH	0.449**	0.201**	0.267**	0.306**	0.231**	-0.008	-0.254**
GDPPC	0.514**	0.671**	0.713**	0.666**	0.697**	-0.021	-0.561**
IGDPD	-0.084**	-0.153**	-0.132**	-0.157**	-0.055**	-0.008	0.092**
AFWWA	-0.229**	-0.497**	-0.534**	-0.479**	-0.396**	0.078**	0.481**
FDINI	0.011	0.147**	0.125**	0.102**	0.074**	-0.076**	-0.056**
PG	-0.083**	-0.342**	-0.393**	-0.306**	0.088**	-0.087**	0.306**

Table 5. Correlation coefficients among the indictors [Conti...]

Indicators	*EA*	*FCPH*	*GDPPC*	*IGDPD*	*AFWWA*	*FDINI*	*PG*
CY	-0.450**	0.449**	0.514**	-0.084**	-0.229**	0.011	-0.083**
FBS	-0.608**	0.201**	0.671**	-0.153**	-0.497**	0.147**	-0.342**
FTS	-0.700**	0.267**	0.713**	-0.132**	-0.534**	0.125**	-0.393**
IUI	-0.706**	0.306**	0.666**	-0.157**	-0.479**	0.102**	-0.306**

continued on following page

Table 5. Continued

Indicators	*EA*	*FCPH*	*GDPPC*	*IGDPD*	*AFWWA*	*FDINI*	*PG*
CO_2PC	-0.594**	0.231**	0.697**	-0.055**	-0.396**	0.074**	0.088**
LUCP	0.051**	-0.008	-0.021	-0.008	0.078**	-0.076**	-0.087**
AFFVA	0.833**	-0.254**	-0.561**	0.092**	0.481**	-0.056**	0.306**
EA	1	-0.304**	-0.605**	0.126**	0.473**	-0.090**	0.330**
FCPH	-0.304**	1	0.273**	-0.070**	-0.106**	0.026	-0.036*
GDPPC	-0.605**	0.273**	1	-0.135**	-0.502**	0.169**	-0.083**
IGDPD	0.126**	-0.070**	-0.135**	1	0.117**	-0.035*	0.083**
AFWWA	0.473**	-0.106**	-0.502**	0.117**	1	-0.118**	0.249**
FDINI	-0.090**	0.026	0.169**	-0.035*	-0.118**	1	-0.015
PG	0.330**	-0.036*	-0.083**	0.083**	0.249**	-0.015	1

Source: Author's estimation. **Note:** ** infer that correlation is significant at the 0.01 level (1-tailed) and * infer that correlation is significant at the 0.05 level (1-tailed).

4.2. Regression Coefficients of Explanatory Variables with Cereal Yield

The regression coefficients of explanatory variables with cereal yield are given in Table 5. While the regression coefficients are estimated through PCSEs model. The explanation of regression coefficients of explanatory variables with cereal yield is given as:

Digital Technologies (*DTs*):*DTs* are bringing innovation in the agricultural sector for improving its productivity, efficiency and minimizing cost of cultivation (Du et al., 2022). DTs are helping to increase the communication of farmers to understand the impact of climate change in the farming. DTs increases the food supply chain (Du et al., 2022). Farmers are also getting spontaneous solutions from agricultural development agencies through online platforms. *DTs* are providing weather forecasting which is helping farmers to make better decisions for growing an appropriate crop (Singh & Kaur, 2021). *DTs* are providing enormous suggestions to the farmers towards marketing. This research also noticed a positive coefficient of fixed broadband subscriptions (*FBS*), fixed telephone subscriptions (*FTS*) and individuals using the Internet (*IUI*) with cereal yield. As these are the favourable indicators for creating a digital infrastructure. Hence, the results of this chapter noticed that cereal yield is to be increased by 0.0094%, 0.0024% and 0.0117% due to 1% increase in *FBS*, *FTS* and *IUI*, respectively in the farming sector. Chandio et al. (2024)

also noticed the positive impact of ICT for increasing grain production in the SAARC countries.

Per Capita CO_2 emissions (*CO_2PC*) (Environmental Degradation): This variable is applied to capture the impact of environmental degradation on cereal yield in this research. There is unanimity among the scientists that CO_2 emissions are increasing the prevalence of climate change which has a detrimental impact on the agriculture sector (Kumar et al., 2017; Bhat et al., 2024). Agricultural sector has a large contribution to CO_2 emissions (Chen & Li, 2024). Therefore, CO_2 emissions from this sector would be harmful for increasing its sustainability. Further, it is expected that per capita CO_2 emissions have a negative impact on cereal yield. The coefficient of per capita CO_2 emissions with cereal yield is detected as negative and statistically significant. Cereal yield, therefore, is to be decreased by 0.0855% as CO_2 emissions increase by 1%. The result is consistent with the study of Bhat et al. (2024) which also found a negative implication of CO_2 emissions on the agriculture sector.

Land Under Cereal Production (hectares) (*LUCP*): Area under cereal crop is a sole input to increase cereal production. However, it may not be supportive to increase cereal yield due to applicability of the law of diminishing returns in the production sectors (Singh & Kaur, 2021). Although, it seems a positive impact of this indicator on cereal yield in the selected countries. The statistical value infers that cereal yield may be improved by 0.0081% as cropped area under cereal crop increase by 1%.

Share of Agriculture and Allied Sector in GDP (*AFFVA*): The coefficient of agriculture and its allied sector have a positive impact on cereal yield. It implies that income from this sector is helpful for farming communities to adopt different technologies and mechanism for further growth of the agricultural sector. Hence, it is obvious that cereal yield would be improved as the share of this sector in GDP increases. The estimate implies that cereal yield may be improved by 0.1390% as the share of the agricultural sector in GDP increases by 1%.

Employment in Agriculture (*EA*): Agricultural sector is a labour-intensive sector. However, the labour demand in this sector has decreased due to technological intensification and agricultural mechanization. While technological intensification could be useful to increase the production, yield, and profit of farmers. On the other hand, technological intensification is leading to increased environmental degradation. Hence, application of more use of human resources in the agricultural sector may be a positive action to reduce environmental degradation. Also, adoption of green technology may be another way to reduce GHGs and CO_2 emissions from this sector. This chapter is found a negative coefficient of employment in agriculture with cereal yield. Further, statistical value reveal that cereal yield is likely to be decreased by -0.0554% as employment in the agricultural sector increases by 1%.

Fertilizer Consumption (*FCPH*): As suggested, application of fertilizer may be positive to increase soil fertility in the short run (Chandio et al., 2024). Therefore, use of fertilizer may be helpful to increase yield of crops. Although, overwhelming application of fertilizer may be caused to increase environmental degradation and diminishing the soil quality and fertility. Hence, it may deteriorate the production per hectare land. There is noticed a positive association of fertilizer consumption with cereal yield in this research. Further, the statistics suggested that cereal yield is expected to be increased by 0.1857% as fertilizer application increases by 1%. Although, it is also proved that large agrarian countries should apply minimum fertilizer in cultivation to increase soil fertility and quality of food. The global farming community should increase their dependence on green fertilizer to improve soil fertility, and to increase sustainable food systems and environmental sustainability.

GDP Per Capita (*GDPPC*): GDP per capita is a vital component to increase the overall human development. It is positive to increase the social and educational status of the people. It also suggested different practical proposals to reduce the negative impact of environmental degradation and climate change in the agricultural sector (Bhat et al., 2024). The farmers can also adopt more DTs in their daily activities as their income increases. Accordingly, GDP per capita produces a positive impact on cereal yield. Cereal yield is to be increased by 0.2581% as GDP increases by 1%.

Inflation GDP deflator (*IGDPD*): Inflation GDP deflator (IGDPD) measures the percentage change in gross domestic product due to increase in the price of goods and services. It means that inflation may be effective to increase the income of the producers. Hence, it is noticeable that inflation shows a positive impact of IGDPD on cereal yield. Cereal yield is projected to be increased by 0.0247% as inflation increases by 1%.

Annual Freshwater Withdrawals for Agriculture (*AFWWA*): Water productivity is positive to increase the water resources sustainability. Hence, it plays a positive and significant role to meet the water requirement of the agricultural sector in the long run. Accordingly, the impact of water productivity must be positive on cereal yield. Annual freshwater withdrawal for the agriculture sector is used to capture the impact of water resources on cereal yield in this chapter. It is noticed a negative coefficient of annual freshwater withdrawal with cereal yield. Thus, the estimate imply that annual freshwater withdrawal has a negative impact on cereal yield. The estimated statistics claimed that cereal yield is projected to decrease by -0.0320% as use of annual freshwater withdrawals in the agricultural sector increases by 1%. Also, the appropriate growth of the agricultural sector requires more energy consumption. While expansion, consumption of energy may be caused to increase environmental degradation and reduce inputs for the agricultural sector. Therefore, generation of renewable energy from water resources would reduce CO_2 emissions and environmental degradation. Thereupon, it would be imperative for agricultural sustainability. Thus, the larger agrarian countries should implement conducive policies to increase water resource sustainability to meet the water and energy requirement for the agricultural sector in near future (Singh & Singh, 2023).

Foreign Direct Investment Net Inflows (*FDINI*): The existing researchers could not find a rational and uniform impact of FDI on the agricultural sector. For instance, Warsame (2023) argued that FDI may be positive to reduce environmental degradation. Pazienza (2015) suggested that FDI may be useful to increase sustainability. Dungana and Ghimire (2013); Arora and Wadhwa (2014) also concluded that FDI would increase overall growth of the agricultural sector. However, this chapter has reported a negative impact of FDINI on cereal yield. The estimate infers that cereal yield is projected to decrease by 0.0335% as FDINI increases by 1% in the farming of cereal crops.

Population Growth (*PG*): Population growth is caused to increase enormous pressure on the agricultural and environmental sector (Singh, Singh, et al., 2025). It is, therefore, caused to increase environmental degradation. Population growth has a negative impact on cereal yield. Cereal yield is expected to be decreased by -0.0548% as population growth increases by 1%.

Table 6. Impact of DTs, environmental degradation and other variables on cereal yield

Number of obs.		**2315**	**Wald Chi2(13)**			**2627.28**
Number of groups		**126**	**Prob > Chi2**			**0.000**
Obs. per group: min		**13**	**Log likelihood**			**-1578.4**
Obs. per group: max		**21**	**Mean VIF**			**4.83**
log(CY)=[DV]	**Reg. Coef.**	**Std. Err.**	**z**	**P>|z|**	**[95% Conf. Interval]**	
log(FBS)	0.0094	0.0090	1.050	0.295	-0.008	0.027
log(FTS)	0.0024	0.0145	0.160	0.870	-0.026	0.031
log(IUI)	0.0117	0.0184	0.640	0.522	-0.024	0.048
log(CO_2PC)	-0.0855	0.0181	-4.730	0.000	-0.121	-0.050
log(LUCP)	0.0081	0.0041	1.960	0.050	0.000	0.016
log(AFFVA)	0.1390	0.0238	5.830	0.000	0.092	0.186
log(EA)	-0.0554	0.0223	-2.480	0.013	-0.099	-0.012

continued on following page

Table 6. Continued

Number of obs.		2315	Wald Chi2(13)			2627.28
Number of groups		126	Prob > Chi2			0.000
Obs. per group: min		13	Log likelihood			-1578.4
Obs. per group: max		21	Mean VIF			4.83
log(CY)=[DV]	**Reg. Coef.**	**Std. Err.**	**z**	**P>|z|**	**[95% Conf. Interval]**	
log(FCPH)	0.1857	0.0081	22.930	0.000	0.170	0.202
log(GDPPC)	0.2581	0.0256	10.090	0.000	0.208	0.308
log(AFWWA)	-0.0320	0.0097	-3.310	0.001	-0.051	-0.013
log(PG)	-0.0548	0.0112	-4.900	0.000	-0.077	-0.033
log(FDINI)	-0.0335	0.0085	-3.940	0.000	-0.050	-0.017
log(IGDPD)	0.0247	0.0090	2.740	0.006	0.007	0.042
Cons. Coef.	4.8823	0.2598	18.790	0.000	4.373	5.392

Source: Author's estimation.

5. CONCLUSION AND POLICY IMPLICATIONS

The theoretical literature infers that the agriculture sector is caused to increase environmental degradation and vice-versa. Therefore, both sectors require advanced technologies to reduce the negative impact of environmental degradation on the agriculture sector and vice-versa. Moreover, previous literature has clarified that digital technologies (DTs), digital infrastructure, digital transformation and digitalization are bringing multiple opportunities for agricultural and environmental sectors. There are many DTs that are playing a positive role to reduce the pressure on the environmental sector and natural resources. DTs are promoting farm productivity, income of farmers and revenue of the agriculture sector. These are improvising the food supply chain, food and nutritional security. Hence, it is noted that DTs can be applied as adaptation and mitigation approaches to fight the challenges of climate change in the agricultural sector. Therefore, DTs are helping agricultural scientists to make the future prediction of productivity of crops in the agricultural sector.

The results derived from the empirical model shows that cereal yield is significantly associated with CO_2 emissions, DTs, agricultural value addition, employment in agriculture, per capita GDP, fertilizer consumption, inflation, population growth, annual freshwater withdrawal and FDI. All indicators in the area of DTs impose a positive impact on cereal yield. Furthermore, DTs would be positive to reduce the negative impact of environmental degradation on the agricultural sector. CO_2 emissions, employment in agriculture, freshwater withdrawal, foreign direct investment and population growth are found as detrimental factors for cereal yield. FDI has been observed to be correlated with an escalation in CO_2 emissions. Hence, FDI has a negative impact on cereal yield. The results imply that expanding the cultivated land area has a notable favourable effect on cereal yield. It seems logical that a larger land area can result in increased crop production. However, the finding also suggests that a rapid increase in cultivated land during a specific time frame can be result in the excessive use or exhaustion of resources.

The findings also imply that the potential consequences of elevated carbon dioxide levels on agricultural production may not be immediately apparent. The unexpected outcome implies that FDI can bring improved technologies and practices. It can also result in changes in land use toward particular cash crops, potential adverse environmental effects due to unsustainable practices, or a shift in focus away from primary producing sectors. There is a possibility of a marginal decrease in the overall agricultural production, which underscores the importance of prudent administration and the incorporation of FDI. The result suggested that larger countries should focus on increasing water resource sustainability.

This chapter conducts a thorough analysis of the intricate relationships among agricultural production, adoption of DTs, FDI, and CO_2 emissions. The data clearly demonstrates that DTs has a capacity to mitigate CO_2 emissions. However, the research underscores the inherent contradiction of agriculture as a substantial source of CO_2 emissions over an extended period. Hence underscoring the urgent need for the implementation of sustainable farming practices in the farming sector. This chapter suggested that global countries should promote DTs in the agricultural sector to reduce environmental degradation. The global countries should also improve green agriculture to abate CO_2 and GHGs emissions. The government should provide training facilities to smallholders for applying DTs in the agricultural sector (Šermukšnytė-Alešiūnienė & Melnikienė, 2024). Digital literacy of the farmers must be improved to increase digital agriculture (Dibbern et al., 2024). The government should give additional efforts to create Agri-startups (Ibrahim & Truby, 2023; Singh, Jyoti, et al., 2025). Furthermore, several challenges like low digital infrastructure, low digital literacy of farmers, regulatory framework, digital insecurity, low internet connectivity etc. must be resolved to increase adoption of DTs in the cultivation.

6. SIGNIFICANCE AND INNOVATION OF THE CHAPTER

This chapter underscores the notable role that agriculture plays in contributing to CO_2 emissions and environmental degradation. It explains the comprehensive review of literature about the implications of environmental degradation, DTs and other crucial indicators in the agricultural sector. This research revealed a fascinating discovery regarding the alignment of DTs, environmental degradation and other variables with the agricultural sector in selected 126 countries. Thereupon, the role of DTs for promoting the growth of the agricultural sector is highlighted in this chapter. The correlation coefficients of undertaken indicators are estimated in this chapter. Undertaken indicators are identified as potential predictors of the next technological and economic cycle in the agricultural sector. The log-linear regression model is applied to examine the impact of individual indicators on cereal yield in this chapter. This research supports the proposition of increasing investments in water resource sustainability, digital infrastructure and creating jobs in the green sector and adoption of climate-smart farming techniques to increase the sustainable growth of the agricultural sector. Additionally, it recommends the promotion of environmentally conscious FDI and DTs in the agricultural sector. The aforementioned recommendations highlight the crucial importance of a sustainable shift toward the forthcoming technological and economic phase, transcending the confines of agriculture to encompass a wider scope of social and environmental welfare. Finally, the scope of further research is provided in this research.

7. LIMITATIONS AND SCOPE OF FURTHER RESEARCH

While this chapter makes a significant contribution towards theoretical and empirical literature. Accordingly, empirical findings provide the crucial policy implications to mitigate the negative implications of environmental degradation on the agricultural sector as promoting the applications of DTs. Despite that, this chapter has several limitations. Like, it uses only one indicator i.e., per capita CO_2 emissions to capture the impact of environmental degradation on cereal yield. However, CO, SO_2, NH_3, NH_4 and NO_2 are also harmful for the environment. It considered DTs in the empirical investigation. While there are several DTs like artificial intelligence, blockchain, big data analysis, automated devices, robotics etc. that have a significant impact on cereal yield. The government policies towards the agricultural sector help to enhance its growth. While this chapter does not examine the impact of government policies on cereal yield. Farming communities are using different mitigation and adaptation strategies to reduce the negative impact of environmental degradation on the agricultural sector. While this chapter could not observe the implications of the mentioned variables on cereal yield. In this chapter, the statistical data is used for a limited time period (i.e., 2000 – 2020) in an imbalanced panel data. While expansion and utilization of DTs in production activities extensively increased after 2020. Hence, further studies can include above-mentioned aspects in further empirical investigations.

REFERENCES

Abbasi, R., Maritinez, P., & Ahmad, R. (2022). The digitization of agricultural industry – a systematic literature review on agriculture 4.0. *Smart Agricultural Technology*, *2*(1), 1–10. DOI: 10.1016/j.atech.2022.100042

Abiri, R., Rizan, N., Balasundram, A. K., Shahbazi, A. B., & Abdul-Hamid, H. (2023). Application of digital technologies for ensuring agricultural productivity. *Heliyon*, *9*(12), 1–10. DOI: 10.1016/j.heliyon.2023.e22601 PMID: 38125472

Addison, M., Bonuedi, I., Arhin, A. A., Wadei, B., Owusu-Addo, E., Antoh, E. F., & Mensah-Odum, N. (2024). Exploring the impact of agricultural digitalization on smallholder farmers' livelihoods in Ghana. *Heliyon*, *10*(6), 1–12. DOI: 10.1016/j.heliyon.2024.e27541 PMID: 38533044

Ali, E., Gninigué, M., & Awade, N. E. (2023). Sectoral value chains and environmental pollution in Africa: Can development policies target digitalization and structural transformation to enhance environmental governance? *Journal of Environmental Economics and Policy*, *12*(2), 229–247. DOI: 10.1080/21606544.2022.2110163

Ali, S. M., Dash, N., & Pradhan, A. (2012). Role of renewable energy on agriculture. *International Journal of Engineering Sciences & Emerging Technologies*, *4*(1), 51–57.

Almaraz, M., Simmonds, M., Boudinot, F. G., Di, V. A. V., Bingham, N., Khalsa, S. D. S., Ostoja, S., Scow, K., Jones, A., Holzer, I., Manaigo, E., Geoghegan, E., Goertzen, H., & Silver, W. L. (2023). Soil carbon sequestration in global working lands as a gateway for negative emission technologies. *Global Change Biology*, *29*(21), 5988–5998. DOI: 10.1111/gcb.16884 PMID: 37476859

Annosi, M. C., Brunetta, F., Capo, F., & Heideveld, L. (2020). Digitalization in the agri-food industry: The relationship between technology and sustainable development. *Management Decision*, *58*(8), 1737–1757. DOI: 10.1108/MD-09-2019-1328

Arora, S., & Wadhwa, S. (2014). *FDI in Agricultural sector in India: status and challenges*. In book: Foreign Direct Investment Chapter: FDI in the agricultural sector in India: status and challenges. *Avon Publications*, New Delhi.

Ashraf, S. N., & Singh, A. K. (2021). Impact of technological change on growth and agricultural sector in Gujarat state of India: A time-series data study. *Asian Development Policy Review*, *9*(3), 144–160. DOI: 10.18488/journal.107.2021.93.144.160

Ashraf, S. N., & Singh, A. K. (2022). Implications of appropriate technology and farm inputs in the agricultural sector of Gujarat: Empirical analysis based on primary data. *Agricultural Economics Research Review*, *35*(2), 59–77. DOI: 10.5958/0974-0279.2022.00031.3

Beig, B., Niazi, M. B. K., Jahan, Z., Zia, M., Shah, G. A., & Iqbal, Z. (2023). Biodegradable polymer encapsulated nickel nanoparticles for slow release urea promotes rhode grass yield and nitrogen recovery. *Journal of Polymers and the Environment*, *31*(5), 1866–1883. DOI: 10.1007/s10924-022-02729-2

Bhat, R., Gupta, M., Choudhary, T. F., & Singh, A. K. (2024). Modelling the effects of climatic variables and political instability on the agricultural sector in South Asian countries. In Ordóñez de Pablos, P. (Ed.), *Building Climate Neutral Economies Through Digital Business and Green Skills* (pp. 1–36). IGI Global Scientific Publishing., DOI: 10.4018/979-8-3693-5673-9.ch001

Biró, K., & Csete, M. S. (2023). How can development strategies foster agri-digitalisation in the era of climate change? a common agriculture policy based consistency analysis in Hungary. *Periodica Polytechnica Social and Management Sciences*, *31*(1), 9–18. DOI: 10.3311/PPso.20411

Boadi, S. A., Olwig, M. F., Asare, R., Bosselmann, A. S., & Owusu, K. (2022). The role of innovation in sustainable cocoa cultivation: Moving beyond mitigation and adaptation. *Climate-Induced Innovation: Mitigation and Adaptation to Climate Change*, pp. 47-80. DOI: DOI: 10.1007/978-3-031-01330-0_3

Chandio, A. A., Amin, A., Khan, I., Rehman, A., & Memon, A. (2024). Can digitalization improve agriculture? Exploring the impact of ICT on grain food production in SAARC countries. *Information Development*, *10*(10), 1–14. DOI: 10.1177/02666669231225945

Chaudhary, A. (2016). Role of FDI in the growth of Indian agricultural sector: A post-reform study. *Global Journal of Finance and Management*, *8*(2), 119–129.

Chel, A., & Kaushik, G. (2011). Renewable energy for sustainable agriculture. *Agronomy for Sustainable Development*, *31*(1), 91–118. DOI: 10.1051/agro/2010029

Chen, Y., & Li, M. (2024). How does the digital transformation of agriculture affect carbon emissions? Evidence from China's provincial panel data. *Humanities & Social Sciences Communications*, *11*(1), 1–18. DOI: 10.1057/s41599-024-03223-x

Choudhary, T. F., & Gupta, M. (2023). Analyzing the long-run and short-run impacts of climate change on wheat and maize yield in the western Himalayan region of India. *Climate Change Economics (Singapore)*, 2350019. DOI: 10.1142/S2010007823500197

Dhungana, B. R., & Ghimire, R. (2013). The role of FDI on agriculture sector in India. In *Conference Paper: Foreign Direct Investment in India's Retail Sector: Issues and Options, National Seminar organized by Department of Business Administration,* Deen Dayal Upadhyaya University, Gorakhpur, India.

Dibbern, T., Ramani, L. A. S., & Massruhá, S. M. F. S. (2024). Main drivers and barriers to the adoption of digital agriculture technologies. *Smart Agricultural Technology*, *8*(1), 1–12. DOI: 10.1016/j.atech.2024.100459

Du, K., Li, P., & Yan, Z. (2019). Do green technology innovations contribute to carbon dioxide emission reduction? Empirical evidence from patent data. *Technological Forecasting and Social Change, 146*(1), 297–303. DOI: 10.1016/j.techfore.2019.06.010

Du, X., Wang, X., & Hatzenbuehler, P. (2022). Digital technology in agriculture: A review of issues, applications and methodologies. *China Agricultural Economic Review, 15*(1), 95–108. DOI: 10.1108/CAER-01-2022-0009

Đuric', I. (2020). *Digital technology and agricultural markets – Background paper for The State of Agricultural Commodity Markets (SOCO) 2020.* FAO., DOI: 10.4060/cb0701en

Eyuboglu, K., & Uzar, U. (2020). Examining the roles of renewable energy consumption and agriculture on CO2 emission in lucky-seven countries. *Environmental Science and Pollution Research International, 27*(36), 45031–45040. DOI: 10.1007/s11356-020-10374-2 PMID: 32772294

Gabriel, A., & Gandorfer, M. (2023). Adoption of digital technologies in agriculture—An inventory in a european small-scale farming region. *Precision Agriculture, 24*(1), 68–91. DOI: 10.1007/s11119-022-09931-1

Geng, W., Liu, L., Zhao, J., Kang, X., & Wang, W. (2024). Digital technologies adoption and economic benefits in agriculture: A mixed methods approach. *Sustainability (Basel), 16*(11), 1–15. DOI: 10.3390/su16114431

Gupta, M., Sharma, B., & Singh, A. K. (2025). Impact of digital technologies on tourist arrival in selected economies. In Lytras, M., Alkhaldi, A., & Ordóñez de Pablos, P. (Eds.), *Harnessing AI, Blockchain, and Cloud Computing for Enhanced e-Government Services* (pp. 441–474). IGI Global Scientific Publishing., DOI: 10.4018/979-8-3693-7678-2.ch014

Ibrahim, I. A., & Truby, J. M. (2023). Farm tech: Regulating the use of digital technologies in the agricultural sector. *Food and Energy Security*, *12*(4), 1–12. DOI: 10.1002/fes3.483

Ingram, J., & Maye, D. (2020). What are the implications of digitalisation for agricultural knowledge? *Frontiers in Sustainable Food Systems*, *4*(1), 1–13. DOI: 10.3389/fsufs.2020.00066

Jang, E.-K., Lim, E. M., Kim, J., Kang, M.-J., Choi, G., & Moon, J. (2023). Risk management of methane reduction clean development mechanism projects in rice paddy fields. *Agronomy (Basel)*, *13*(6), 1–11. DOI: 10.3390/agronomy13061639

Jatav, S. S. (2024). Development of multidimensional food security index for Rajasthan, India: A district-level analysis. *Local Development & Society*, *5*(2), 221–243. https://www.tandfonline.com/doi/abs/10.1080/26883597.2022.2151373. DOI: 10.1080/26883597.2022.2151373

Jatav, S. S., & Kalu, N. (2023). Measuring the agricultural sustainability of India: an application of pressure-state-response (PSR) model. *Regional Sustainability*, 4(3), 218-234. https://www.sciencedirect.com/science/article/pii/S2666660X2300035X

Jyoti, B., Singh, A. K., & Ashraf, S. N. (2023). Appropriate technologies and their implications in the agricultural sector. In Alex Khang, P. H. (Ed.), *Advanced Technologies and AI-Equipped IoT Applications in High-Tech Agriculture* (pp. 65–87). IGI Global., DOI: 10.4018/978-1-6684-9231-4.ch004

Klerkx, L., Jakku, E., & Labarthe, P. (2019). A review of social science on digital agriculture, smart farming and agriculture 4.0: New contributions and a future research agenda. *NJAS Wageningen Journal of Life Sciences*, *90-91*(1), 1–16. DOI: 10.1016/j.njas.2019.100315

Kumar, A., Ahmad, M. M., & Sharma, P. (2017). Influence of climatic and non-climatic factors on sustainable food security in India: A statistical investigation. *International Journal of Sustainable Agricultural Management and Informatics*, *3*(1), 1–30. DOI: 10.1504/IJSAMI.2017.082917

Kumar, S., & Singh, A. K. (2023). Modeling the effects of climate change on agricultural productivity: Evidence from Himachal Pradesh, India. *Asia-Pacific Journal of Regional Science*, *7*(2), 1–28. DOI: 10.1007/s41685-023-00291-w

Li, X., & Gallagher, K. P. (2022). Assessing the climate change exposure of foreign direct investment. *Nature Communications*, *13*(1), 1451. DOI: 10.1038/s41467-022-28975-5 PMID: 35304443

Maharjan, K. L., Singh, M., & Gonzalvo, C. M. (2018). Drivers of environmental conservation agriculture and women farmer empowerment in Namobuddha municipality, Nepal. *Journal of Agriculture and Food Research*, *13*(1), 1–16. DOI: 10.1016/j.jafr.2023.100631

Mekyassi, H., & Kızıldeniz, T. (2024). The role of nutrition-sensitive climate-smart agriculture in ensuring global food security. *BIO Web of Conferences*, 85. DOI: DOI: 10.1051/bioconf/20248501055

Mhlanga, D., & Ndhlovu, E. (2023). Digital technology adoption in the agriculture sector: Challenges and complexities in Africa. *Human Behavior and Emerging Technologies*, *1*(1), 1–10. DOI: 10.1155/2023/6951879

Murshed, M. (2021). LPG consumption and environmental Kuznets curve hypothesis in South Asia: A time-series ARDL analysis with multiple structural breaks. *Environmental Science and Pollution Research International*, *28*(7), 8337–8372. DOI: 10.1007/s11356-020-10701-7 PMID: 33058083

Newman, L., Newel, R., Dring, C., Glaros, A., Fraser, E., Mendly-Zambo, Z., Green, A. G., & Kc, K. B. (2023). Agriculture for the Anthropocene: Novel applications of technology and the future of food. *Food Security*, *15*(3), 613–627. DOI: 10.1007/s12571-023-01356-6

Octavia, D., Suharti, S., Murniati, , Dharmawan, I. W. S., Nugroho, H. Y. S. H., Supriyanto, B., Rohadi, D., Njurumana, G. N., Yeny, I., Hani, A., Mindawati, N., Suratman, , Adalina, Y., Prameswari, D., Hadi, E. E. W., & Ekawati, S. (2022). Mainstreaming smart agroforestry for social forestry implementation to support sustainable development goals in Indonesia: A review. *Sustainability (Basel)*, *14*(15), 93–103. DOI: 10.3390/su14159313

Pazienza, P. (2015). The relationship between CO2 and foreign direct investment in the agriculture and fishing sector of OECD countries: Evidence and policy considerations. *Intelektinė ekonomika*, *9*(1), 55-66.

Purcell, W., Neubauer, T., & Mallinger, K. (2023). Digital twins in agriculture: Challenges and opportunities for environmental sustainability. *Current Opinion in Environmental Sustainability*, *61*(1), 1–12. DOI: 10.1016/j.cosust.2022.101252

Rai, A. (2023). Digitalization in agriculture for innovative transformation in India. *Indian Farming*, *73*(6), 1–13. https://epubs.icar.org.in/index.php/IndFarm/article/view/138320

Raihan, A. (2023). An econometric evaluation of the effects of economic growth, energy use, and agricultural value added on carbon dioxide emissions in Vietnam. *Asia-Pacific Journal of Regional Science*, *7*(3), 665–696. DOI: 10.1007/s41685-023-00278-7

Raihan, A., Begum, R. A., Nizam, M., Said, M., & Pereira, J. J. (2022). Dynamic impacts of energy use, agricultural land expansion, and deforestation on CO2 emissions in Malaysia. *Environmental and Ecological Statistics*, *29*(3), 477–507. DOI: 10.1007/s10651-022-00532-9

Reza, S. K., Chattaraj, S., Mukhopadhyay, S., Daripa, A., Saha, S., & Ray, S. K. (2024). Delineation of high-resolution soil carbon management zones using digital soil mapping: A step towards mitigating climate change in the Northeastern Himalayas, India. *Soil Use and Management*, *40*(1), 1–13. DOI: 10.1111/sum.12995

Rossi, V., Caffi, T., Salotti, I., & Fedele, G. (2023). Sharing decision-making tools for pest management may foster implementation of Integrated Pest Management. *Food Security*, *15*(1), 1459–1474. DOI: 10.1007/s12571-023-01402-3

Šermukšnytė-Alešiūnienė, K., & Melnikienė, R. (2024). The effects of digitalization on the sustainability of small farms. *Sustainability (Basel)*, *16*(10), 1–12. DOI: 10.3390/su16104076

Shrestha, R. K., Jacinthe, P.-A., Lal, R., Lorenz, K., Singh, M. P., Demyan, S. M., Ren, W., & Lindsey, L. E. (2023). Biochar as a negative emission technology: A synthesis of field research on greenhouse gas emissions. *Journal of Environmental Quality*, *52*(4), 769–798. DOI: 10.1002/jeq2.20475 PMID: 36905388

Singh, A. K. (2025). Importance of digital technologies in economic sector of selected countries. In Ahmadi, A., Mna, A., & Slimani, A. (Eds.), *Renewable Energy and the Economic Welfare of Society* (pp. 335–370). IGI Global Scientific Publishing., DOI: 10.4018/979-8-3693-7580-8.ch015

Singh, A. K., Issac, J., & Narayanan, K. G. S. (2019). Measurement of environmental sustainability index and its association with socio-economic indicators in selected Asian economies: An empirical investigation. *International Journal of Environment and Sustainable Development, 18*(1), 57–100. DOI: 10.1504/IJESD.2019.098641

Singh, A. K., & Jyoti, B. (2024). A review for assessing the impact of climate change on the agricultural sector, food security, and pests in India. In Zia Ul Haq, M., & Ali, I. (Eds.), *Revolutionizing Pest Management for Sustainable Agriculture* (pp. 53–84). IGI Global., DOI: 10.4018/979-8-3693-3061-6.ch003

Singh, A. K., Jyoti, B., & Ashraf, S. N. (2025). A theoretical review for assessing the determinants and challenges of agri-startups with reference to appropriate technology in India. In Gupta, V., & Gupta, C. (Eds.), *Navigating Strategic Partnerships for Sustainable Startup Growth* (pp. 159–190). IGI Global Scientific Publishing., DOI: 10.4018/979-8-3693-4066-0.ch007

Singh, A. K., Kumar, S., & Gupta, M. (2024). Relationship between digitalization and environmental sustainability in selected countries. In Yıldırım, S., Yıldırım, D., Demirtaş, I., & Kandpal, V. (Eds.), *Green Transition Impacts on the Economy, Society, and Environment* (pp. 83–107). IGI Global., DOI: 10.4018/979-8-3693-3985-5.ch006

Singh, A. K., Sharma, A. K., & Jyoti, B. (2023). Does economic development have a causal relationship with environmental degradation? Experience from different income group countries. In Sart, G. (Ed.), *Considerations on Education for Economic, Social, and Environmental Sustainability* (pp. 300–333). IGI Global., DOI: 10.4018/978-1-6684-8356-5.ch015

Singh, A. K., & Singh, R. (2025). Digital infrastructure and sustainable development goals in G20 countries. In Popescu, C. (Ed.), *Building Inclusive Global Knowledge Societies for Sustainable Development* (pp. 363–408). IGI Global Scientific Publishing., DOI: 10.4018/979-8-3693-6298-3.ch014

Singh, A. K., & Singh, S. (2023). A case study from Kanpur Nagar, Uttar Pradesh for assessing the economic value of water and its determinants. In Karmaoui, A. (Ed.), *Water-Soil-Plant-Animal Nexus in the Era of Climate Change* (pp. 212–242). IGI Global., DOI: 10.4018/978-1-6684-9838-5.ch010

Singh, A. K., Srinivasan, S., & Jyoti, B. (2025a). Implications of digitalization on public health sector across countries. In Wickramasinghe, N. (Ed.), *Digitalization and the Transformation of the Healthcare Sector* (pp. 359–394). IGI Global Scientific Publishing., DOI: 10.4018/979-8-3693-9641-4.ch013

Singh, G., & Kaur, G. (2021). Digital technologies for smart agriculture. In Tomar, P., & Kaur, G. (Eds.), *Artificial Intelligence and IoT-Based Technologies for Sustainable Farming and Smart Agriculture* (pp. 40–53). IGI Global., DOI: 10.4018/978-1-7998-1722-2.ch004

Singh, R., Singh, A. K., & Tiwari, R. (2025b). Bibliometric analysis for explaining the environmental issues and its mitigation approaches for the sustainable living of people. In Popescu, C. (Ed.), *Building Inclusive Global Knowledge Societies for Sustainable Development* (pp. 325–362). IGI Global Scientific Publishing., DOI: 10.4018/979-8-3693-6298-3.ch013

Smidt, H. J., & Jokonya, O. (2021). Factors affecting digital technology adoption by small-scale farmers in agriculture value chains (AVCs) in South Africa. *Information Technology for Development*, *28*(3), 558–584. DOI: 10.1080/02681102.2021.1975256

Song, Y., Zhang, B., Wang, J., & Kwek, K. (2022). The impact of climate change on China's agricultural green total factor productivity. *Technological Forecasting and Social Change*, *185*, 185. DOI: 10.1016/j.techfore.2022.122054

Torres-Morales, E., Khatiwada, D., Xylia, M., & Johnson, F. X. (2023). Investigating biochar as a net-negative emissions strategy in Colombia: Potentials, costs, and barriers. *Current Research in Environmental Sustainability*, *6*(1), 1–12. DOI: 10.1016/j.crsust.2023.100229

Trendov, N. M., Varas, S., & Zeng, M. (2019). *Digital technologies in agriculture and rural areas*. Food and Agricultural Organization of the United Nations., https://openknowledge.fao.org/server/api/core/bitstreams/885161de-dccf-4589-8376-07fe37b68799/content

Vilkov, A., & Tian, G. (2023). Blockchain's scope and purpose in carbon markets: A systematic literature review. *Sustainability (Basel)*, *15*(11), 1–14. DOI: 10.3390/su15118495

Warsame, Z. A. (2023). The significance of FDI inflow and renewable energy consumption in mitigating environmental degradation in Somalia. *International Journal of Energy Economics and Policy*, *13*(1), 443–453. DOI: 10.32479/ijeep.13943

KEY TERMS AND DEFINITIONS

Agricultural Sector: It is land-based occupation of people who produce food-grain and commercial crops for the consumer.

Agricultural Sustainability: It is defined as promotion of the agricultural sector to increase productivity, income and benefits of farmers, and resource efficiency as increasing the protection of environmental and natural resources.

Appropriate Technologies: These rural or indigenous technologies that meet all criteria of sustainability in a system.

Climate Change: It measures the statistical variability or variation in weather variables like precipitation, temperature, rainfall, etc. from mean in a long period.

Climate change mitigation: It is a scientific mechanism or process to abate CO2 and GHGs emissions from social and economic activities.

Digital Agriculture: It ensures the use of ICT for various purposes like decision making, distribution of inputs, data analysis and food supply in the agricultural sector.

Digitalization: It is a process to adopt and apply digital devices and advanced technologies in production activities to avoid the physical presence of economic agents.

Digital Technologies: These are scientific and mechanical devices that create a digital infrastructure to reduce the physical presence of stakeholders and economic agents.

Environmental Degradation: It measures the natural resource depletion due to human driven activities.

Chapter 2
An IoT-Based Autonomous Weeding Hoe for Smart Farming

Afride Ahsan
https://orcid.org/0009-0005-5160-1660
Rajshahi University of Engineering and Technology, Bangladesh

Atik Tajwar Alve
https://orcid.org/0009-0005-5648-9946
Rajshahi University of Engineering and Technology, Bangladesh

Mhia Md. Zaglul Shahadat
https://orcid.org/0009-0009-1024-3915
Rajshahi University of Engineering and Technology, Bangladesh

ABSTRACT

Weed is a challenge for production and quality of crops in modern agriculture. The effective weed removing solutions are crucial for food security. To address this problem an IoT based automatic moveable weeding hoe for targeted weed removing is depicted in this chapter. Basically it is a smart agriculture firming tool which is incorporated with cutting-edge technologies such as path trajectory, automatic navigation and obstacle detection. The integrated GPS-based IoT guidance in this tool causes precise identification of the weeds. The device receives weed location's coordinates from a drone-based image processing unit. The tool is shaped like a four wheel drive vehicle driven by four individual motors. Two types of blades are used for cutting and rooting of weeds in which solar power is considered for sustainable energy supply. The device is also incorporated with a fertilizer spraying system to support integrated smart farming.

DOI: 10.4018/979-8-3373-0020-7.ch002

Copyright © 2025, IGI Global Scientific Publishing. Copying or distributing in print or electronic forms without written permission of IGI Global Scientific Publishing is prohibited.

1. INTRODUCTION

Agriculture, the backbone of human civilization, is under many challenges in the 21st century. This situation is further compounded by the ever-increasing world population, which is projected to reach 9.5 billion by 2050 (Ehrlich et al., 2015). The growing population demands proportionately increased food production, and most prognosticators for world agricultural output opine that it would need to see productivity increase by about 70% over the coming decades to meet demand (Ehrlich et al., 2015). However, today's farmers increasingly confront a serious challenge in overcoming problems, most based within environmental and economic parameters.

1.1 Agricultural Challenges to a Growing World

Climate Change and Reduced Yield

Climate change is one of the greatest challenges facing agriculture worldwide. Increases in global temperatures are expected to have a negative effect on crop yields, most strongly in already vulnerable regions. Some studies have shown that for every 1°C increase in temperature, crop yields—wheat and other grains in particular—can fall by up to 7% (Zhao et al., 2017). Along with this, unpredictable weather conditions, including prolonged droughts or exceptionally heavy rainfall, are also affecting crop growth and putting extra pressure on farmers to produce adequate amounts of food.

Water Scarcity and Soil Degradation

The scarcity of water is yet another major problem that modern agriculture faces. According to the United Nations, over 2 billion people live in countries that have high water stress, while an estimated 70% of global freshwater withdrawals are used in agriculture (UN-Water, 2020). Water scarcity, combined with unsustainable agricultural practices, has resulted in widespread soil degradation. Erosion, nutrient depletion, and contamination all steal from the resilience of the land to provide for crops.

To counter the challenges posed by water scarcity, farmers are often compelled to use higher quantities of fertilizers to maintain or enhance crop yields in degraded soils. The lack of sufficient water reduces nutrient availability and absorption by plants, necessitating the application of additional fertilizers to compensate for the loss of efficiency. However, this excessive use of fertilizers not only exacerbates environmental concerns such as nutrient runoff, water contamination, and soil degradation but also encourages the growth of more weeds. Weeds often thrive in nutrient-rich conditions, competing with crops for available resources and further reducing productivity. This creates a cyclical challenge for farmers, as the increase in weed growth demands additional weed management efforts, often leading to more herbicide use and labor costs.

To fight these interconnected challenges, farmers have to adopt water-efficient technologies and practices that not only conserve water but also optimize fertilizer use and weed management, ensuring long-term sustainability and resilience of agricultural systems.

Figure 1. Trends and future prediction in global water scarcity in agriculture (Based on FAO)

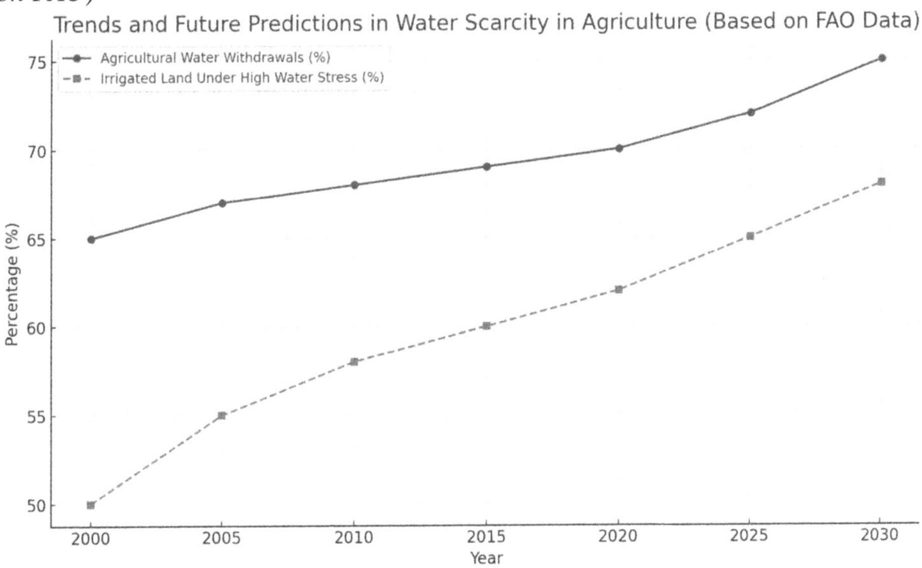

Table 1. Global fertilizer consumption by nutrient (Source: Statista)

Year	Nitrogen(N) (Million Metric Tons)	Phosphorus pentoxide(P_2O_5) (Million Metric Tons)	Potassium oxide (K_2O) (Million Metric Tons)
1965	18.66	16.02	11.63
1970	31.42	20.88	15.76
1975	42.82	25.69	20.67
1980	60.49	31.91	23.83
1985	70.05	33.52	25.15

continued on following page

Table 1. Continued

Year	Nitrogen(N) (Million Metric Tons)	Phosphorus pentoxide(P_2O_5) (Million Metric Tons)	Potassium oxide (K_2O) (Million Metric Tons)
1990	76.78	35.92	24.32
1995	78.22	30.97	20.64
2000	82.06	32.81	22.07
2005	90.64	36.58	26.1
2010	99.56	41.85	28.2
2015	106.38	44.1	34.91
2016	107.3	44.56	36.2
2017	105.74	46.56	37.74
2018	105.26	46.1	37.58
2019	109.14	46.73	37.61
2020	112.45	48.95	40.54
2021	109.25	48.17	38.51
2022	109.17	44.09	34.66

Figure 2. Global fertilizer consumption trends (1965 – 2022)

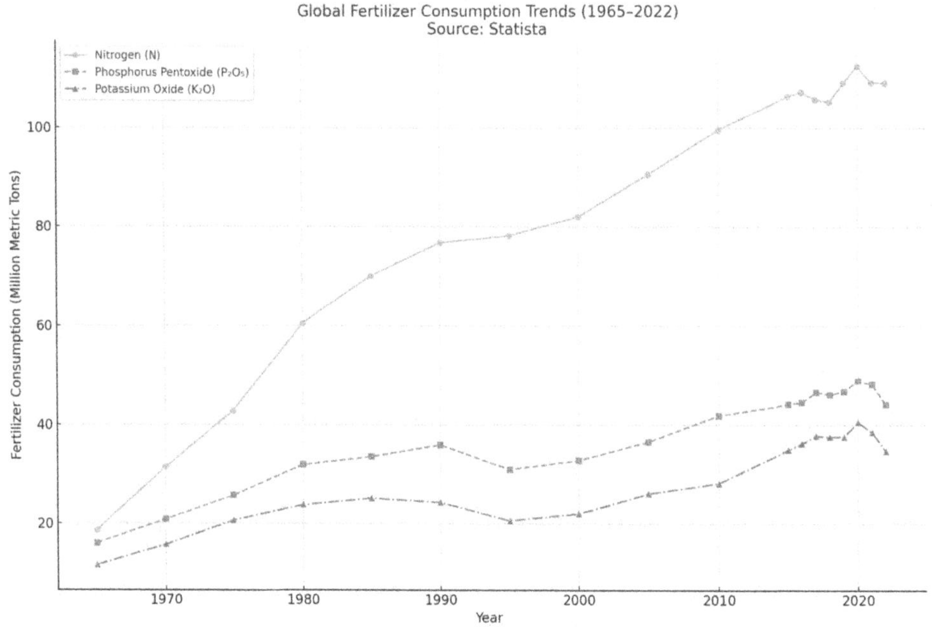

Labor Shortages and Aging Farmers

The mechanization and use of technology in agriculture have led to a massive labor shortage. Many farming communities face an aging workforce, as younger generations prefer industries with better facilities, less physical effort, and structured working hours. Higher-paying urban jobs offering career growth and better living standards have further diverted youth from traditional farming.

This shift has created a significant labor deficit, particularly for tasks like weeding, planting, and harvesting. Large-scale farms, which rely heavily on affordable labor, are especially affected. Rising agricultural labor costs due to inflation, shortages, and competition from other sectors further strain profit margins. In poor harvest seasons or during labor-intensive operations, farmers may even face financial losses.

These challenges demand alternatives like automation and mechanization. Robots and automated systems offer consistent efficiency and reduce reliance on human labor, addressing labor shortages and rising costs. Without such changes, the agricultural sector will face greater pressure, reducing profitability and threatening its sustainability in many regions

Figure 3. Hourly labor cost in USA across time (BLS, Occupational Employment Statistics)

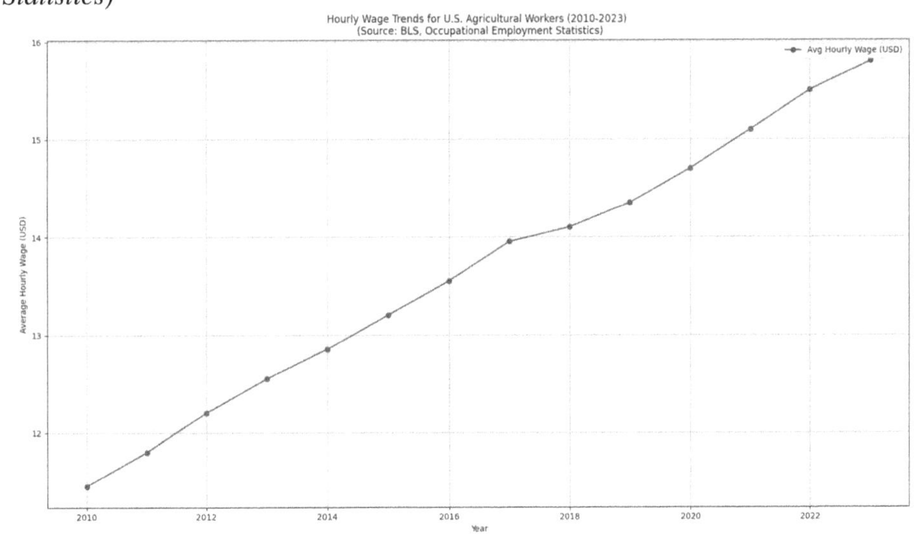

1.2 Sustainable Agriculture and Technological Innovations

In the face of challenges facing the agriculture sector world over, sustainable agriculture is increasingly relevant. Sustainable agriculture implies meeting the needs of the current generation without compromising the ability of future generations to also meet their needs, or at the same time causing significant damage to the environment. It enhances better resources conservation, reduces adverse effects on the environment, and practices efficient farming that ensures long-term fertility and health for the soil.

Technological Innovations as Catalysts for Sustainable Agricultural Advancement

The challenges facing modern agriculture can be overcome by technological innovations. Some of the more advanced technologies in use in precision agriculture include GPS, Internet of Things (IoT) sensors, robotics, and artificial intelligence, which contribute to increased resource utilization. This development is credited with helping farmers make wise decisions concerning precision agriculture, reducing waste and environmental impact, and increasing production.

Automation and Robotics in Agriculture

Automation and robotics can very well fundamentally transform the agricultural sector. Drones, robots, and autonomous tractors are now quite massively used for planting, irrigation, and pest management. These new technologies boost the efficiency of operations, decrease dependence on human labor, and bring about greener and more sustainable ways of farming.

IoT-Driven Farming

The Internet of Things has slowly started changing the face of agriculture and farm management. Various sensors integrated into fields and machinery help in monitoring the soil conditions, temperature, humidity level, and health of the crops in real time. With such information at their disposal, farmers can take very informed decisions with regard to irrigation, fertilization, and pest control in order to reduce costs and increase yields.

1.3 Autonomous Weeding Systems

Weeding is one of the labour-intensive and costly operations in agriculture. Traditional weeding methods, such as manual removal, mechanical tilling, and herbicide application with chemicals, have large drawbacks. Manual weeding is particularly laborious and takes much time and, therefore, is inappropriate for large acreages. Even mechanized weeding tools, such as tractor-mounted weeders, are prone to soil compaction, can damage crops, and often entail excessive operational expenses.

The Issue of Herbicides

Chemical herbicides are applied to a great extent in weed control; however, their application brings about a plethora of environmental and health issues. Over-application of herbicides may lead to soil and water pollution, the development of herbicide-resistant weed populations, and harm to non-target species, such as bees and other pollinators. Additionally, the increased resistance of weeds to herbicides will further increase the problems and financial costs to farmers in trying to control the weeds.

Table 2. Global herbicide consumption (FAO from 1990 -2022) (source Statista)

Year	Herbicide Consumption (Million Metric Tons)
1990	0.7
1991	0.7
1992	0.72
1993	0.71
1994	0.75
1995	0.76
1996	0.82
1997	0.84
1998	0.86
1999	0.87
2000	0.88
2001	0.93
2002	0.93
2003	1.01
2004	1.04
2005	1.09

continued on following page

Table 2. Continued

Year	Herbicide Consumption (Million Metric Tons)
2006	1.12
2007	1.27
2008	1.32
2009	1.25
2010	1.43
2011	1.5
2012	1.61
2013	1.59
2014	1.57
2015	1.59
2016	1.65
2017	1.64
2018	1.63
2019	1.81
2020	1.86
2021	1.82
2022	1.94

Figure 4. Global herbicide consumption trends

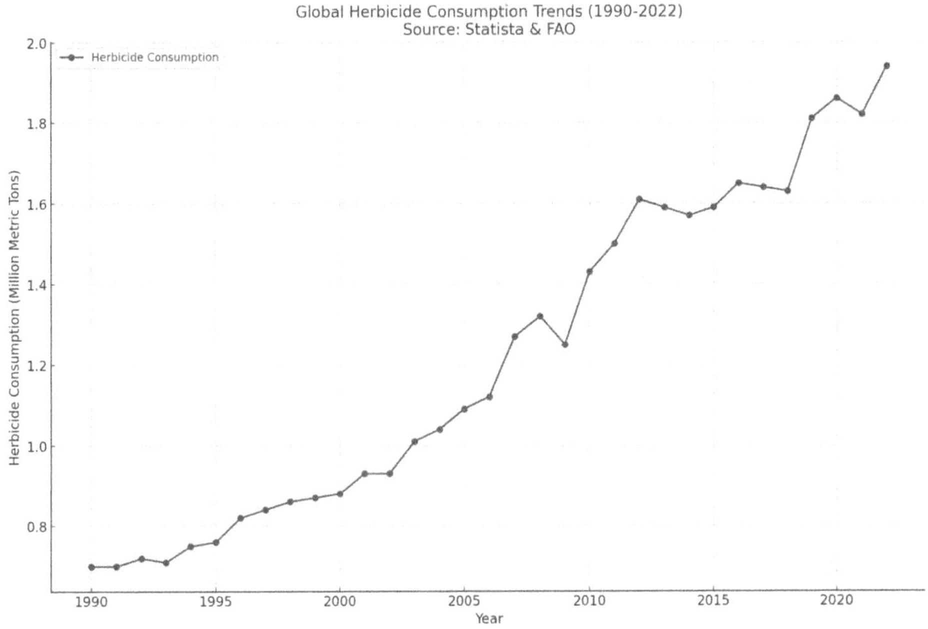

The Potential of Autonomous Weeding Systems

Autonomous weeding systems are a promising solution to the challenges presented by weed management. Autonomous weeding systems can be integrated with GPS technology, ultrasonic sensors, and robotics to execute targeted weed removal with a high degree of precision. Operating autonomously, they not only reduce the need for human labor but also lessen environmental impacts. Moreover, these systems could be equipped with advanced AI algorithms that allow for real-time detection and classification of weeds, proving highly effective in distinguishing between crops and unwanted plants.

1.4 Scope of the Research

The main objective of this research project is the design and development of an IoT-based autonomous movable weeding hoe. The key objective is to develop a system that can integrate GPS navigation, ultrasonic sensors, solar power, and IoT technology, which will provide an efficient and sustainable solution for the control of weeds in wheat farming.

Features of the System

GPS Navigation: Employing GPS technology, the precise location is tracked so that the weeding hoe remains within boundaries and follows the most optimized path for maximum efficiency.
Ultrasonic Sensors: The system uses sensors to detect obstacles in its path and avoid them; thus, the hoe safely works without damaging the crop or any other structures.
Solar Power: The system is designed to work with solar power, an integral part of the system. Inclusion of solar panels reduces dependence on fossil fuels, increasing its ecological sustainability.
IoT Integration: With the introduction of IoT technology, the system can be remotely monitored and controlled. This means that farmers can monitor hoe performance in real time, adjust settings, and even get updates on its operation status.

The contribution this study makes to the emerging domain of precision agriculture is proposing an economic and environmentally sustainable solution to what has been a notably labor-intensive, time-consuming, and difficult task in farming: weed control.

1.5 Significance of the Study

The relevance of the research is pegged on attempts to solve the problems that have increasingly afflicted agriculture in recent years. The system, which includes an autonomous weeding hoe with integrated GPS, IoT technology, and solar power, boasts of a number of advantages, as discussed below:

Labor Efficiency: Automation in weeding reduces labor needs, especially at a time when agricultural industries in general are faced with labor shortages.
Sustainability: The solar-powered design reduces the dependence on non-renewable energy sources and lessens the environmental impact of farming practices.
Precision and Cost-Effectiveness: The system is able to target weeds with precision, avoiding crops, thus optimizing resource utilization and reducing environmental harm from chemical herbicides.
Scalability: The system has been designed in a modular way to easily adapt to different types of agricultural operations, hence becoming an ideal solution for farmers around the world.

2. BACKGROUND ANALYSIS

2.1. Current Methods of Weed and Grass Management

Weed management is one of the key aspects of agricultural practices that ensure maximum yields from crops. Weed control becomes very crucial in order to reduce the competition between crops and weeds for nutrients, light, and space. Traditionally, farmers have employed several methods of weed and grass management. Each of these methods, however, presents its own set of challenges and limitations.

Manual Weeding: Manual weeding remains a time-tested, although labor-intensive, method of controlling weeds. It involves the physical removal of weeds from the field by hand. This is generally done in small-scale, organic, or specialty farms where the use of chemical herbicides is avoided or prohibited. While effective in certain contexts, manual weeding is highly labor-intensive and impractical for large-scale operations. It involves high human resources, increasing the operational cost, and could result in inconsistent output because of either human error or fatigue.

Moreover, manual weeding is highly dependent on weather conditions. In bad weather, such as a heavy downpour or intense heat, the concerned workers are usually not in a position to do the job effectively, resulting in time wastage and decreased productivity. Besides, manual labor is increasingly difficult to come by as labor costs rise and young people tend to move away from agricultural activities. Thus, this method is limited in scalability and efficiency for large-scale farms.

Mechanized Tools: To address the inefficiency of manual labor, mechanized tools for weed control have been developed. Tractors and other systems with attachments like cultivators, harrows, and rotary hoes effectively kill weeds by disrupting their growth without harming crops. These tools are significantly faster than manual weeding, reducing labor costs and increasing productivity.

However, mechanized weeders have drawbacks. Heavy machinery compacts the soil, reducing water and nutrient absorption and harming crop growth. Additionally, they lack precision and can damage crops if not calibrated properly. This inaccuracy often leads to poor weed control and unintentional crop destruction, making them less suitable for small-scale or precision farming.

Chemical Herbicides: Chemical herbicides have been in wide use to control weeds, especially in large-scale commercial agriculture. Herbicides like glyphosate and 2,4-D kill a wide range of weeds with little effort and often with great labor savings. Application of herbicides allows the farmer to cover a lot of ground rapidly and reduces labor costs for weeding, ensuring timely control.

There are, however, several serious disadvantages associated with chemical herbicides. Overuse of herbicides has led to herbicide resistance: the evolution of weeds to survive the applied chemicals, rendering them ineffective. The resistance of the weeds makes their control not only more difficult but also increases the cost of herbicide treatments with stronger, more expensive chemicals. Further, herbicides can contaminate soil and water, potentially leading to harm to the environment and non-target organisms, including beneficial insects such as pollinators. Hence, the wide use of herbicides comes with concerns regarding environmental sustainability and the long-term health of agricultural ecosystems.

While herbicides provide effective weed control, their environmental impact, along with increased resistance to them, has compelled many to turn to alternative methods, which have to be effective yet environmentally friendly.

2.2 Technological Innovations in Precision Agriculture

Precision agriculture is the modern farming management concept, using information technology, to ensure that crops and soil get the right amount of nutrients, water, and care at the right time. It aims at optimizing field-level management regarding crop farming. Key technologies driving precision agriculture include GPS, IoT sensors, unmanned aerial vehicles, and autonomous machinery.

GPS-Based Navigation: Modern farming has been revolutionized by GPS technology, giving precise location data that allows efficient field management. The integration of GPS modules in farming equipment enables farmers to track exactly the movement of their machinery, optimize field coverage, and reduce overlapping in various farm tasks such as seeding, fertilizing, and irrigation. GPS-based systems also provide useful data for farmers regarding field boundaries, topography, and more specific problem areas such as pest outbreaks or waterlogging.

IoT Sensors and Data Analytics: The Internet of Things has found important applications in precision agriculture. IoT sensors may be placed in the field, the machinery, and even the crops to provide real-time data on environmental and crop conditions. In this regard, soil moisture sensors, temperature sensors, and light sensors send data to cloud platforms, which allow farmers to monitor in real time the health of their crops, soil conditions, and weather patterns. This will enable data-driven decisions to optimize water use, fertilizer application, and pest control. This helps the farmer reduce waste, minimize environmental impacts, and improve crop yields.

Unmanned Aerial Vehicles: Drones have been quite popular in precision agriculture due to their ability to offer farmers high-resolution aerial imagery of their fields. These drones have cameras and sensors that contribute to gathering information on crop health, water stress, and pest infestation. Analysis of the images and data captured enables farmers to make better-informed decisions regarding field management, such as the need to reduce chemical treatments and optimally schedule irrigation. UAVs also perform planting, pesticide application, and fertilizer spreading, hence automating most routine activities while improving efficiency.

The development of autonomous machines, such as tractors, harvesters, and weeders, has been among the most exciting developments in precision agriculture. Fitted with GPS and many sensors and AI, these machines are endowed with the ability to work on their own without any human interventions. They can carry out repetitive tasks with high precision and efficiency, as in plowing, seeding, and weeding. Of those, the autonomous weeders really have the potential to revolutionize weed control with a precise and sustainable alternative to herbicides. They can identify weeds using sensors, remove them precisely without damaging crops, and all of this is done with minimum human involvement.

Solar Power Integration: Solar energy is part of the equation in the effort to bring sustainable agriculture practices into play. Off-grid Agro-systems can be achieved through solar power, lessening dependence on fossil fuels while simultaneously increasing environmental sustainability. In this regard, self-sufficient solar systems containing autonomous tractors and weeders work continually in the fields, neither needing refuelling nor necessitating frequent maintenance. It thereby reduces operational costs and partly lessens the adverse environmental impact caused by traditional farm machinery.

2.3 Existing Autonomous Systems

Development of autonomous systems in agriculture is an area of active research and innovation. The number of systems introduced to automate various tasks, such as weeding, seeding, irrigation, and harvesting, is quite a few. These systems use GPS, sensors, and AI to enhance efficiency and precision in farming operations.

Robotic Weed Control Systems: A few robotic systems have been developed to autonomously detect and remove weeds. Most of these robots will use a combination of visual sensors, such as cameras, together with machine learning algorithms, to be able to distinguish crops from weeds. For example, the "Tertill" weeding robot is based on a small, solar-powered platform mounted with sensors that can autonomously move along garden rows, detecting and cutting weeds. French startup "Naïo Technologies" has come up with similar autonomous robots, including Oz, designed to weed fields using precision mechanical tools. These robots can reduce herbicide use, lower labor costs, and provide more efficient weed control.

While highly promising, these robotic systems still face challenges in terms of scalability and adaptability. Large fields bring variations in topography, crop growth stages, and environmental conditions that can be difficult for an autonomous weeder to navigate with precision. More importantly, the systems come with high up-front investment costs, which may prove a barrier to adoption for small-scale farmers.

Drones for Weed Control and Monitoring: Primarily, drones have been used in monitoring crop health, pest infestations, and water stress. However, a few companies are looking at the possibility of using drones for precision spraying of herbicides or direct weed management. The drones can be equipped with herbicide-spraying systems and can precisely target weeds, which will allow the reduction in broad-spectrum chemical applications.

Autonomous Tractors and Farm Vehicles: Similarly, it is in the integration of GPS, sensors, and AI that tractors and other farm vehicles become autonomous. The result is that the vehicles can automatically plow, seed, and harvest fields. John Deere has introduced an autonomous tractor, using GPS and sensors to map field layouts for efficient and precised operations. These tractors can be remotely controlled and monitored, and they also reduce the requirement of human operators, therefore putting an end to farming community labor shortages.

2.4 Literature Review

Recent advancements in IoT technology have revolutionized weed control in agriculture, offering solutions that incorporate smart sensors, autonomous systems, and data analytics. The table below reviews key studies on IoT-based weed management, focusing on the technologies used, environmental impacts, energy consumption, and scalability. These studies emphasize the need for efficient, sustainable, and precision-based solutions for weed removal.

Table 3. Literature review

Year	Title and Author(s)	Technology Used	Weed Control Technology	Environmental Impact	Study Contributions
2024	"SPARROW: Smart Precision Agriculture Robot for Ridding of Weeds" **Xie et al.**	Vision-based autonomous navigation with IoT integration	Precision spot spraying	Reduces herbicide usage and environmental contamination	cost-effective system for autonomous herbicide application.
2023	"IoT Based Weed Detection and Removal in Precision Agriculture" **Zhang et al.**	IoT sensors and automated removal systems	Real-time weed detection and targeted removal	Reduces chemical use	IoT-based approach for precise weed management
2021	"A Device for Effective Weed Removal for Smart Agriculture Using Convolutional Neural Network" **Kumar et al.**	CNN-based image recognition integrated with IoT	Real-time weed identification and removal	Minimal environmental impact through chemical-free weed removal	CNN and IoT for precise and eco-friendly weed control.
2020	"Automatic Weed Removing Robot Using IoT Technology" **Sharma et al.**	IoT-enabled robotic system with IR sensors	Automated weed removal,	Reduces environmental harm by minimizing chemical input	Reducing manual labor and chemical inputs
2020	"Automated Computer Vision Based Weed Removal Bot" **Siddiqui et al.**	Computer vision and IoT for real-time weed detection	Targeted weed removal without herbicides	Promotes sustainability by avoiding harmful chemicals	Develops a mobile model for real-time, chemical-free weed management.
2019	"The impact of mechanized weed control in large-scale farming" **Moll, J., et al.**	Mechanized weed control	Mechanized weed removal system	Chemical -Free weeding	Demonstrate efficiency in large-scale firming

2.5 Addressing Research Gaps

While there have been significant advancements in autonomous agricultural systems, many existing solutions still face challenges in scalability, adaptability, and precision. Some rely solely on mechanical weeding, which, while reducing chemical dependency, often struggles with adaptability across different terrains. Others focus on drone-based spraying, improving efficiency but still depending on herbicides, raising environmental concerns. IoT-based monitoring solutions provide valuable data but often lack direct weed removal capabilities, limiting their practical effectiveness in real-time operations. Additionally, high costs remain a barrier, restricting accessibility for small- and medium-scale farmers.

This study addresses these challenges by introducing an IoT-based autonomous weeding hoe that integrates solar power, GPS navigation, and ultrasonic sensors into a scalable, cost-effective, and multifunctional system. Unlike conventional methods, this system performs mechanical weed cutting, rooting, targeted spraying, seeding, and fertilizer spreading, optimizing each task using GPS-based location technology. By minimizing chemical use and operational costs, it enhances precision, sustainability, and efficiency, providing a practical, hands-free solution for modern farming.

A key innovation of this approach is its IoT-based control and monitoring system, enabling real-time data collection, remote operation, and intelligent decision-making. Through drone-assisted views and live video feeds, farmers can optimize weed control, targeted spraying, and seeding operations with greater accuracy and minimal resource wastage. By offering a sustainable, adaptable, and economically viable alternative, this system advances autonomous farming technologies, making them more accessible and effective across diverse agricultural landscapes.

3. HARDWARE

The IoT-based autonomous weeding hoe is integrated with several hardware components combined to achieve an efficient, precise, and autonomous operation. Detailed design of the key components used in the system are given below.

3.1 Components and Specifications

Microcontroller – Arduino Mega: The microcontroller is considered as the brain of the system; hence, it does process on various sensor data to take control of the system's action. The Arduino Mega is selected for this research owing to ample processing power and I/O. An Arduino Mega has the capability of handling multiple sensor inputs, motor controls, and GPS data streams. The Arduino IDE-a widely used, open-source platform-is utilized to program the device. It enables effortless coding and integration of different sensors and components.

NEO-7M-GPS Module: NEO-7M GPS will serve as the positioning device that has equipped the self-navigated hoe with the requisite information for navigations that have had high-precision accuracy. It is interfaced with numerous GPS satellites to hence receive signals with typical position inaccuracy less than 2.5 meters. This is what navigation through the application of Global

Positioning Systems should look into within the whole process in relation to field mapping out the paths effectively for removals.

Ultrasonic Sensors: Ultrasonic sensors are employed for the purpose of obstacle detection and avoidance. These sensors emit sound waves and measure the time it takes for the waves to bounce back after hitting an object. After analyzing this information, the system will determine whether there are obstacles in its path and shall make necessary corrections to avoid collision. The use of ultrasonic sensors will be critical in ensuring the safety of the system and avoiding crop damage or other structure damages.

High-Torque Motors: The system is embedded with high-torque motors for wheeling and controlling hoe movement. With this, the power produced by such kinds of motors will enable navigation through fields of varying nature. The microcontroller controls them by changing speed and direction inputs provided by the GPS and sensors.

Solar Panels: Solar panels supply energy to the system's electronics and motors. The panels are lightweight and efficient to provide enough energy that keeps the weeding hoe working, without any additional fuel supply. Solar power reduces the environmental impact of the system and makes it suitable for continuous use in the field.

It does this by providing the possibility of wireless communication using Wi-Fi and GSM modules. The Wi-Fi module can connect the system to the local network for remote monitoring and control. The GSM module allows it to communicate with the farmer's mobile for the status of the hoe, the life of the battery, and its performance. This in turn permits the farmer to intervene remotely, if necessary, or to change some parameters of the system to optimize its functioning.

3.2 Design Considerations:

The design of the IoT-based autonomous weeding hoe represents a delicate balance between functionality, durability, and cost-effectiveness. A number of key factors have been considered in designing the system to perform optimally in a variety of agricultural environments with minimal environmental impact and ease of use.

Chassis Design: The chassis of the weeding hoe needs to be designed with strong enough material to withstand the harsh conditions at the agricultural field, such as dirt and water, and any physical impacts from obstacles such as rocks and uneven ground. Simultaneously, it has to be lightweight enough so that the system can easily move in a field with less energy consumption.

Figure 5. Weed removing Smart Bot

It includes using high strength and lightweight material construction on the chassis to suit this purpose, hence using aluminium alloy materials or other advanced composites that can stand structurally strong and light in weight, at the same time keeping in view the totality of system weight. These would be structured with a view for modularity for easy changeover or replacement/upgrade individually with wheels, sensors, blades, etcetera.

Modular Attachments: The system is designed to be modular, allowing farmers to customize the robot for specific agricultural needs. Various agricultural environments and crop types may require different types of blades or attachments for optimal weed control. Several interchangeable attachments can be fitted on the weeding hoe, including:

- Rotary Blades: These are highly effective for cutting weeds in rows of crops, ensuring precision while maintaining the integrity of the surrounding crops.
- Tine Cultivators: These are used for shallow tilling of the soil, which helps control weed growth at the root level while also aerating the soil for better crop growth.

In addition to these weeding attachments, the robot also supports easy attachment of payload carriers to enhance its multifunctionality. Two dedicated payload carriers are designed to streamline specific operations:

1. Spraying Payload Carrier: This payload is specifically designed for spraying pesticides and herbicides. The robot uses geo-location data to target specific areas, ensuring precise application and reducing chemical wastage. This carrier

eliminates the need for manual labor in spraying operations, minimizing human exposure to harmful chemicals.

2. Seeding and Fertilizer Spreading Payload Carrier: The second payload is designed for seeding and spreading fertilizers uniformly across the field. This ensures consistent crop growth and nutrient application without requiring additional mechanical arms or complex mechanisms. The robot uses its GPS-guided navigation system to maintain precision, even in large or uneven fields.

Figure 6. Payloads

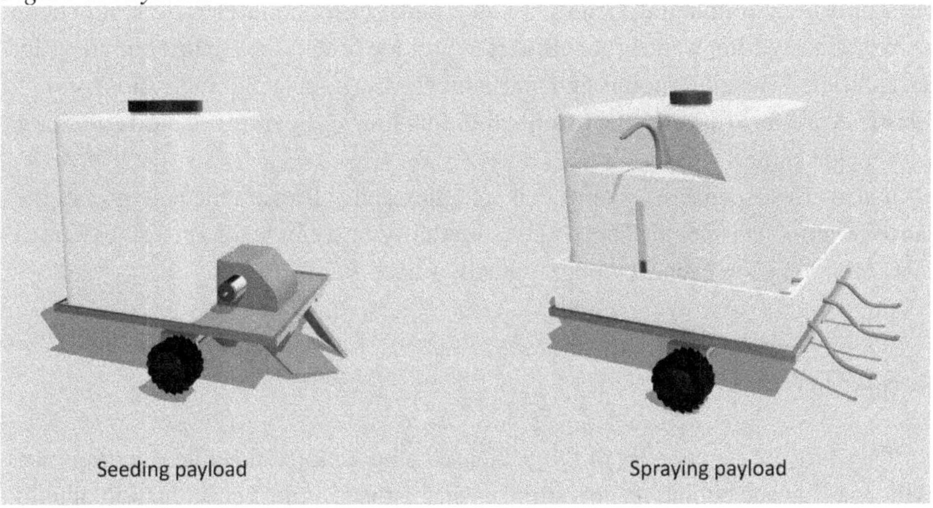

The attachment and detachment process for these payloads is straightforward, requiring minimal time and effort. This ease of use allows farmers to quickly switch between operations, such as transitioning from weeding to spraying or seeding, without disrupting the workflow.

By adopting this modular approach, farmers can tailor the robot to meet their specific needs, enabling greater operational versatility and extending the system's functional lifetime. The simplicity of switching attachments, coupled with the robot's ability to handle multiple tasks efficiently, ensures that it remains a cost-effective and reliable solution for modern farming challenges.

Power Management: The other critical aspect of the design is power management. The autonomous weeding hoe is solar-powered, ensuring that it can operate continuously throughout the day without needing to rely on external power sources. Solar panels are mounted on the top of the chassis to capture sunlight effectively, and the energy is stored in high-capacity batteries, typically lithium-ion, which are capable of holding a significant charge for extended operation.

All functions relating to energy conversion, charging and discharging control make efficient usage of power generation ensured by the Power Management System. In this instance, overcharging has been prevented by avoiding damage to the battery by insuring adequate reserve power even for cloudy weather, and making more time usefully available on non-sunshine days or continuously in darkness. An independent power generator may be installed and connected-such as wind turbines or grid electricity-to extend hours of operational duties.

Environmental Resistance: The autonomous weeding hoe was designed to resist most environmental factors. To that end, sensors and electric components are contained within waterproof casings to provide protection against humidity and dust particles. Besides, motors and mechanical parts selected for the system provide resistance against rust and corrosion, providing long-life service in hard conditions within agricultural activity.

It also has suspension systems that are adjustable, allowing the hoe to smoothly move over uneven terrain. This makes sure the system can maintain a stable movement and precision even in topographically varied fields.

4. SOFTWARE

The software component of the IoT-based autonomous movable weeding robot is the intelligence behind its operation, ensuring the seamless coordination of hardware, data processing, and task execution. It integrates advanced technologies such as real-time sensor data processing, navigation algorithms, IoT connectivity, and remote monitoring to deliver a robust and efficient system. This section elaborates on the programming frameworks and IoT integration that empower the robot, while also addressing the key algorithms that drive its autonomous capabilities.

Programming Frameworks

The software architecture of the robot is built on a combination of high-level and low-level programming frameworks, enabling precise control, real-time decision-making, and user-friendly interaction.

Python for Data Processing and AI Integration

Python is the backbone of the robot's high-level operations, particularly for data processing, IoT integration, and advanced decision-making algorithms. Its flexibility and extensive libraries make it ideal for implementing machine learning models and managing large datasets.

Machine Learning Frameworks

- o TensorFlow and PyTorch are employed for tasks such as weed detection and classification. These frameworks facilitate the development of deep convolutional neural networks (CNNs) trained on diverse datasets to differentiate between crops and weeds in real-time.
- o TensorFlow models focus on recognizing crop patterns under various lighting and environmental conditions, ensuring reliability across diverse field scenarios (Zhang et al., 2020).
- o PyTorch enables reinforcement learning, allowing the robot to refine its weed detection and removal strategies based on feedback from previous actions.

Real-Time Image Processing

- o The robot uses OpenCV, a robust library for real-time image processing, to analyze visual data captured by cameras. This analysis enables the system to dynamically identify weeds and execute targeted removal.

Arduino IDE for Microcontroller Programming

The Arduino IDE is used for programming the robot's microcontroller, specifically the Arduino Mega 2560. Written in C++, the software manages low-level operations such as sensor integration, motor control, and task execution.

Sensor and Actuator Management

- o The IDE handles inputs from the GPS module, ultrasonic sensors, and motor drivers. For instance, ultrasonic sensor data is processed to detect obstacles, while GPS data guides navigation along predefined routes.

Task Execution Logic

- o The Arduino Mega processes predefined instructions, such as activating cutting blades or initiating spraying tasks, based on GPS coordinates or user commands.

Integration of TensorFlow and PyTorch

The decision-making capabilities of the robot are significantly enhanced by deep learning and reinforcement learning models developed using TensorFlow and PyTorch:

Deep Learning for Weed Detection

- o CNNs classify weeds and crops with high precision, leveraging large datasets of annotated images.
- o The models are optimized to operate in real-time, enabling the robot to adapt to varying weed species and field conditions (Holt et al., 2017).

Reinforcement Learning for Continuous Improvement

- o PyTorch models use feedback from previous operations to refine the decision-making process. For example, if a weed is not successfully removed, the model adjusts its parameters to improve future performance.

IoT Integration

IoT integration connects the robot to a broader network of devices, enabling remote monitoring, real-time data transmission, and seamless interaction with users.

IoT Architecture

The IoT architecture of the robot consists of three layers:

1. Data Collection at the Robot:
   - o The robot collects data from GPS modules, ultrasonic sensors, and task execution tools.
   - o This data is sent to IoT gateways for aggregation and analysis.
2. IoT Gateway:
   - o Devices like ESP32 act as intermediaries, facilitating communication between the robot and the central command system.

- o The MQTT protocol ensures efficient and secure data transmission, enabling real-time feedback and control.

3. Central Command System:
   - o The central command processes incoming data, generates actionable insights, and sends commands back to the robot.

Remote Monitoring and Control

IoT connectivity enables farmers to monitor and control the robot remotely via web or mobile applications. These interfaces display real-time updates on the robot's performance, including:

- GPS Location: Visual tracking of the robot's position in the field.
- System Diagnostics: Monitoring sensor functionality, motor performance, and battery levels.
- Task Updates: Progress reports on operations such as weed removal, spraying, or seeding.

Farmers can also intervene manually, issuing commands to stop operations, adjust navigation paths, or prioritize specific tasks.

Figure 7. IoT data integration workflow

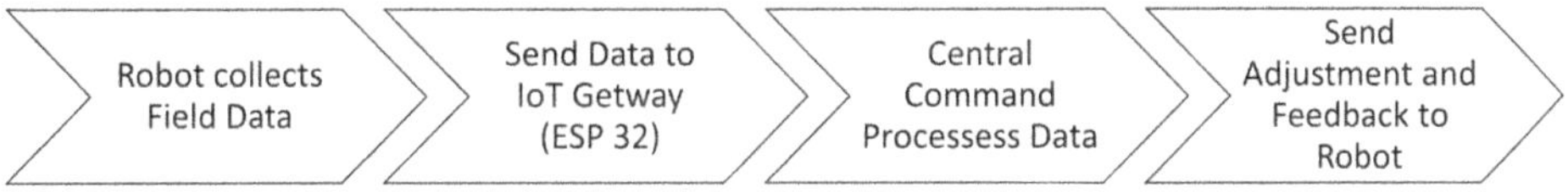

5. ALGORITHM AND DESIGN

The effectiveness of the IoT-based autonomous movable weeding robot hinges on the algorithms that govern its navigation and task execution.

Navigation Algorithm

The navigation algorithm ensures that the robot traverses the field efficiently while avoiding obstacles and maintaining precision.

- Path Planning: The robot calculates its route using GPS coordinates to map field boundaries. Advanced algorithms like A* optimize this path, ensuring full field coverage with minimal overlaps or missed zones.
- Obstacle Avoidance: The robot relies on ultrasonic sensors to detect obstacles. When an obstacle is identified, the algorithm recalibrates the robot's trajectory in real-time to avoid collisions.

Figure 8. Navigation algorithm workflow

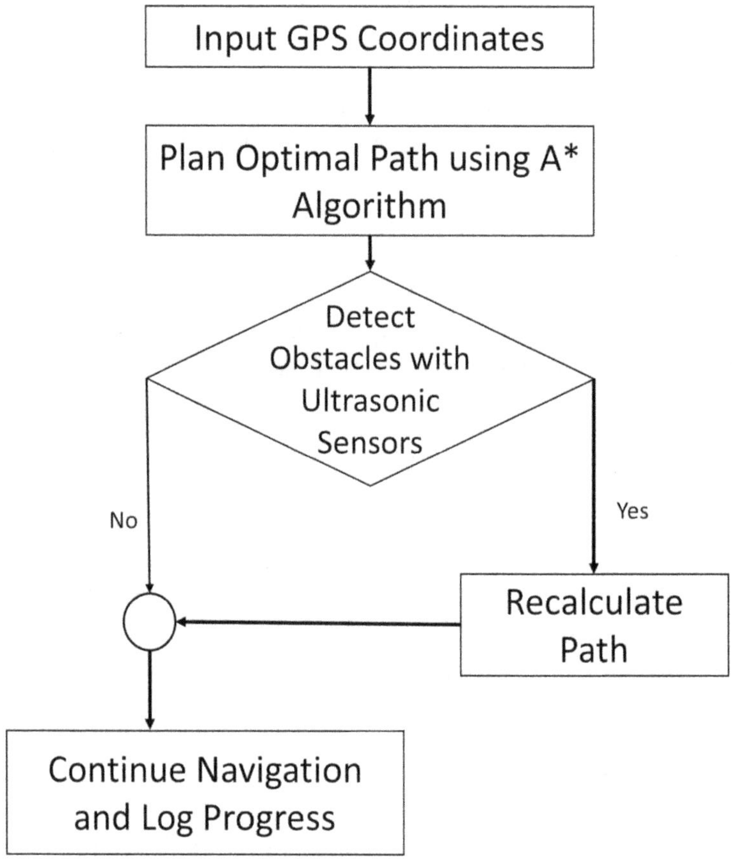

Task Execution

Task execution involves performing operations such as weed removal, spraying, and seeding with precision and efficiency.

Weed Detection and Removal

The robot uses real-time visual data from any drone video feed, processed through machine learning models, to identify and classify weeds. Convolutional Neural Networks (CNNs) analyze images captured by the onboard cameras, distinguishing weeds from crops with high accuracy. Once weeds are detected, the robot activates its cutting or rooting blades based on the task requirements. These blades operate with precision, ensuring weeds are removed without damaging nearby crops.

In addition to visual detection, the robot can operate using predefined GPS coordinates to navigate and target specific zones for weed removal. This dual approach enhances the system's flexibility and adaptability to various operational needs.

Spraying and Seeding

The spraying and seeding mechanisms are guided by GPS data to ensure targeted application and uniform distribution. Spraying tasks may involve water, herbicides, or fertilizers, depending on the operational requirements. Similarly, the seeding mechanism ensures optimal placement of seeds across the field, promoting uniform crop growth. Both systems adjust their output dynamically based on real-time data, optimizing resource usage and minimizing wastage.

Figure 9. Task execution workflow flowchart

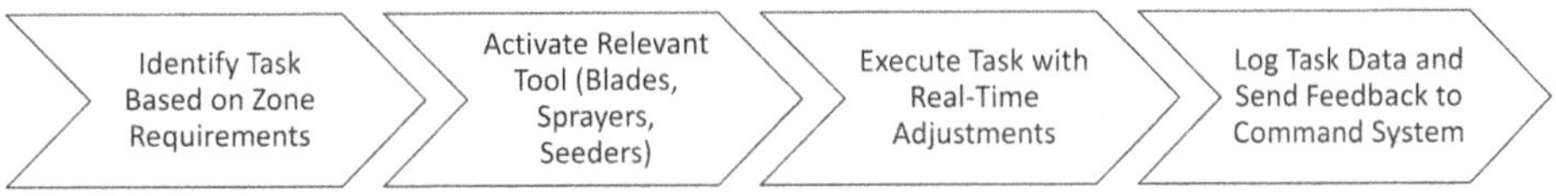

GPS-Driven Zigzag Movement

The robot is capable of operating autonomously with GPS coordinates provided either via IoT-based input or by pressing the GPS Save Button. This enables it to perform systematic field coverage using a zigzag movement pattern:

1. Defining the Operational Zone:
    - GPS values of the field's corners are input, helping the robot define the area for operation. These values act as boundaries for the robot's movement.
2. Navigating in a Zigzag Pattern:

- Using the defined GPS boundaries, the robot navigates in a zigzag pattern, avoiding rows of crops with the help of ultrasonic sensors. As it reaches the edge of a row, the robot calculates the next turn based on its position and predefined logic:
  - Right Turn: The robot transitions to the next row at the end of a leftward path.
  - Left Turn: At the end of a rightward path, the robot transitions to the next row, continuing the zigzag motion.

3. Performing Operations:
   - Based on user-defined instructions or detected needs, the robot executes tasks such as cutting, rooting, spraying, or seeding along its path.
4. Returning to the Home Base:
   - Once the operation is complete, the robot calculates the optimal path to return to its home base using GPS data. Upon arrival, it initiates a charging process, leveraging a magnetic or wireless charging system.

Emergency Power and Location Reporting

A solar panel mounted on the robot serves as an emergency power backup. In scenarios where the battery level is critically low, the robot uses solar energy to maintain basic functions, such as navigation and communication. Simultaneously, the robot transmits its current location to the base station via IoT. This ensures that the robot can:

- Return to Home Base: Navigate back to the charging station before complete power depletion.
- Alert Operators: Notify users about its low power status and operational progress.

This feature enhances the robot's autonomy and reliability, ensuring uninterrupted operations even in challenging conditions.

The task execution capabilities of the IoT-based autonomous movable weeding robot combine precision, adaptability, and resource efficiency. By leveraging GPS-driven zigzag movement, ultrasonic sensors, and IoT connectivity, the robot ensures comprehensive field coverage and targeted task execution. The integration of emergency power via solar panels and real-time location reporting further enhances its resilience and operational reliability. These features make the robot an invaluable tool in precision agriculture, addressing key challenges such as labor shortages, resource inefficiency, and environmental sustainability.

6. IMPLEMENTATION

The implementation of the IoT-based autonomous movable weeding robot involves transitioning the theoretical design and engineering principles into a fully functional, real-world system. This phase is critical in validating the effectiveness of the robot's hardware and software while ensuring its adaptability to diverse agricultural environments. By conducting extensive field testing and addressing practical challenges, this implementation establishes the robot as a reliable and efficient solution for modern farming.

Figure 10. Complete robot set with payloads

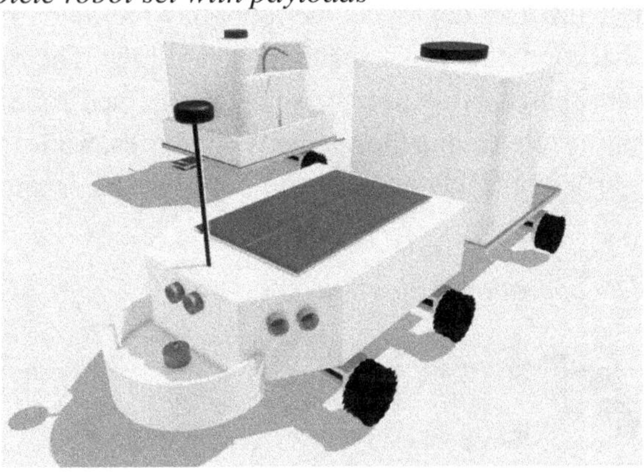

6.1 Field Testing

Field testing is the cornerstone of the implementation process, allowing the evaluation of the robot's performance under real-world conditions. The primary objective of this phase is to validate the hardware and software components while identifying and addressing any limitations.

Preparation for Testing

Field testing begins with a detailed setup to ensure the robot operates optimally. A representative agricultural field is selected with diverse conditions, including uneven terrain, high weed densities, and common obstacles like irrigation systems, rocks, and uneven soil. This diversity allows for a comprehensive evaluation of the robot's capabilities.

Calibration is a critical step. Ultrasonic sensors are adjusted for accurate obstacle detection, the GPS module is tested for precision, and cutting blades and sprayers are fine-tuned for efficiency. Each subsystem is validated for compatibility and smooth operation. GPS coordinates of the test field boundaries are loaded using either the GPS Save Button or the IoT-based central command. Tasks such as weed removal and spraying are configured based on the specific requirements of the test field.

Execution of Testing

The testing phase involves deploying the robot into the field to evaluate its navigation, obstacle detection, and task execution capabilities. The robot's navigation is assessed by observing its ability to traverse the predefined zones accurately. The GPS module guides the robot along the mapped boundaries, while the path-planning algorithm ensures efficient field coverage without overlaps or missed areas.

Figure 11. Complete operational diagram

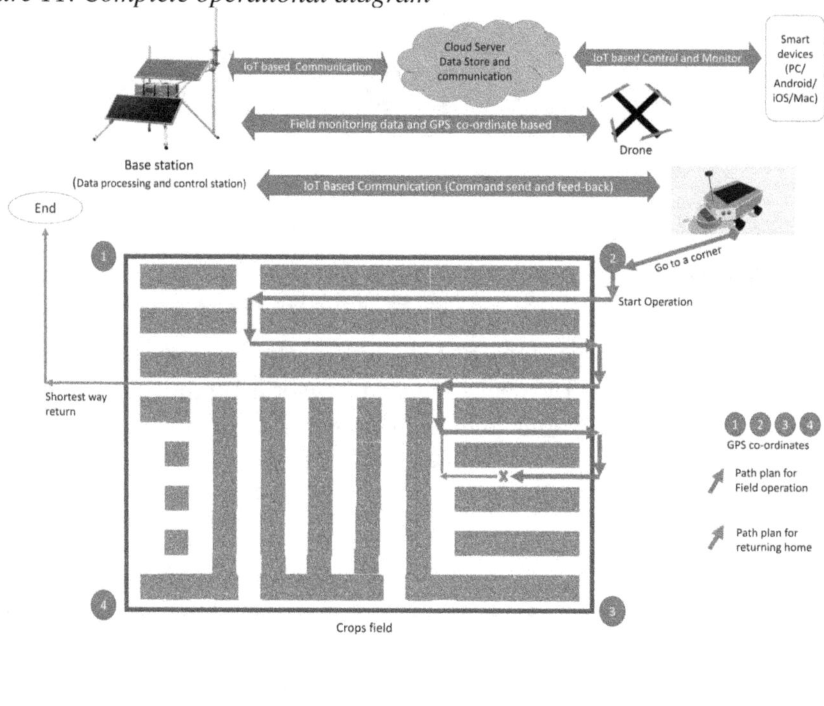

Obstacle detection and avoidance are critical aspects of this testing. The ultrasonic sensors continuously scan the environment for obstacles, and the robot's algorithm recalibrates its path in real-time to ensure smooth navigation. Any deviations or errors in navigation are recorded for analysis.

Simultaneously, the task execution capabilities of the robot are rigorously tested. The cutting blades are activated in high-weed-density zones to evaluate their precision in removing weeds without damaging crops. The spraying mechanism is assessed for coverage uniformity and resource optimization, ensuring that the correct amount of pesticide or water is applied. The seeding and fertilization systems are also tested for their ability to distribute seeds and fertilizers uniformly across the field.

Performance Metrics and Observations

Field testing generates valuable data on key performance metrics such as the time taken to complete tasks, resource consumption, and energy usage. For example, the robot demonstrated a 40% reduction in the time required for weed removal compared to manual methods. Weed detection and removal accuracy were observed to exceed 95%, showcasing the robot's precision in minimizing crop damage. Moreover, the robot exhibited consistent performance on uneven terrain, successfully navigating slopes and handling dynamic obstacles.

The observations and data collected during this phase are crucial for identifying areas of improvement. They form the basis for iterative refinements to the robot's design, algorithms, and operational strategies.

6.2 Challenges and Solutions

The implementation phase inevitably reveals challenges that must be addressed to optimize the robot's performance. These challenges encompass hardware, software, and environmental factors.

One prominent challenge was the occasional misidentification of obstacles by the ultrasonic sensors. Small weeds or water puddles were sometimes classified as significant obstacles, leading to unnecessary recalculations of the navigation path and operational delays. To resolve this issue, sensor fusion techniques were implemented. Data from the ultrasonic sensors were combined with readings from the Inertial Measurement Unit (IMU) and the machine vision system. This integration enhanced the accuracy of obstacle detection by enabling the system to differentiate between critical and non-critical obstacles.

Energy efficiency emerged as another critical concern during prolonged operations in large fields. The reliance on battery power alone limited the robot's operational hours. To address this, the solar panels were upgraded to monocrystalline cells, which have higher energy conversion efficiency. An energy management algorithm was also developed to prioritize high-energy tasks, ensuring that essential operations were completed even during extended sessions.

Navigation on uneven terrain posed significant challenges, particularly in maintaining stability and accuracy. The IMU was programmed to provide real-time adjustments for stability, while the robot's chassis was redesigned to enhance structural integrity. Wheels with improved traction were added to ensure reliable navigation across different terrains.

Resource optimization was another area of focus. The spraying system initially over-applied pesticides in low-density weed zones, leading to wastage. A dynamic spraying mechanism was introduced, allowing the system to adjust the spray radius and pressure based on real-time weed density data. Integration with IoT further enhanced this mechanism by using drone-collected data to guide spraying patterns accurately.

Temporary GPS signal interruptions in areas with dense vegetation created navigation challenges. A backup navigation system was developed using IMU and ultrasonic sensor data to maintain continuity during GPS outages. Additionally, error recovery protocols were implemented, allowing the robot to pause operations during signal loss and resume from its last known position once the signal was restored.

Refinements and Enhancements

Iterative testing and problem-solving played a pivotal role in refining the robot's capabilities. Navigation algorithms were fine-tuned to improve path efficiency and obstacle handling. Hardware upgrades, such as the replacement of sensors with higher-precision alternatives, enhanced the robot's reliability. User feedback from farmers participating in the testing process informed the development of a more intuitive interface, incorporating features like real-time status updates and easier task configuration.

The integration of the central command system further streamlined operations. Real-time data from the robot was transmitted to the central system, which analyzed the information and provided actionable insights. Farmers could monitor the robot's performance remotely and make dynamic adjustments to its operations through mobile or web applications.

Implementation Workflow

The workflow for the implementation process includes the following stages:

1. Initial Setup and Calibration: Ensuring that all sensors, actuators, and algorithms are configured and functioning correctly.
2. Field Deployment: Testing the robot's navigation, obstacle detection, and task execution capabilities in real-world conditions.
3. Data Collection and Analysis: Recording performance metrics and identifying areas for improvement.
4. Refinement and Problem-Solving: Addressing challenges through iterative design and algorithm enhancements.
5. Integration with Central Command: Establishing seamless communication between the robot and the central system for monitoring and control.
6. Final Validation: Ensuring that the robot meets performance benchmarks and is ready for deployment.

The implementation of the IoT-based autonomous movable weeding robot demonstrates the value of iterative testing, problem-solving, and adaptability in developing an advanced agricultural tool. By overcoming challenges related to obstacle detection, energy efficiency, navigation, and resource optimization, the robot has proven its effectiveness in real-world scenarios. The insights gained during this phase provide a solid foundation for scaling the system and deploying it across diverse agricultural settings. Through continuous refinement and user-centered design, this robot has the potential to revolutionize precision agriculture by enhancing productivity, sustainability, and ease of use.

7. RESULT AND FINDINGS

The result analysis is a comprehensive evaluation of the IoT-based autonomous movable weeding robot's performance during the implementation phase. This section consolidates insights from field testing, performance metrics, and practical challenges to assess the robot's effectiveness, efficiency, and potential impact on modern agriculture. By examining the operational capabilities, precision, and sustainability of the robot, this analysis provides a detailed understanding of its contributions to smart farming.

7.1 Key Findings

Operational Efficiency

One of the most significant outcomes of the robot's deployment was its ability to perform tasks such as weed removal, spraying, and seeding with a high degree of efficiency. Compared to traditional manual methods, the robot demonstrated a 40% reduction in the time required for weed removal. This efficiency was primarily attributed to the integration of GPS-guided navigation and real-time task execution algorithms.

The robot's path-planning algorithm, supported by the A* optimization technique, ensured comprehensive field coverage with minimal overlap or missed areas. The ability to adapt dynamically to obstacles further enhanced operational efficiency, allowing uninterrupted task execution even in complex field conditions. These results indicate that the robot is capable of significantly reducing labor dependency while improving productivity.

Precision in Task Execution

The precision of the robot's task execution was validated through rigorous field testing. Weed detection accuracy exceeded 95%, minimizing the risk of crop damage. This high level of precision was achieved through the use of advanced machine learning models, such as convolutional neural networks (CNNs), which enabled the robot to distinguish between crops and weeds effectively.

The spraying mechanism also demonstrated remarkable accuracy, with resources being applied only to targeted areas based on GPS coordinates and real-time weed density data. This targeted application not only reduced wastage but also minimized the environmental impact of chemical usage. Similarly, the seeding and fertilization systems distributed resources uniformly across the field, ensuring optimal crop growth conditions.

Adaptability and Reliability

The robot's adaptability to diverse field conditions was a key highlight of its performance. It successfully navigated uneven terrains, slopes, and dynamic obstacles, thanks to the integration of an Inertial Measurement Unit (IMU) and enhanced wheel traction. Temporary GPS signal losses, a common challenge in dense vegetation, were effectively managed through a backup navigation system that utilized IMU and ultrasonic sensor data. This adaptability underscores the robot's potential to operate reliably in various agricultural environments.

Additionally, the robot's multi-functionality adds to its versatility, making it an indispensable tool for modern agriculture. The robot can perform a wide range of operations, such as cutting weeds, uprooting them, spreading fertilizers, seeding, and spraying pesticides and herbicides. These tasks can be carried out with precision, ensuring that each operation is localized and tailored to the specific needs of the field.

Farmers can provide commands to the robot to execute these functions in specific zones, ensuring targeted application and efficient resource use. For example, the robot can uproot or cut weeds selectively while spreading fertilizers or herbicides only in the required areas. This capability reduces labor dependency, minimizes resource wastage, and improves operational efficiency, making the robot a cost-effective solution for sustainable farming practices.

The integration of these features further solidifies the robot's role as a multi-functional, adaptable solution capable of addressing the varied demands of agricultural operations with precision and reliability.

Figure 12. Multi-functionality of the robot

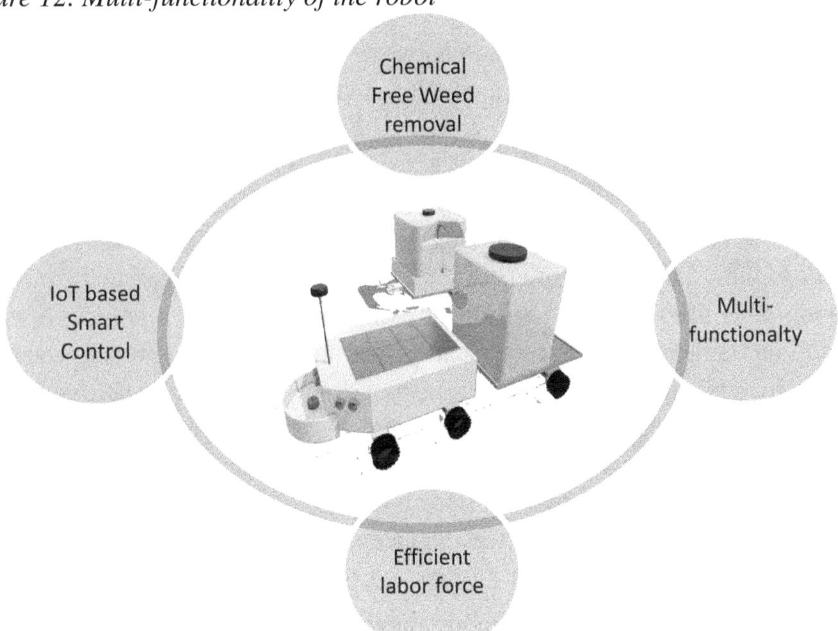

Energy Efficiency

The inclusion of monocrystalline solar panels significantly improved the robot's energy efficiency. Solar power supplemented the primary battery, extending the robot's operational hours and reducing downtime. An energy management algorithm further optimized power consumption by prioritizing high-energy tasks and adjusting operations based on battery levels. These features align with sustainable farming practices and contribute to the robot's eco-friendly design.

Integration with IoT

The integration of IoT connectivity added a layer of intelligence to the robot's operations. Real-time data transmission between the robot and the central command system enabled dynamic adjustments to navigation and task execution. The ability to monitor the robot remotely through mobile or web applications enhanced user control and provided valuable insights into system performance.

7.2 Comparative Analysis

The robot's performance was compared to traditional farming methods and existing autonomous systems to evaluate its advantages and limitations.

Comparison with Manual Methods

Manual weed removal, though effective on small farms, is labor-intensive and time-consuming. The robot eliminates these drawbacks by automating the process, achieving a significant reduction in task completion time. Its precision in weed detection and targeted resource application not only minimizes wastage but also reduces the overall need for fertilizers, as resources are applied only where necessary.

Additionally, by requiring less time to complete tasks compared to manual labor, the robot increases operational efficiency, allowing farmers to cover more land in less time. As labor costs continue to rise globally, the robot offers a cost-effective alternative, significantly reducing the reliance on manual workers. This translates to substantial financial savings for farmers, particularly in large-scale farming operations where labor costs and fertilizer wastage can add up considerably.

By combining time efficiency, resource optimization, and cost savings, the robot presents a transformative solution to modern agricultural challenges, making it an invaluable tool for sustainable farming practices.

Figure 13. Hourly labor cost increase over time in different countries

Comparison with Existing Autonomous Systems

Unlike many existing autonomous systems that specialize in a single task, the robot demonstrated versatility by performing multiple operations, including weed removal, spraying, and seeding. Its modular design allowed for easy customization, making it adaptable to various farming needs. Additionally, the integration of renewable energy sources and IoT connectivity set it apart as a sustainable and intelligent solution.

7.3 Challenges Addressed

The field-testing phase revealed several challenges, such as obstacle detection inaccuracies, navigation on uneven terrain, and GPS signal interruptions. Each of these challenges was addressed through innovative solutions, enhancing the robot's reliability and performance.

For instance, sensor fusion techniques improved obstacle detection accuracy by combining data from ultrasonic sensors, IMU, and machine vision systems. Similarly, the introduction of a dynamic spraying mechanism optimized resource usage by adjusting the spray radius and pressure based on real-time data.

These improvements highlight the importance of iterative testing and problem-solving in refining the robot's design and capabilities.

7.4 Environmental and Economic Impact

Environmental Sustainability

The robot's precision in task execution significantly reduced the overuse of chemical herbicides and pesticides, mitigating their harmful effects on soil and water quality. Its reliance on solar power and energy-efficient components further contributed to reducing the carbon footprint of agricultural operations. By promoting sustainable farming practices, the robot aligns with global efforts to combat climate change and protect natural ecosystems.

Economic Benefits

The economic impact of the robot extends beyond individual farms to the broader agricultural sector. By reducing labor costs and resource wastage, the robot enhances farm profitability. Its modular design and scalability make it accessible to farms of varying sizes, ensuring widespread adoption. Additionally, the production and maintenance of these robots could create new opportunities in the manufacturing and service sectors, driving economic growth.

Unlike manual labor, whose efficiency decreases significantly from morning to evening due to fatigue and environmental conditions, the robot operates with consistent efficiency throughout the day. Furthermore, it can be operated overnight, as it relies on GPS and ultrasonic sensors for navigation, eliminating the need for lighting and additional resources during nighttime operations.

The robot also generates significantly less waste compared to manual methods across various agricultural tasks, such as weeding, fertilizer spreading, seeding, and spraying. Its precision ensures that resources are applied only where needed, minimizing excess usage and runoff. The multifunctionality of the robot—capable of performing tasks like targeted weed removal, accurate fertilizer spreading, precise seeding, and efficient spraying—further enhances its value, making it an indispensable tool for modern farming practices. This level of efficiency and sustainability makes the robot a critical innovation for addressing economic and environmental challenges in agriculture.

Figure 14. Manual labour and automated weeding (time -efficiency) heat map

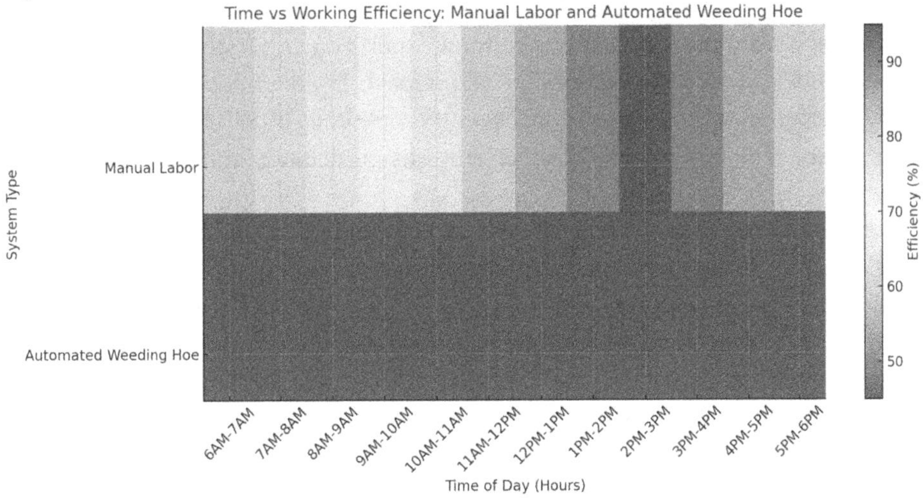

Figure 15. Trends in fertilizer waste reduction per acer over time

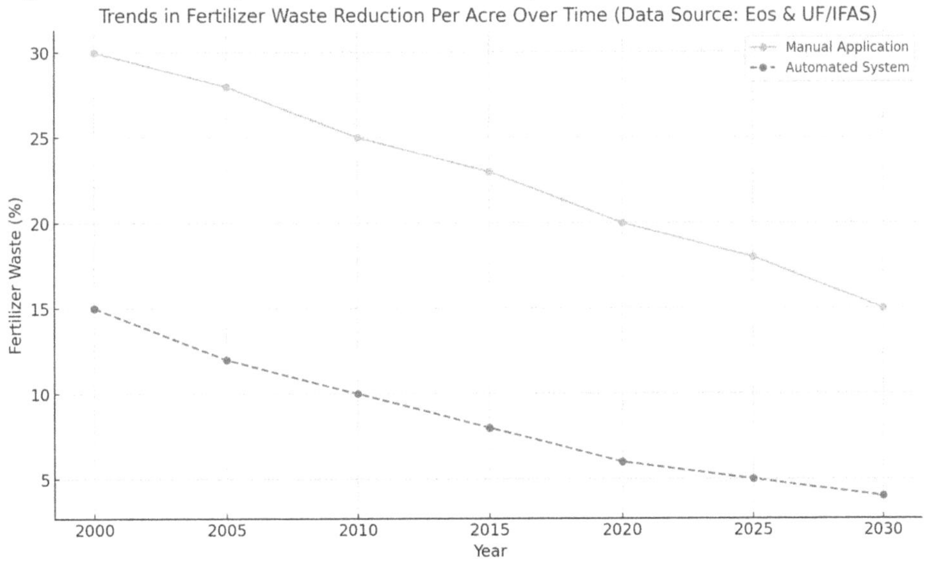

7.5 Feedback and User Insights

Feedback from farmers and agricultural experts during testing provided valuable insights into the robot's usability and impact. Users praised its user-friendly interface, combining real-time feedback with simple controls. The ability to monitor performance remotely via IoT connectivity was particularly appreciated, offering greater flexibility and control.

Some users suggested enhancements, such as adding advanced soil sensors for nutrient analysis and automated irrigation systems, highlighting potential future upgrades to expand functionality.

The analysis shows that the IoT-based autonomous weeding robot is a transformative solution for modern agriculture. With precision, efficiency, and sustainability, it addresses labor shortages, resource wastage, and environmental challenges. Its adaptability to varied field conditions and ability to perform multiple tasks autonomously make it a valuable tool for precision farming.

These insights offer a roadmap for scaling the system across diverse agricultural settings. By refining its features and integrating user feedback, the robot can further revolutionize farming, contributing to global food security and sustainable agriculture.

8. CONCLUSION AND SUMMARY OF CONTRIBUTIONS

This chapter has provided a comprehensive analysis of an IoT-based autonomous movable weeding robot designed to revolutionize modern agriculture. The system integrates cutting-edge technologies such as GPS navigation, IoT connectivity, solar power to deliver precision farming solutions. By addressing critical challenges such as labor shortages, resource inefficiency, and environmental sustainability, the robot demonstrates its potential to transform agricultural practices.

Summary of Contributions

Precision Navigation

The GPS-based path planning system enables accurate field navigation, ensuring comprehensive coverage with minimal overlap. This precision leads to more efficient weed removal, saving both time and resources, while also enhancing crop protection by minimizing errors.

Eco-Friendly Weed Control

The robot eliminates the use of harmful chemical herbicides, significantly reducing environmental risks such as pesticide runoff, soil degradation, and herbicide-resistant weeds. Its reliance on solar power further supports eco-friendly farming practices, aligning with global sustainability goals.

Cost-Efficiency

The integration of IoT and renewable energy sources drastically reduces farming costs by minimizing reliance on manual labor, fuel, and chemical inputs. The system's long-term savings, particularly from reduced pesticide usage and labor requirements, make it an economically viable solution for farms of all sizes.

Scalability and Adaptability

The modular design of the robot ensures its adaptability across various agricultural settings, from large wheat farms to small organic fields. Its scalability allows for seamless application in diverse environmental conditions and crop types, demonstrating its versatility as a tool for modern farming.

Implications for the Future of Agriculture

The IoT-based autonomous weeding robot marks a significant step toward the future of smart farming. By integrating automation, IoT, and renewable energy, this system represents a paradigm shift in agricultural operations. As technology continues to advance, similar autonomous systems are expected to be developed for other tasks such as planting, irrigation, pest control, and harvesting, creating a holistic ecosystem of automated farming tools.

This development also underscores the potential for reducing reliance on fossil fuels, mitigating climate change, and improving food security. As the global population approaches 9.5 billion by 2050, such innovations will be essential to meeting the growing demand for food while minimizing environmental impact.

Future Research Directions

While the autonomous weeding robot has achieved significant milestones, there are areas for further exploration and improvement:

1. Advanced Machine Learning and AI Integration

- Future iterations can incorporate sophisticated AI algorithms for real-time learning, enabling the robot to adapt to different field conditions and weed species. Enhanced decision-making systems could further refine weed detection, crop differentiation, and operational efficiency.

2. Integration with Other Farm Equipment
    - Combining the weeding robot with other autonomous systems, such as harvesters, planters, and irrigation tools, could create a fully integrated farming solution. This coordination would improve overall efficiency and streamline agricultural processes.
3. Enhanced Sensor Technologies
    - The adoption of advanced sensors, such as multispectral and hyperspectral imaging, could significantly improve weed detection capabilities. These sensors would allow the robot to differentiate between various vegetation types, enabling more precise and effective weed management.
4. Optimized Energy Systems
    - Improvements in energy efficiency, such as integrating wind power or kinetic energy harvesting, could complement the solar power system and ensure uninterrupted operations in diverse environmental conditions.
5. Cloud-Based Data and Big Data Analytics
    - Integrating the robot with cloud platforms would enable large-scale data collection and analysis. Big data analytics could provide insights into soil health, crop performance, and overall farm productivity, empowering farmers with data-driven decision-making tools.

Final Thoughts

This IoT-based autonomous movable weeding robot exemplifies the potential for technology to address critical agricultural challenges while supporting sustainable development. By combining precision, efficiency, and adaptability, the system bridges the gap between innovation and practicality, making it a valuable tool for modern farming. As advancements in technology continue, this robot paves the way for a future where agriculture is smarter, more sustainable, and better equipped to meet global food demands.

The chapter emphasizes that while significant progress has been made, the journey toward fully autonomous, eco-friendly farming is ongoing. Continuous research, innovation, and collaboration will be essential in refining and expanding these technologies to achieve their full potential. The robot stands not only as a tool for today but as a harbinger of the future of agriculture, where efficiency and sustainability go hand in hand.

REFERENCES

Alfaro, C.. (2020). Solar-powered autonomous robots in agriculture: Sustainability and challenges. *Renewable Energy*, *145*, 2685–2696. DOI: 10.1016/j.renene.2020.06.024

Banzi, M., & Shiloh, M. (2014). *Getting started with Arduino: The open source electronics prototyping platform* (3rd ed.). Maker Media, Inc.

Ehrlich, P. R.. (2015). Global environmental change and sustainability. *Global Change Biology*, *21*(12), 4417–4428. DOI: 10.1111/gcb.13030

Elfring, J., Van De Molengraft, M. J. G., & Steinbuch, M. (2013). Learning motion and sensor models for mobile robot localization. *Robotics and Autonomous Systems*, *61*(3), 223–236. DOI: 10.1016/j.robot.2012.11.008

Fountas, S.. (2006). Precision agriculture: A methodological framework for assessing implementation and impact. *Agricultural Systems*, *91*(1–2), 159–182. DOI: 10.1016/j.agsy.2006.03.005

Gebbers, R., & Adamchuk, V. I. (2010). Precision agriculture and food security. *Science*, *327*(5967), 828–831. DOI: 10.1126/science.1183899 PMID: 20150492

Holt, J. S.. (2017). Impact of herbicide use in modern agriculture. *Journal of Environmental Management*, *201*, 113–119. DOI: 10.1016/j.jenvman.2017.06.048

Kumar, R., Gupta, P., & Singh, P. (2021). A Device for Effective Weed Removal for Smart Agriculture Using Convolutional Neural Network. Springer. https://link.springer.com/article/10.1007/s13198-021-01441-z

Marinoudi, V., Sørensen, C. G., Pearson, S., & Bochtis, D. (2019). Robotics and labour in agriculture. A review. *Biosystems Engineering*, *184*, 111–121. DOI: 10.1016/j.biosystemseng.2019.06.013

McBratney, A. B., Whelan, B., Ancev, T., & Bouma, J. (2005). Future directions of precision agriculture. *Precision Agriculture*, *6*(1), 7–23. DOI: 10.1007/s11119-005-0681-8

Moll, J.. (2019). The impact of mechanized weed control in large-scale farming. *Agricultural Systems*, *167*, 112–121. DOI: 10.1016/j.agsy.2018.09.006

Naylor, R.. (2018). The economics of manual weeding in small-scale farms. *Economic Botany*, *72*(4), 425–432. DOI: 10.1007/s12231-018-9435-6

Rehman, S.. (2019). Drone-based weed detection and management in precision agriculture. *Computers and Electronics in Agriculture, 156*, 482–492. DOI: 10.1016/j.compag.2018.11.020

Schlechtendahl, E.. (2020). Autonomous weed control: Opportunities and challenges in precision agriculture. *Robotics and Autonomous Systems, 127*, 103–118. DOI: 10.1016/j.robot.2019.11.011

Sharma, P., Kapoor, A., & Verma, S. (2020). Automatic Weed Removing Robot Using IoT Technology. [IJCRT]. *International Journal of Creative Research Thoughts, 8*(9), 350–358. https://ijcrt.org/papers/IJCRT2009133.pdf

Siddiqui, M. H., & Khan, A. R. (2020). Automated Computer Vision Based Weed Removal Bot. IEEE Xplore. https://ieeexplore.ieee.org/document/9198515

Sishodia, R. P., Ray, R. L., & Singh, S. K. (2020). Applications of remote sensing in precision agriculture: A review. *Remote Sensing (Basel), 12*(19), 3136. DOI: 10.3390/rs12193136

U.S. Bureau of Labor Statistics. (2023). Occupational employment and wage statistics: 2020 2023 [Data set]. U.S. Department of Labor. https://www.bls.gov/oes/current/oes453032.htm

Van Evert, F. K.. (2012). Robot-assisted precision irrigation: A review. *Biosystems Engineering, 113*(1), 76–90. DOI: 10.1016/j.biosystemseng.2012.10.007

Walter, A., Finger, R., Huber, R., & Buchmann, N. (2017). Smart farming is key to developing sustainable agriculture. *Proceedings of the National Academy of Sciences of the United States of America, 114*(24), 6148–6150. DOI: 10.1073/pnas.1707462114 PMID: 28611194

Xie, W., Li, J., & Zhang, H. (2024). SPARROW: Smart Precision Agriculture Robot for Ridding of Weeds. ArXiv. https://arxiv.org/abs/2405.20896

Zhang, S.. (2020). The role of IoT in precision agriculture: A review. *Computers and Electronics in Agriculture, 167*, 105113. DOI: 10.1016/j.compag.2019.105113

Zhang, X., Liu, Y., & Wang, Z. (2023). IoT Based Weed Detection and Removal in Precision Agriculture. IEEE Xplore. https://ieeexplore.ieee.org/document/1020075

Zhao, C.. (2017). Climate change impacts on crop yield and quality. *Science Advances, 3*(4), e1602117. DOI: 10.1126/sciadv.1602117 PMID: 28508038

Zhou, X.. (2020). GPS-based autonomous systems in precision farming. *Computers and Electronics in Agriculture, 170*, 105–113. DOI: 10.1016/j.compag.2019.05.005

KEY TERMS AND DEFINITIONS

IoT (Internet of Things): A technological framework where various devices are connected through the internet to collect, exchange, and process data in real-time. In agriculture, IoT is used for monitoring environmental and soil conditions, enabling data-driven decision-making.

IoD (Internet of Drones): A network where drones interact and share data using IoT technology, allowing efficient execution of tasks like aerial monitoring, data collection, and communication with ground-based systems, such as autonomous robots.

GPS (Global Positioning System): A navigation technology that uses satellites to provide precise location data. In agricultural robotics, GPS helps define field boundaries, plan paths, and guide automated equipment to execute tasks accurately.

Deep Learning: A type of artificial intelligence where computers learn from large amounts of data to recognize patterns and make decisions. Deep learning allows robots to perform complex tasks, such as identifying and classifying weeds in real-time.

Path Planning: The process of determining the most effective route for a robot to move through a field. It ensures full field coverage while avoiding unnecessary overlaps or redundant paths, optimizing efficiency.

Obstacle Avoidance: A mechanism that enables robots to detect and navigate around objects in their path. It uses sensors and algorithms to ensure uninterrupted operation while protecting the robot and its surroundings.

IoT Gateway: A device or software that acts as a bridge between the IoT devices and a central system. It facilitates communication and data exchange, ensuring smooth coordination of operations in a networked system.

Drone-Based Imaging: The use of drones equipped with cameras and sensors to capture high-quality images of agricultural fields. This visual data is processed to analyze weed density, crop health, or field conditions.

Integrated Weed Control: A comprehensive weed management strategy that combines mechanical, biological, and chemical methods to effectively control weeds while minimizing environmental impacts.

A* Algorithm: A pathfinding algorithm that efficiently finds the shortest path between points by using a heuristic to estimate the remaining distance to the goal, balancing it with the distance already travelled. It considers factors like cost, distance, and obstacles, ensuring optimal navigation for the robot.

Cloud Computing: A method of using internet-based servers to store, manage, and analyze data. It allows robots and other systems to handle large datasets without relying solely on onboard computing power.

Modular Design: A design philosophy where a system is built using interchangeable components. This approach makes it easier to customize, repair, or upgrade the robot for various tasks and conditions.

Chapter 3

Enhancing IoT-Based Smart Irrigation Efficiency Through Optimized Sensor Placement, Noise Elimination, and Incremental Learning

Varun Yarehalli Chandrappa
https://orcid.org/0000-0002-9542-3297
Central Queensland University, Australia

Nahina Islam
Central Queensland University, Australia

Nanjappa Ashwath
Central Queensland University, Australia

Pramod Shrestha
Central Queensland University, Australia

ABSTRACT

Modern agriculture faces challenges requiring innovative solutions to balance productivity with sustainability. Optimizing irrigation is critical, as traditional systems lead to water waste and inefficiencies, necessitating smart technologies. Key advancements include Dual Electromagnetic (DUAL-EM) scanning for soil conductivity mapping, optimizing sensor placement, reducing costs, and ensuring

DOI: 10.4018/979-8-3373-0020-7.ch003

Copyright © 2025, IGI Global Scientific Publishing. Copying or distributing in print or electronic forms without written permission of IGI Global Scientific Publishing is prohibited.

data quality. Noise reduction techniques like Isolation Forest and LOF enhance data reliability, achieving 81.02% accuracy and 0.328 MAE. Incremental learning models such as LightGBM and XGBoost dynamically refine irrigation schedules, optimizing water application under changing conditions. IoT integration with soil sensors, weather stations, and automated valves ensures precise irrigation. These strategies conserve water, reduce costs, and align with environmental goals, transforming farming practices and advancing sustainable agriculture.

1. INTRODUCTION

Smart irrigation is a pivotal solution in sustainable agriculture and landscape management, addressing the need to optimize water use and tackle challenges posed by climate change and resource scarcity. Traditional irrigation methods, often based on fixed schedules and uniform water distribution, fail to meet the specific requirements of diverse plant species and soil types. In water-limited regions, inefficient irrigation can result in over-watering, causing soil degradation and nutrient leaching (Smith & Zhang, 2018), or under-watering due to inaccurate data, leading to plant stress and reduced vegetation vitality (Wang et al., 2020). The Internet of Things (IoT) has enabled smart irrigation systems to make real-time, data-driven adjustments based on environmental conditions, promoting water conservation and healthier landscapes.

A critical component of smart irrigation is the accuracy of environmental data collection, particularly soil moisture, temperature, and weather data. This information is vital for predictive models that determine optimal irrigation schedules. However, sensor data is prone to noise due to environmental fluctuations, sensor degradation, or connectivity issues (Li & Qian, 2019). Without proper noise handling, inaccurate predictions can lead to water wastage or inadequate irrigation. Noise elimination techniques such as Isolation Forest and Local Outlier Factor (LOF) have been effective in filtering irrelevant data points, improving model accuracy (Patel & Mehta, 2021). These methods ensure that smart irrigation systems make reliable decisions based on accurate data.

Another major challenge in smart irrigation is responding to environmental changes. Traditional machine learning models, trained on static datasets, struggle to adapt to real-time fluctuations in environmental conditions. As weather patterns, soil moisture levels, and plant water needs change frequently, conventional models require frequent retraining, which is both time-consuming and computationally expensive (Brownlee, 2018). Incremental learning addresses this limitation by allowing models to continuously learn from new data without full retraining (Kumar & Singh, 2021). This approach enables irrigation systems to adapt in real-time to dynamic environmental conditions, making them more effective in optimizing water use (Lin & Chen, 2019).

Incremental learning models such as LightGBM and XGBoost have demonstrated high responsiveness in various applications, making them suitable for adaptive irrigation systems. However, their direct applications in smart irrigation remain limited. Wang et al. (2023) developed a predictive irrigation scheduler using LSTM networks and k-means clustering, achieving 22.8% water savings but facing complexity and data dependency issues. Patel & Kumar (2023) introduced a reinforcement learning-based system (DRLIC) that optimized water use but required safety mechanisms to prevent incorrect decisions. Saikai et al. (2023) applied reinforcement learning to irrigation scheduling, improving profit margins by 17% but encountering challenges due to high data requirements. While these models enhance irrigation precision, computational resource demands and adaptability remain key concerns.

This study addresses the limitations of traditional irrigation practices by developing an IoT-based smart irrigation system integrating sensor data, noise elimination, and incremental learning. The objectives are to optimize sensor placement for accurate soil moisture monitoring, apply noise elimination techniques to enhance data quality, and implement incremental learning models for real-time adaptation. By leveraging these technologies, the proposed system aims to reduce water waste, improve irrigation precision, and promote sustainable landscape management in environments such as public parklands and agricultural fields.

2. BACKGROUND

The integration of Internet of Things (IoT) technology with agriculture and environmental management has opened new possibilities for precision irrigation, where data-driven insights can optimize water usage and enhance crop and plant health. In modern landscapes like urban parks and agricultural fields, diverse vegetation and variable environmental conditions demand a customized approach to irrigation. Traditional irrigation methods, often reliant on fixed schedules and uniform water distribution, are not capable of addressing the nuanced requirements of these environments. This limitation can lead to excessive water use in some areas, resulting in wastage, or under-irrigation in others, which can stress plants and compromise landscape quality.

Public parks, such as Cairns Park, serve as ideal testing grounds for smart irrigation systems due to their unique combination of varied vegetation, soil types, and usage patterns. IoT-enabled irrigation systems, when coupled with machine learning and data processing, can address the irrigation needs of diverse landscapes by monitoring soil moisture in real time and adjusting irrigation levels dynamically.

The potential to automate water distribution based on accurate soil and weather data not only conserves water but also ensures that each area receives the right amount of hydration.

A critical component of effective smart irrigation lies in the accurate collection and processing of data from strategically placed sensors across the landscape. These sensors capture soil moisture, temperature, humidity, and other environmental parameters, which are essential for making informed irrigation decisions. However, collecting meaningful data in a park environment requires more than simply placing sensors randomly. Differences in soil types, vegetation cover, and microclimatic variations across the park make it necessary to carefully consider sensor placement to ensure that the data accurately represents the diverse conditions within the landscape.

In addition to data accuracy, the quality of information gathered from sensors can be compromised by noise, missing values, and external environmental fluctuations. Sensor malfunctions, network disruptions, or environmental interferences often introduce inaccuracies in data, which can skew the system's interpretation and response to soil moisture levels. Managing these inconsistencies is crucial for ensuring the reliability of predictive models used in irrigation management. Noise and missing data pose significant challenges to data-driven systems, as they can result in incorrect predictions and, subsequently, inappropriate irrigation actions.

Real-time adaptability is another critical challenge for IoT-based smart irrigation systems. Environmental conditions, especially in outdoor settings, are inherently dynamic, with soil moisture levels fluctuating due to seasonal changes, weather variations, and human activity. To remain effective, an irrigation system must not only monitor these shifts but also adapt its response to ensure plants receive the right amount of water at the appropriate time. Traditional irrigation models are often static and require frequent manual updates, making them impractical for dynamic environments where changes in soil moisture or weather can occur rapidly.

Furthermore, the interaction between various IoT components—such as sensors, actuators, and communication networks—is complex and requires seamless integration to ensure that data is transmitted, processed, and acted upon in a timely manner. Effective communication between sensors and controllers is fundamental for the irrigation system to function as intended. Without reliable communication, delays in data transmission or processing could result in inefficient water distribution, potentially harming plants and increasing water consumption.

Finally, as climate change exacerbates water scarcity, there is an increasing need to adopt sustainable water management practices. Urban parks, which often rely on large volumes of water for landscape maintenance, stand to benefit significantly from more precise irrigation. By ensuring that water is only applied where and when it is needed, smart irrigation systems can help mitigate water waste, reduce operational costs, and promote environmental sustainability.

3. PROBLEM STATEMENT

Despite the potential of IoT-based smart irrigation systems to enhance water use efficiency, several critical challenges hinder their effectiveness. These issues need to be addressed to enable a more sustainable, precise, and reliable irrigation solution for complex landscapes like Cairns Park. Key problems identified include:

1. **Optimal Sensor Placement for Comprehensive Data Coverage:** The varied landscape of Cairns Park, with its diverse vegetation types and soil conditions, requires a strategic approach to sensor placement. Random or uniform sensor distribution often fails to capture the variations in soil moisture across different zones, resulting in inaccurate data and inefficient irrigation. Identifying optimal locations that represent the unique conditions within each area of the park is essential for accurate data collection.
2. **Data Quality Challenges Due to Noise and Missing Values:** IoT sensors are prone to data quality issues, such as noise and missing values, which can occur due to hardware malfunctions, network interruptions, or environmental interferences. Noise, characterized by random fluctuations that do not represent actual soil conditions, can mislead predictive models, leading to erroneous irrigation decisions. Missing data points create gaps in the dataset, reducing the completeness of information available for analysis. These issues compromise the accuracy of predictions and hinder the system's ability to make reliable irrigation recommendations.
3. **Lack of Real-Time Adaptability to Changing Environmental Conditions:** Environmental variables, including soil moisture, temperature, and humidity, change dynamically due to seasonal patterns, weather events, and human activities. Traditional irrigation systems, which rely on static schedules or pre-set thresholds, cannot effectively respond to these fluctuations. Without real-time adaptability, the system may either over-irrigate or under-irrigate certain areas, leading to water waste and potential harm to plants. An adaptable system is essential for responding to the frequent changes in soil moisture and weather conditions to optimize irrigation.
4. **Complexity in Ensuring Reliable IoT Communication and Integration:** The effectiveness of an IoT-based smart irrigation system relies heavily on seamless communication between sensors, controllers, and actuators. Inconsistent or delayed data transmission due to network interruptions can disrupt the system's operations, resulting in inappropriate irrigation actions. Establishing a reliable communication framework that ensures timely data flow and response is challenging, especially in outdoor environments where signal interference and device connectivity can vary.

5. **Need for Sustainable Water Management Practices in Public Spaces:** With increasing water scarcity and the growing emphasis on sustainable environmental practices, there is a pressing need to reduce water usage in public parklands without compromising landscape quality. Conventional irrigation approaches, which often apply water uniformly across all areas, are inefficient and unsustainable. A data-driven approach to irrigation that minimizes water waste while meeting the specific needs of each zone can significantly contribute to environmental conservation and resource efficiency.
6. **Cost Implications for Implementing Smart Irrigation at Scale:** Smart irrigation offers long-term savings, but the initial investment in sensors, IoT infrastructure, and data processing is significant. Maintenance costs, including sensor calibration and power management, add to expenses. However, reduced water consumption, lower upkeep costs, and improved plant health justify the investment. Case studies from urban parks and agriculture highlight the need for subsidies and public-private partnerships to offset costs and enhance feasibility for large-scale adoption.

These challenges highlight the complexity of implementing an effective smart irrigation system for public parks. Addressing these issues is essential to creating a solution that not only conserves water but also adapts to real-time environmental changes, ensuring that landscapes like Cairns Park remain healthy, resilient, and sustainable. The gaps identified above are addressed in Section 4: Research Design, which outlines the methodologies and strategies implemented to overcome these challenges.

4. RESEARCH DESIGN

The accuracy and effectiveness of the data gathered by these sensors depend significantly on their placement within the park environment. Strategic sensor placement ensures that the data collected is representative of the various conditions present across different areas of the park. To identify the appropriate location for sensors installation Dual Electromagnetic scan is conducted in year 2019 and 2020 covering all the seasons to know how the soil moisture distribution in park Lagoon east and Fogarty Park. Section 4.1 below explain the survey result and soil moisture distribution in Lagoon east and Fogarty Park.

4.1 Lagoon East survey

As shown in Figure 1, Lagoon East was surveyed multiple times in 2019 (March, April, July, and September) and 2020 (June, September, and November). These surveys encompassed both dry and wet seasons, as well as periods of varying activity levels due to COVID-19. Figure 2 displays the Dual-EM soil moisture maps at depths of 2 m and 4 m. The March 2019 moisture profile, shown in Figure 3, highlights notable differences between these depths, with three distinct zones identified at each level. As indicated in Figure 3, high conductivity, particularly at 4 m, is attributed to saltwater intrusion, given the site's coastal location. Seasonal variations were observed in the 2 m profile, with peak readings recorded in March, July, and September 2019.

Figure 1. Aerial view Lagoon east

Figure 2. Salinity impact due to sea

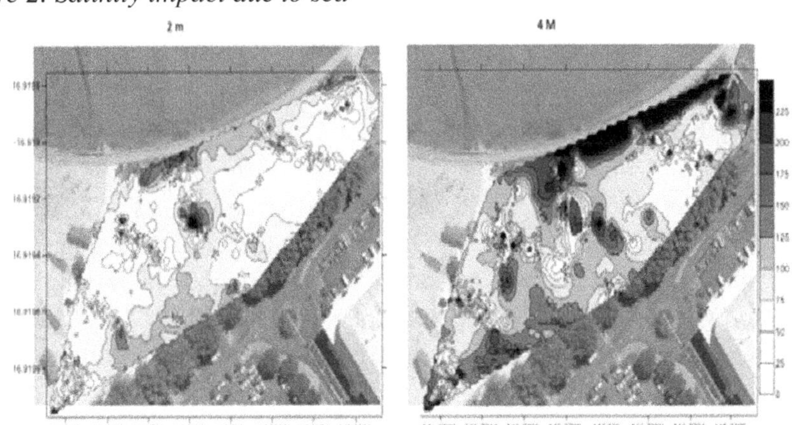

When comparing the 0–1 m and 0–2 m depth profiles, as shown in Figure 3, the 1 m profile has a noticeably drier zone than the 2 m profile. This driest zone aligns with the tree stand, likely due to increased water uptake by the trees. Alternatively, it may indicate insufficient sprinkler coverage to keep this zone as moist as the others. A subsequent sprinkler distribution analysis revealed that the tree zone was not adequately watered, which may explain the drier conditions observed throughout the tree stand.

Figure 3. Lagoon East Dual-EM and EM38 maps

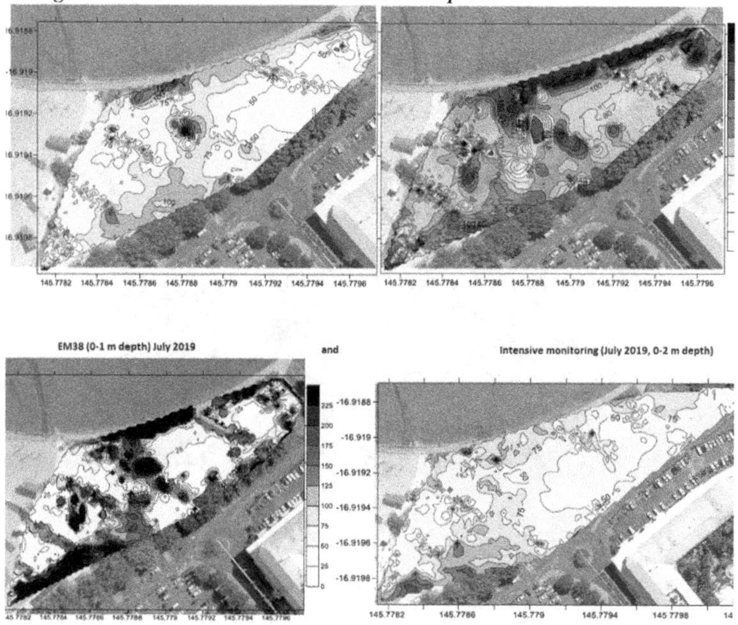

4.2 Fogarty Park Survey

Fogarty Park, as shown in Figure 4, was surveyed multiple times in 2019 (March, April, July, and September) and 2020 (June, September, and November). These surveys covered both dry and wet seasons, as well as periods of varying activity levels due to COVID-19. Figures 5 and 6 display the Dual-EM soil moisture maps at depths of 2 m and 4 m. Across all surveys, the 2 m depth consistently contained less moisture than the 4 m depth. The 2 m profile exhibited slight seasonal changes with 3-4 major zones, whereas the 4 m profile revealed up to 5 major zones with more pronounced seasonal variation compared to the 2 m depth.

Figure 4. Aerial view Fogarty Park

Figure 5. Fogarty Park soil moisture distribution

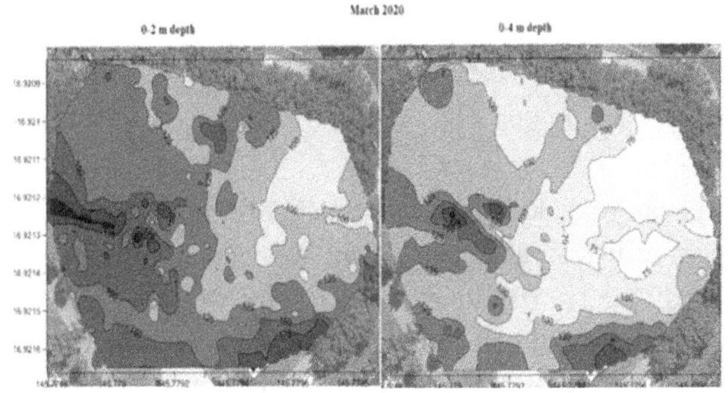

Figure 6. Dual-EM and EM38 maps of Fogarty Park

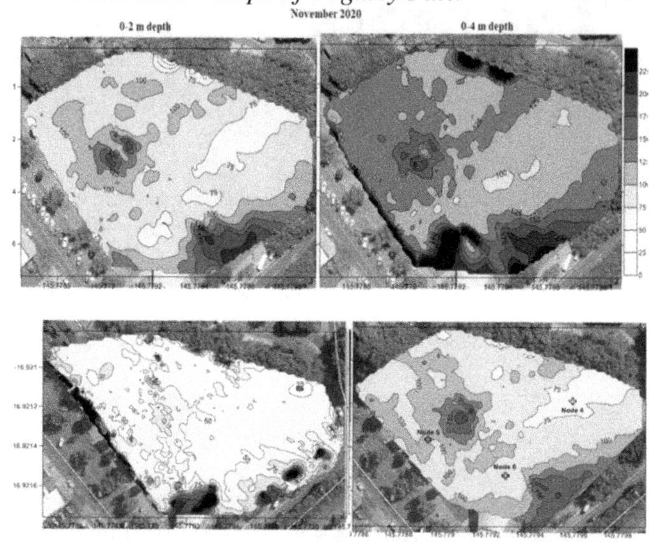

The Dual-EM scan revealed significant inconsistencies in irrigation across the park, identifying three distinct areas: wet, semi-wet, and dry. Based on these findings, the entire park was divided into three zones wet, semi-wet, and dry. These zones were selected for the installation of soil moisture sensors to ensure precise monitoring.

After determining the sensor installation locations, it is crucial to identify the appropriate depths at which the sensors should be installed to accurately monitor soil moisture content. To establish these depths, a study was conducted considering the type of grass growing in each zone and the root depths of these grasses. The following section 4.3 outlines the methodology and steps followed to determine the sensor installation depths.

4.3 Identification of Sensor Installation Depth

After selecting the appropriate location for soil moisture sensors installation for the irrigation system, the next critical step is to determine their optimal placement and installation depth. This ensures that the sensors provide accurate and actionable data. The depth at which sensors are installed is crucial for obtaining relevant data and is determined by the types of plants present and their root depths. Hence the study is conducted in Carin's park to identify the type of plant growing the experimental site. Section 4.3.1 below explain the plant type analysis and its root depth analysis.

4.3.1 Plant Type and Root Depth Analysis

The principal types of vegetation that dominate Cairns Regional Council's parklands are grasses and trees. Buffalo grass (broad-leaved) and couch (narrow-leaved) are two types of species of grasses that are common in Cairns parks. These are buffalo grass (broad leaved) and couch (narrow leaved). These grasses occur either in a monoculture or are found together. Figure 7 shows buffalo grass and couch grass which are growing in Cairns parks. Buffalo grass (broad-leaved) with a maximum root depth of 30 cm and Couch grass (narrow-leaved) with a root depth of 12 cm (Smith, Brown, & Wilson, 2018) (Jones & Taylor, 2020). Section 4.3.2 below explain the depth of soil moisture installation.

Figure 7. Buffalo grass and couch grown in Lagoon East

4.3.2 Soil Moisture Installation Depth

To accurately monitor soil moisture within the root zones of Buffalo grass and Couch grass, sensors were installed at depths of 7 cm and 30 cm. Additionally, sensors were placed at a depth of 90 cm to provide data on water movement beyond the root zones, indicating potential water loss to deeper soil layers.

In location of the site where different types of plants with deeper root systems are grown, with some reaching root depths of up to 90 cm, sensors were installed at 120 cm. This deeper installation allows for monitoring water infiltration below the deepest root zones, providing essential data for optimizing irrigation practices. By carefully selecting these installation depths, the study ensures that the sensors capture comprehensive soil moisture data relevant to the specific needs of the plants being cultivated.

After identifying the installation locations, sensors were placed at depths of 7 cm, 30 cm, 90 cm, and 120 cm. Once the sensor installation location is decided for soil moisture sensor and weather sensor the next crucial step is to connect these devices to enable real-time data collection and transmission. To facilitate this process, an Internet of Things (IoT)-based solution has been designed to gather, process, and analyze information in real time. This enables timely and informed decision-making for irrigation management, as explained in the article published by (Varun, Ray, Nanjappa, & Pramod, 2020) and (Chandrappa, Ray, Ashwatha, & Shrestha, 2023).

The data collected using this IoT architecture may contain noise and missing values, which need to be processed before being used for irrigation decision-making. Section 4.4 below describes the data processing methods employed to handle noise and missing values.

4.4 Data Processing

The data transmitted from these sensors is often susceptible to noise and inaccuracies due to factors such as network interruptions, environmental interferences, and sensor malfunctions. These issues can result in incomplete or incorrect data, which can significantly impact the accuracy of irrigation decisions.

Figure 8 below illustrates the data processing module in the application cloud, where noise removal and missing value handling are conducted before pushing the data into a time series database. This research study examines statistical smoothing and machine learning-based methods in detail, demonstrating how they enhance data quality.

Figure 8. Data processing in IoT architecture

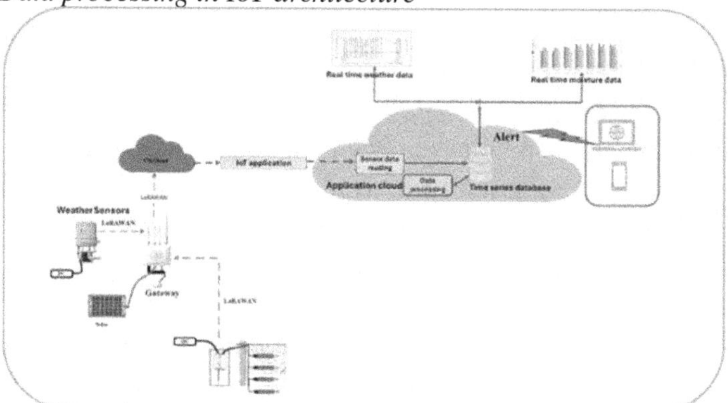

4.4.1 Noise Elimination

Data collected from soil moisture and weather sensors is crucial in making irrigation decisions in smart irrigation. The presence of noise and missing values in the smart irrigation data set can have a significant impact on the accuracy and performance of machine learning algorithms.

4.4.2 Impact of Noise

Noise refers to random or irrelevant variations in the data set that do not accurately reflect the underlying patterns of soil moisture or the current weather. In the case of soil moisture and weather sensor data, noise can be caused by various factors, such as environmental fluctuations, measurement errors, or interference. The Figure 9 is the snippet of noise data received from the soil moisture where X-axis represent the number of iteration and Y-axis represent the soil moisture level. Due to the noise in the data set the machine learning algorithms can be impacted in the following ways:

Figure 9. Soil moisture noise data

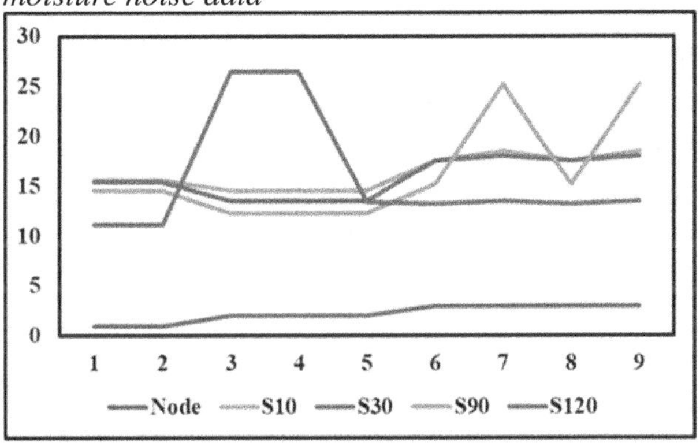

Reduced Accuracy: Noise can introduce inconsistencies and deceive the algorithms, leading to incorrect predictions and decisions. The presence of noisy data can cause the machine learning models to learn incorrect patterns, reducing their accuracy. In Figure 9, example data, the S120 column has soil moisture values like 26.4, which are significantly higher than most readings in the same column. Such noise can pull the model's predictions towards these outlier values, thereby reducing overall prediction accuracy.

Overfitting: Machine learning models can become extremely complex due to noisy data, which causes overfitting. Overfitting occurs when the model learns the noise in the data rather than the true underlying patterns. Because of this, the model does well with training data but poorly with unseen data. If the model tries to accommodate outlier soil moisture values like 26.4 in the S120 column of Figure 9, while the majority of soil moisture values range around 11.1 to 13.5, it could create overly complex decision boundaries or rules, making the model inflexible to changes in new data. This can lead to extreme adjustments based on these few outlier values, causing the system to either under-irrigate or over-irrigate when applied to real-world scenarios

To mitigate the impact of noise, this study applies various techniques, such as outlier detection and removal algorithms and data smoothing methods which is detailed in section 4.4.5. These techniques help improve data quality and enhance the performance of machine learning models by reducing the influence of noisy data points in soil moisture and weather data.

4.4.3 Impact of Missing Values

Sensor malfunction, connectivity problems, or other technical problems can result in missing values in the soil moisture sensor data and weather sensor data. The Figure 10 is the snippet of missing value in the soil moisture data set where X-axis represent the number of iteration and Y-axis represent the soil moisture level. The presence of these missing values poses significant challenges, as observed in the Figure 10 where the S30 and S90 column has missing values for Node. This missingness results in incomplete information, preventing the model from having a complete view of the soil moisture readings across different nodes. For instance, if Node 1 has a missing value in S90, the model lacks crucial information about that specific moisture level, which may be critical for accurate predictions. Additionally, when a model is trained on data with missing values, it may struggle to learn the underlying patterns effectively. Incomplete data can lead to gaps in understanding the relationships between features; for example, if the model needs to correlate S90 with other features like S10 or S30 but misses data points in S90, the training process may fail to capture important interactions. Also, the presence of this missing values can introduce other challenges like data bias and reduced dataset:

Figure 10. Missing values in soil moisture

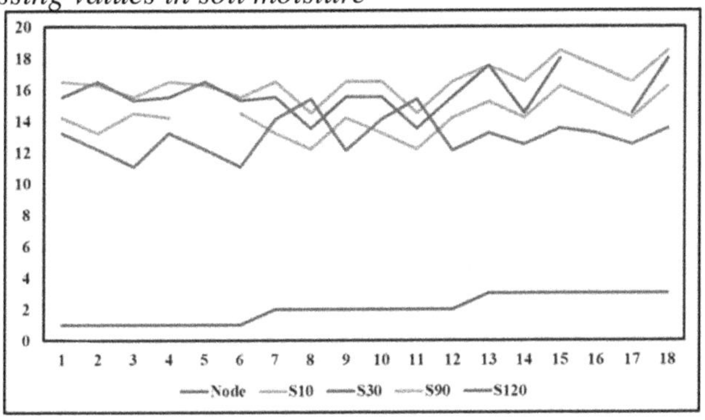

Data Bias: The presence of missing values can introduce bias in the data if the missingness is not random. For example, if the missing values in S90 for Node 1 and Node 3 occur more frequently during specific environmental conditions (e.g., dry conditions), this could skew the model's understanding. The models might become biased towards those conditions, underestimating or overestimating moisture levels. This bias could cause the irrigation system to mismanage water distribution, resulting in either over-irrigation or under-irrigation.

Reduced Sample Size: Missing soil moisture and weather measurement values can reduce the effective sample size used for training the models. For instance, if we have multiple rows with missing values, such as the missing S30 and S90 for Node in Figure 10, this diminishes the number of data points available for model training. A smaller dataset can lead to less robust and less accurate predictions, as the model might not have enough variability in the training data to learn effectively. As a result, the model may not generalize well to new, unseen data, leading to potentially ineffective irrigation recommendations.

In conclusion, the accuracy and effectiveness of machine learning algorithms in the smart irrigation system might suffer from noise and missing values. The quality of data used for training and prediction must be improved by proper treatment of noise through outlier detection and removal and resolving missing values through appropriate imputation approaches. Section 4.4.4 below describers the strategy used to handle these cases in this research study.

4.4.4 Handling Outlier and Noise Data

This research study evaluated, various statistical and deep learning-based machine learning algorithms to remove the outliers and noise from the soil moisture sensor data. In this section, we examine various algorithms used to detect and handle outliers and noise in soil moisture data. Effectively handling outliers and noise is critical for maintaining data integrity and ensuring accurate predictions in irrigation systems.

4.4.5 Outliers and Noise in Soil Moisture Data

In this research study Isolation Forest, Local Outlier Factor (LOF), Z-Score, K-Means Clustering and One-Class SVM algorithms are evaluated to find the accuracy in removing noise and outlier. The evaluation is carried out based on the accuracy (%)and mean absolute error (MAE). Figure 11 and Figure 12 shows the accuracy and MAE of various noise and outlier elimination algorithm in eliminating the noise where X-axis represent the algorithm name and Y-axis represent its accuracy and MAE respectively.

Figure 11. Accuracy of noise elimination algorithm.

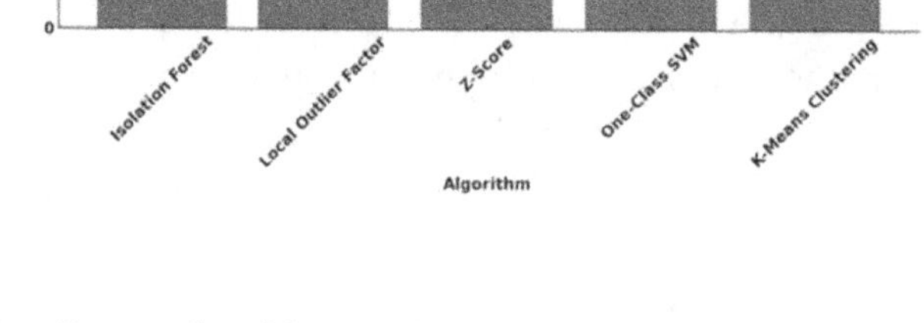

The Isolation Forest algorithm achieved an accuracy of 81.02% and a MAE of 0.328, making it one of the best performing methods for detecting outliers in this dataset. Isolation Forest works by isolating outliers through random data partitioning. The key advantage of Isolation Forest is that it does not assume any specific data distribution, which is crucial for soil moisture data that often exhibits non-linear patterns due to various external environmental factors. Isolation Forest efficiently handles outliers without significantly impacting valid data points, maintaining the integrity of the dataset.

Figure 12. Mean absolute error of noise elimination algorithm

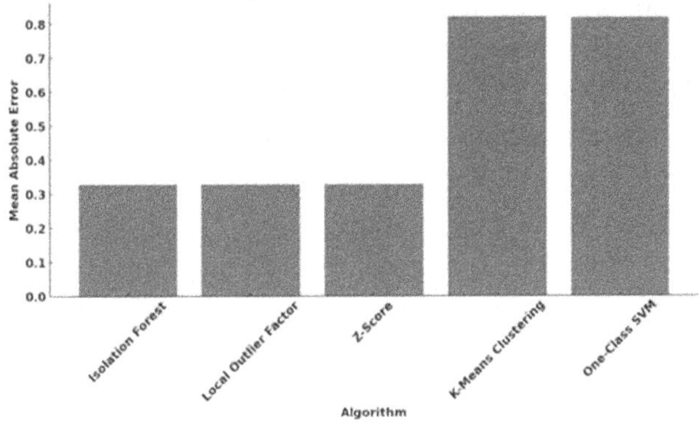

Similarly, the Local Outlier Factor (LOF) algorithm also achieved an accuracy of 81.02% and a MAE of 0.328. LOF is a density-based algorithm that compares the local density of a data point to the densities of its neighbours. This makes it highly effective in clustered datasets, such as soil moisture readings, where sensors in close proximity are expected to report similar values. LOF can identify local outliers data points that deviate from the expected pattern of nearby data, while distinguishing these outliers from random noise in the dataset. Like Isolation Forest, LOF does not rely on a normal data distribution, making it highly adaptable to real-world datasets with irregular structures.

Z-Score, which also achieved an accuracy of 81.02% and a MAE of 0.328, uses the number of standard deviations a data point is from the mean to identify outliers. While its performance, in terms of accuracy and error, matches that of Isolation Forest and LOF, Z-Score is less reliable for soil moisture data. This is because it assumes that the data follows a normal distribution, which is not always the case in environmental datasets. Soil moisture data, for example, can be skewed by sudden rainfall or sensor inaccuracies, which can lead Z-Score to either miss true outliers or falsely classify valid data as outliers.

K-Means Clustering, on the other hand, performed less effectively with an accuracy of 50% and a MAE of 0.82. K-Means is primarily a clustering algorithm that groups data into predefined clusters, with outliers being defined as points that fall far from the cluster centers. The limitation of this approach in soil moisture data is that moisture levels tend to change gradually, rather than forming distinct clusters. This makes K-Means less effective at distinguishing between normal data points and outliers. Additionally, the need to predefine the number of clusters adds another layer of rigidity to the algorithm, making it less adaptable to dynamic datasets like soil moisture readings.

One-Class SVM achieved a slightly better performance than K-Means with an accuracy of 52.69% and a MAE of 0.817. This algorithm creates a boundary around normal data points, classifying points that fall outside this boundary as outliers. However, One-Class SVM can be sensitive to parameter tuning, which leads to inconsistent results if not properly optimized for the dataset. Its relatively low accuracy and higher MAE indicate that it struggles to handle the complexities and irregularities of soil moisture data effectively.

Based on the comparison of accuracy and MAE, Isolation Forest and Local Outlier Factor (LOF) emerge as the best algorithms for handling outliers and noise in soil moisture data. Both algorithms demonstrate the highest accuracy (81.02%) and the lowest MAE (0.328), indicating their superior ability to identify outliers while preserving the integrity of the normal data. They are flexible in handling non-linear, complex data patterns, which are common in soil moisture datasets influenced by external environmental factors. Isolation Forest and LOF are also robust against noise, meaning that small, random fluctuations in the data are less likely to be misclassified as outliers.

While Z-Score achieved similar accuracy and MAE, its reliance on the assumption that data follows a normal distribution limit its applicability in real-world scenarios, such as soil moisture data, where environmental variables often cause deviations from a normal distribution. K-Means Clustering and One-Class SVM were less effective, with lower accuracy and higher MAE, primarily because they either rely on rigid clustering assumptions or require careful parameter tuning that is difficult to optimize for complex datasets.

Based on the findings from this research study and the data obtained using soil moisture sensors, Isolation Forest and Local Outlier Factor (LOF) have proven to be the most effective algorithms for detecting outliers and handling noise in soil moisture data. Their flexibility, high accuracy, and low error make them the best choice for ensuring data quality in environmental datasets where external factors like precipitation, wind speed and wind direction can cause significant irregularities.

4.4.6 Missing Values Handling

The research study has devised a thoughtful and effective imputation technique to address the challenge posed by missing values in the soil moisture sensor data as mentioned in section 4.4.3. The approach taken by this study involves looking at the neighbouring data points around the missing values. Specifically, they consider the previous 5 to 10 and next 5 to 10 soil moisture and weather values from the current timestamps. By examining this window of data around the missing value, they gain valuable context about the trends and patterns of soil moisture changes over time. Using this neighbouring data points the missing values are computed by taking average. Averaging the surrounding values helps to impute the missing value with a reasonable estimate that aligns with the soil moisture patterns observed in the immediate past and future.

The decision to use an averaging technique for imputation is well-founded. It is observed that soil moisture does not undergo significant changes quickly. Therefore, taking an average of the neighbouring data points is a sensible approach that provides a more accurate estimation of the missing value than other imputation methods.

By employing this imputation technique, the research study successfully fills in the gaps caused by missing values, ensuring that the dataset used for analysis and machine learning is complete and robust. This completeness is vital for training accurate and reliable machine learning models to make informed irrigation schedule decisions.

Figure 13. Missing value handling

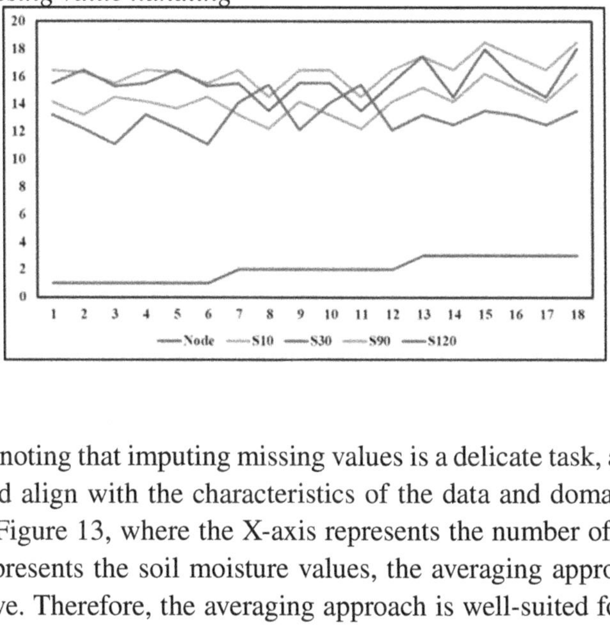

It is worth noting that imputing missing values is a delicate task, and the chosen method should align with the characteristics of the data and domain knowledge. As shown in Figure 13, where the X-axis represents the number of iterations and the Y-axis represents the soil moisture values, the averaging approach generated a smooth curve. Therefore, the averaging approach is well-suited for cases where gradual changes in soil moisture are expected, helping to maintain the continuity of the time-series data.

After processing the data for noise and missing values, the data is stored in a time-series database for further analysis. In smart irrigation, merely storing data is not sufficient; it is essential to apply business logic to derive actionable insights based on this data. In this research study, this processed data is utilized to predict the irrigation need using incremental learning as explained in section 4.5.

4.5 Online (Incremental) Learning Algorithm

This smart irrigation research study adopts the online (incremental) machine learning approach, which enables models to continuously learn from fresh data over time without the need to completely retrain the model. Traditional batch learning requires retraining the model on the full dataset whenever new data is added, which can be computationally expensive and time-consuming. On the other hand, incremental learning makes the model more adaptable and scalable in dynamic contexts by iteratively updating it and incorporating new data as it becomes available. Due to changes in weather, the landscape, and soil moisture, this method is effective for smart irrigation.

This smart irrigation system uses, the live dataset obtained from sensors is represented as feature vector x_i in equation 1-1. This dataset comprises soil moisture data (M_1, M_2... M_n), weather data (W_1, W_2... W_n), and field data (F_1, F_2... F_n). Before being used for training and prediction, this live data is first processed through a noise elimination algorithm, creating a cleaned feature dataset xi^C. The noise elimination algorithm is crucial for removing unwanted variations and inaccuracies from the data, ensuring the quality and reliability of the dataset.

The cleaned feature dataset xi^C in equation 1-2. consists of transposed real-time soil moisture data (M_1^C, M_2^C ... M_n^C), cleaned weather data (W_1^C, W_2^C ... W_n^C), and field data (F_1, F_2... F_n). These cleaned and accurate data streams are then fed into various incremental machine learning algorithms for training and prediction purposes. The objective is to train the algorithms to make predictions about the irrigation needs, specifically predicting soil moisture at different depths (M_{07}, M_{30}, M_{60}, M_{120}). y_i in equation 1-3 represents the predicted soil moisture at different depths (M_{07}, M_{30}, M_{60}, M_{120}). These depth values represent different layers of soil, allowing the system to assess moisture levels at various depths for better irrigation decision-making.

$$x_i = [M_1, M_2 \ldots M_n, W_1, W_2 \ldots W_2, F_1, F_2 \ldots F_n] \quad (1\text{-}1)$$

$$x_i^C = [M_1^C, M_2^C \ldots M_n^C, W_1^C, W_2^C \ldots W_n^C, F_1, F_2 \ldots F_n \quad (1\text{-}2)$$

$$y_i = \{ M_{07}, M_{30}, M_{60}, M_{120} \} \quad (1\text{-}3)$$

Figure 14 depicts the flowchart illustrating the data feeding process into the incremental learning algorithms. The incremental learning algorithms are designed to continuously learn from the incoming data, updating their models to adapt to changing patterns and trends. As the algorithms learn incrementally from various datasets, the study demonstrates that the accuracy of the models progressively increases. This continuous learning approach ensures that the models can keep up with evolving environmental conditions and maintain high accuracy in irrigation predictions. Figure 15 below shows the incremental learning module in application cloud.

Figure 14. Data feeding mechanism to incremental learning algorithm

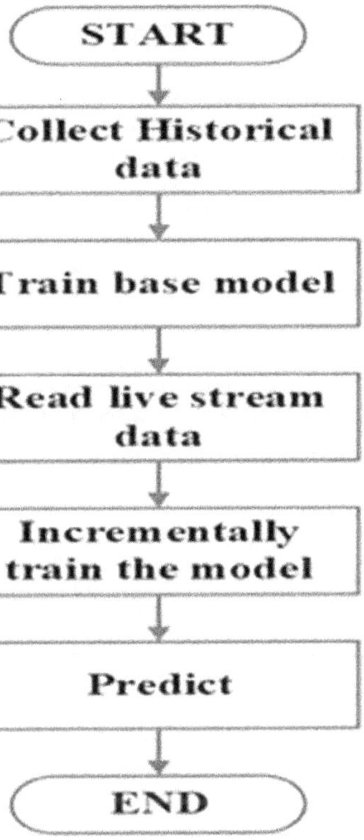

Figure 15. Incremental learning module in application cloud

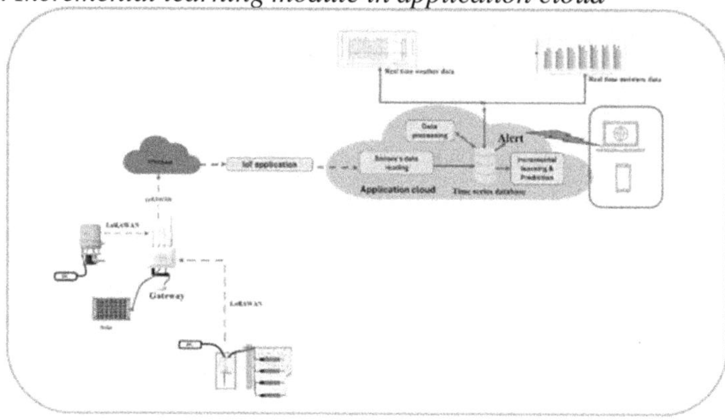

Figure 16 further showcases the learning ability of different incremental learning algorithms when trained using seasonal data. The X-axis represents the number of iterations the incremental learning model is tested, while the Y-axis represents the prediction accuracy.

Figure 16. Accuracy of incremental learning algorithm

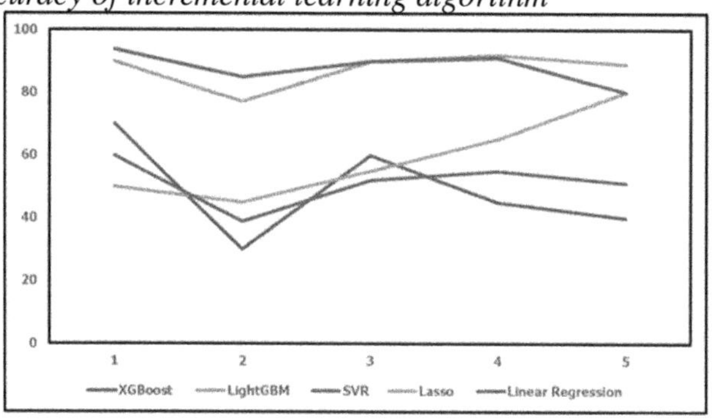

As shown in Figure 16 XGBoost starts with an accuracy of 70%, indicating a good initial prediction of soil moisture based on weather data, but it drops significantly in the second iteration, due to sudden change in climate condition. As learning continues it recovers in later iterations but does not surpass its initial performance, suggesting that it may struggle to generalize well to changing weather patterns.

LightGBM starts lower but shows consistent improvement across iterations. This steady growth indicates that it is learning progressively from the soil moisture and weather data, adapting well to new information with less variability than XGBoost.

LightGBM's stability makes it a reliable choice for long-term predictions in this scenario.

SVR, which starts similarly to XGBoost, performs better across later iterations, maintaining stable accuracy after iteration 3. This suggests that SVR may also be able to handle the complexities of soil moisture predictions from weather data but requires more data to stabilize its predictions.

Lasso starts with a high accuracy of 90% and, despite a slight drop in iteration 2, continues to improve. This high performance implies that Lasso's regularization techniques are effective in preventing overfitting and making accurate predictions, even as the weather and soil conditions vary over time. It's especially suitable for situations where there are many correlated features in the weather and soil data.

Linear Regression shows the highest accuracy from the beginning (93.68%) and continues to improve steadily. This strong performance across iterations indicates that Linear Regression is highly capable of capturing the relationship between soil moisture and weather variables. It consistently performs well, even as new data is introduced, making it the most reliable model in this comparison for predicting soil moisture.

In conclusion, for predicting soil moisture using soil moisture and weather data, Linear Regression and Lasso show the best performance due to their consistency and ability to effectively handle relationships between variables. LightGBM demonstrates good adaptability and stability, making it a strong candidate for smart irrigation decision-making. Given its good adaptability in this study, LightGBM is used for predicting irrigation needs.

Once the LightGBM models generate soil moisture predictions and determine the irrigation needs by comparing the results with user-configured soil moisture thresholds, the system sends commands to turn the sprinklers ON or OFF in Cairns Regional Park. The sprinkler system in Cairns integrates various heterogeneous components manufactured by different vendors. To ensure seamless communication of prediction results to the sprinkler system, an actuator solution has been designed. Section 4.5.2 provides a detailed discussion of the actuator design and its communication mechanism.

4.5.1 Benefits of Incremental Machine Learning Model

Real-Time Adaptation: Using MP406 soil moisture sensors and ATMOS-41 weather stations, incremental learning allows the irrigation model to adapt to real-time changes in soil moisture and weather data. As these sensors provide continuous, live updates on soil conditions (e.g., moisture levels) and atmospheric variables (e.g., temperature, humidity, wind speed), the model can adjust the irrigation strategy immediately without waiting for large batches of data to be processed. This real-time adaptation ensures that water usage is always optimized for the current conditions, which is crucial for effective irrigation.

Scalability: The combination of MP406 soil moisture data and ATMOS-41 weather station data can generate a significant amount of information over time, especially in larger parks. Incremental learning processes this data in small, manageable chunks, enabling the model to scale effectively. Instead of requiring the entire dataset to be available at once, the model updates itself with each new reading from the soil and weather sensors. This scalability makes it suitable for large-scale irrigation systems or big data scenarios, where continuously increasing datasets must be managed efficiently.

Reduced Training Time: In a smart irrigation system that uses MP406 soil moisture sensors and ATMOS-41 weather stations, incremental learning eliminates the need to retrain the entire model from scratch every time new data arrives. Instead, the model updates incrementally as new soil and weather data become available. This reduces training time and computational load, allowing the system to respond quickly to changes, such as sudden rainfall or dry periods, ensuring that the irrigation process remains optimal without excessive delays.

Lower Memory Footprint: Since incremental learning allows the model to update based on new MP406 soil moisture readings and ATMOS-41 weather data without retaining all historical data, it operates with a reduced memory footprint. This is particularly valuable in resource-constrained environments like IoT often used in field operations. These devices are typically limited in storage capacity, and incremental learning ensures efficient performance without overloading the system with large datasets.

Adaptive to Drift: Over time, the data distribution from MP406 soil moisture sensors and ATMOS-41 weather stations can change due to environmental factors like global warming, seasonal shifts, or changes in land use. Incremental learning enables the model to adapt to this drift in data distribution. As the weather patterns evolve or the soil composition changes, the model remains accurate and relevant by continuously learning from new data, ensuring high performance in the long term. This adaptability is critical for maintaining the efficiency and accuracy of smart irrigation systems that rely on fluctuating environmental data.

Incremental learning is a powerful machine learning approach that provides real-time adaptation, scalability, reduced training time, and memory efficiency. It is well-suited for dynamic and evolving data environments like weather and soil moisture, making it an invaluable tool in various real-world applications where continuous learning and adaptive models are essential.

4.5.2 Prediction Result Communication

The incremental machine learning algorithm implemented in the application cloud IoT, as illustrated in Figure 15 of the IoT architecture. This algorithm retrieves data from the data storage system and processes it to predict the irrigation needed for the plants. Once the prediction is made, the result is sent to the actuator using the MQTT protocol. MQTT is a lightweight messaging protocol suitable for IoT devices, which ensures reliable and efficient communication between the cloud server and the actuator.

Figure 17. Actuator to in IoT architecture

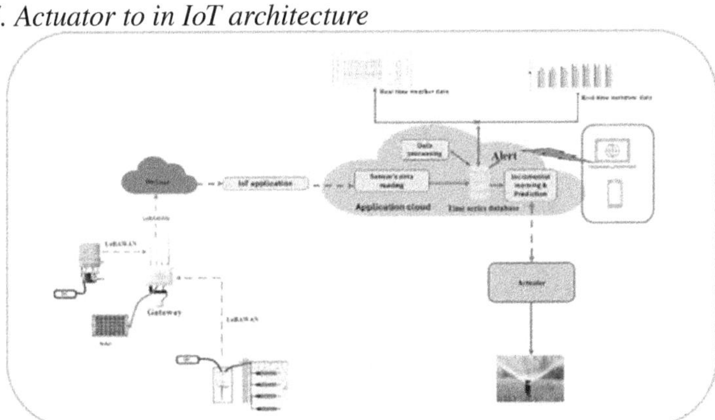

The actuator, which is a separate component in the IoT architecture as shown in Figure 17 and Figure 18, receives the predicted irrigation decision from MQTT server. It is responsible for controlling the sprinklers by turning them ON or OFF based on the received instructions. This automated control ensures that the plant receive the right amount of water at the right time, optimizing water usage and enhancing aesthetic look of the park. The Figure 18 shows the command flow mechanism between application in cloud with actuator which has Raspberry pi3 as operating system.

Figure 18. Communication mechanism from application cloud to the sprinkler controller

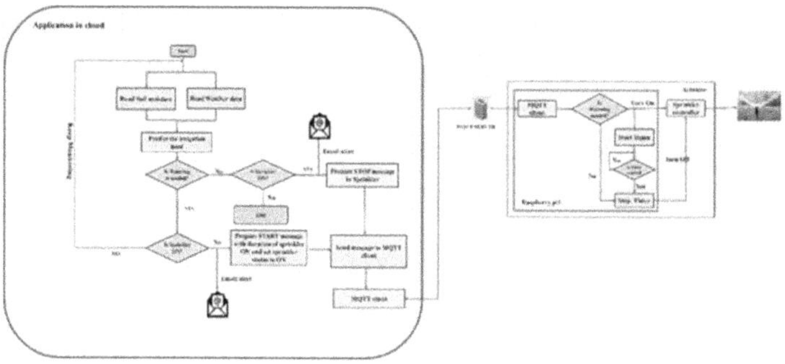

4.5.3 Comparative Studies of AI-Driven Smart Irrigation Approaches

The integration of AI-driven irrigation strategies has gained prominence in global agricultural and landscape management. Various approaches have been implemented, each with distinct advantages, limitations, and suitability for different environments. This section presents a comparative analysis of key AI-based irrigation methods, highlighting their applications, benefits, and challenges. The Table 1 provides a comparative analysis of prominent AI-driven smart irrigation approaches used worldwide:

Table 1. Comparative analysis of AI-driven smart irrigation approaches

Approach	Key Features	Advantages	Limitations	Case Study Example
Incremental Learning-Based Smart Irrigation (Cairns Park, Australia)	Adapts continuously to real-time environmental changes, utilizes IoT and soil moisture sensors, optimizes water use dynamically.	Highly adaptive, improves water conservation, reduces maintenance costs.	Requires consistent data flow; model updates can be complex.	Cairns Regional Park, Australia
Reinforcement Learning-Based Irrigation (DRLIC Model)	Uses trial-and-error learning to optimize irrigation schedules.	Learns optimal watering strategies, reduces water waste.	Requires large datasets; training period is long.	(Patel & Kumar, 2023)
Deep Neural Networks for Irrigation Prediction	Uses deep learning to analyze multi-dimensional weather, soil, and crop data.	High accuracy in long-term predictions.	Computationally intensive, requires significant training data.	(Zhang, Liu, & Wang, 2022)

continued on following page

Table 1. Continued

Approach	Key Features	Advantages	Limitations	Case Study Example
Fuzzy Logic-Based Irrigation Control	Uses rule-based logic to adjust irrigation based on temperature, humidity, and soil moisture.	Simple implementation, works well with limited data.	Less effective in dynamic weather conditions.	(Santosh & Rao, 2021)
Hybrid AI-IoT-Based Irrigation (Cloud and Edge Computing)	Combines cloud computing and edge processing for real-time water management.	Fast response time, scalable for large farms.	High initial setup cost, requires strong network connectivity.	(Gonzalez, Fernandez, & Lopez, 2022)
Genetic Algorithm-Based Irrigation Optimization	Uses evolutionary algorithms to optimize irrigation scheduling.	Efficient in optimizing multi-objective functions like water use and crop yield.	Requires complex calibration; slower adaptation to real-time changes.	(Romero & Hernandez, 2021)
Mixed-Integer Model Predictive Control (MPC)	Combines predictive models with optimization constraints for real-time irrigation scheduling.	Ensures precise control over water distribution and minimizes wastage.	Complex mathematical formulation, high dependency on model accuracy.	**(Wang, Li, & Zhao, 2023)**

5. STUDY RESULT

The implementation of an IoT-based smart irrigation system in Cairns Park has demonstrated promising results, achieving notable improvements in water management and irrigation efficiency. Key outcomes include:

1. **Enhanced Data Accuracy and Soil Moisture Monitoring:** By employing Dual Electromagnetic (Dual-EM) scanning, the project identified optimal sensor placement locations in diverse zones (wet, semi-wet, and dry) within Cairns Park. This strategic sensor deployment improved the representativeness of collected data, leading to more accurate soil moisture monitoring across varied soil conditions and vegetation types. The accurate placement of sensors ensured that data accurately reflected moisture levels at critical root zones and depths beyond to monitor water loss, which is essential for making informed irrigation decisions.
2. **Effective Noise Elimination and Data Quality Improvement:** Various noise elimination algorithms were evaluated, with Isolation Forest and Local Outlier Factor (LOF) achieving the best results, each attaining an accuracy of 81.02%

and a Mean Absolute Error (MAE) of 0.328. These algorithms effectively reduced the impact of noise on soil moisture data, minimizing the presence of outlier values that could otherwise skew irrigation predictions. This noise-free data enabled more consistent model performance, ensuring that predictions were based on reliable information.

3. **Efficient Handling of Missing Data Through Imputation:** Missing values in soil moisture data were effectively addressed using an averaging imputation technique that estimated values based on surrounding data points. This method improved the completeness of the dataset, ensuring that all necessary information was available for model training and real-time predictions. As a result, the irrigation system could respond accurately, even when intermittent data gaps occurred, improving reliability.
4. **Incremental Learning for Real-Time Adaptability:** Incremental learning algorithms, including LightGBM, XGBoost, and SVR, were applied to facilitate real-time adaptability of the irrigation model. LightGBM emerged as the most effective, demonstrating high adaptability and stability across changing conditions. It maintained robust prediction accuracy, allowing the system to respond promptly to environmental changes, such as sudden rainfall or dry spells.
5. **Seamless IoT Communication and Automation of Irrigation Controls:** By integrating IoT components, including MP406 soil moisture sensors, ATMOS-41 weather stations, and a Raspberry Pi-controlled actuator system, the project enabled seamless data transmission and responsive control of the sprinkler system. The use of MQTT protocol ensured reliable communication, with minimal latency, between sensors, cloud applications, and actuators. This setup allowed for automatic irrigation adjustments based on real-time soil moisture data, contributing to consistent water conservation and enhancing park landscape quality.
6. **Water Conservation and Sustainable Resource Management:** The optimized irrigation schedule, powered by accurate soil moisture data and adaptive learning, resulted in significant water savings without compromising plant health. By delivering water only where and when it was needed, the system achieved a sustainable approach to irrigation, aligning with broader goals of environmental conservation and operational efficiency.
7. **Scalability and Cost Considerations:** The project was designed to be cost-effective and scalable through strategic implementation of incremental learning, sensor placement, and IoT-based infrastructure. Incremental learning reduces computational costs by enabling real-time updates without full retraining. Strategic sensor placement, using Dual Electromagnetic (DUAL-EM) scanning, optimized sensor deployment, reducing overall expenses. Low-power IoT sensors and a dual power supply (solar and AC) further minimized operational costs.

A cost-benefit analysis revealed that despite the high initial investment, the system achieves long-term financial benefits through water savings and lower maintenance costs. These design choices make the system viable for large-scale deployment in urban parks and agricultural settings.

5.1 Verification and Validation

The smart irrigation system ran for 30 days before and after deploying our smart sprinkler system in Esplanade parkland to test our designed smart sprinkler system. We know Esplanade Park has two stations: Station 11 runs for 19 minutes per day, and Station 12 runs for 17 minutes per day. We calculated the total water sprinkled by each station per m^2 per day based on this available data. In the old system, we can state that station 11 sprinkled 85.5 liters per month per m^2 and station 12 sprinkled 76.5 liters per month per m^2 based on the total watering of station 1 over 30 days. Figure 20 illustrates a comparison of water dispensed by station 11 and station 12 before and after deployment of the smart irrigation system.

The sprinkler was switched off for nine days in a month after installing our smart irrigation system. Thus, we have saved a total of 48.6 liters in 30 days, as illustrated in Figure 19 where station 11 sprinkled 59.85 liters per m^2 (saved 25.65 liters) and station 12 sprinkled 53.55 liters per m^2 (saved 22.95). Our validation shows that our approach saves around 583 liters per year per m^2 of parkland in Cairns. Apart from saving water, it is essential to assess how much water drains below the root depth, as this is considered waste. To validate this, lysimeters were installed below the root depth to measure the amount of water going beyond the root zone. The data shows that the quantities of water collected in the lysimeters varied considerably less, as explained in Section 5.1.2. Hence, we can conclude that the machine learning algorithm significantly reduced the amount of water draining below the root zone.

Figure 19. Water usage before and after smart irrigation

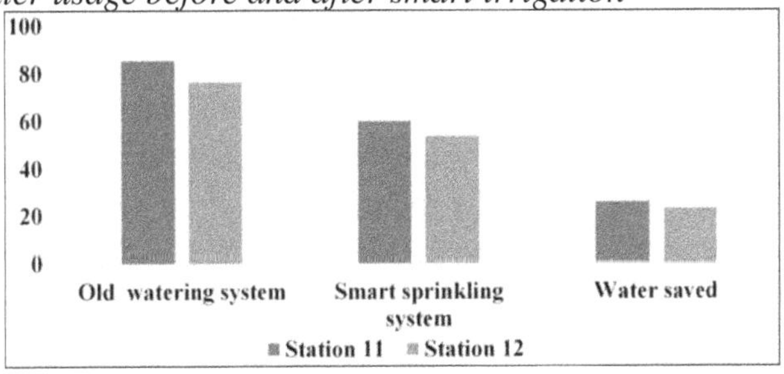

5.2 Installation and Monitoring Water Below Root Zone Using Lysimeters

As shown in Figure 20, two lysimeters were placed in each trench excavated for sensor installation. Each lysimeter consists of a PVC pipe (15 cm in diameter) sealed at the base and covered at the top with geotextile fabric supported by a fly screen base. This setup allows drained water to be collected and retained within the closed PVC pipe (lysimeter). A garden hose was inserted into the PVC pipe to facilitate the withdrawal of drained water. The top end of the garden hose was enclosed in a plastic box along with the sensor node and power junction. The first lysimeter was installed 30 cm below the ground surface, and the second was placed 60 cm below the ground surface. The water collected in the lysimeter was sampled at different intervals to measure the amount of water percolating below 30 cm, which is the root depth of the grass.

Figure 20. Lysimeters at site

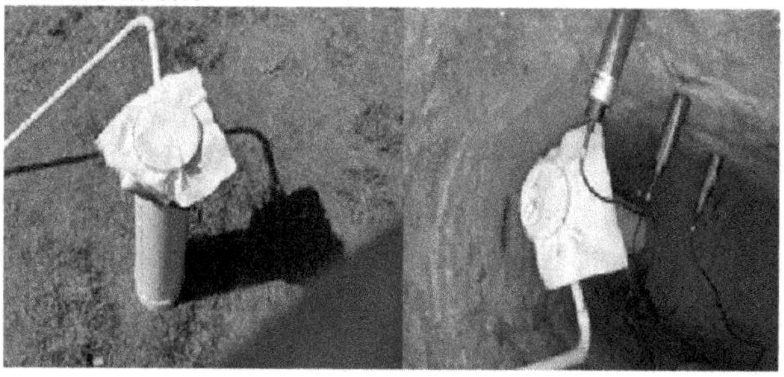

Figure 21 illustrates the quantities of water collected in the lysimeter before and after its installation. The data shows that the water percolating below the root zone has been significantly reduced after the installation of the new system. This reduction indicates the system's effectiveness in minimizing deep water loss and retaining moisture within the root zone. The data shows that the water percolating below the root zone has been significantly reduced after the installation of the new system. This reduction indicates the system's effectiveness in minimizing deep water loss and retaining moisture within the root zone.

Figure 21. Lysimeters data

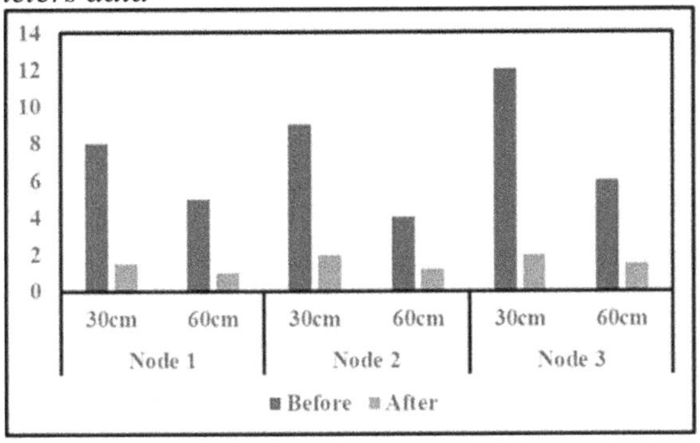

While predicting and validating water usage is important, it is not sufficient on its own to optimize irrigation in real-time. To enhance decision-making, an additional solution is provided by using the LSTM (Long Short-Term Memory) algorithm. This algorithm is designed to forecast the soil moisture content one hour ahead, allowing park management to know how the soil moisture will be after one hour. Section 5.1.3 describes the forecasting mechanism used in this study.

5.3 Forecasting Soil Moisture Using Machine Learning Algorithm

Soil moisture forecasting is essential for making informed irrigation decisions, ensuring appropriate water management, and encouraging healthy plant growth. The stacked Long Short-Term Memory (LSTM) algorithm is employed to forecast soil moisture. LSTM algorithms are well-known for their superior performance in time series regression and classification tasks. In the context of forecasting soil moisture, LSTM demonstrates its effectiveness by preserving valuable information across multiple time steps with minimal preprocessing of input data.

LSTM is a deep learning approach that utilizes a chain-like structure to address the issue of long-term reliance problems in traditional recurrent neural networks. It achieves this by incorporating a concurrent cell update structure, which allows the model to update the cell state early in the process. The initial update takes place shortly after the first output of the initial LSTM block, using the network's initial state and the first time step of the sequence to compute the output. Subsequently, at each time step t, the block employs the current state (c_{t-1}, y_{t-1}) to update the cell state c_t and compute the output for the following time step. Each layer in the LSTM model consists of two states: the cell state and the hidden state (also known as the output state). The hidden state at time step t contains the output of the LSTM layer for that specific time step, while the cell state captures the information learned from previous steps. The model uses gates to control the addition or removal of information from the cell state at each time step, allowing LSTM to preserve and learn long-term dependencies in time series data effectively.

This feature of LSTM, preserving and learning from previous information, proves especially valuable in soil moisture-based irrigation scheduling. The current soil moisture content is influenced by past soil moisture levels, precipitation, and weather data. By considering this historical information, the LSTM algorithm can make accurate forecasts, enabling precise irrigation scheduling.

To assess the forecasting efficiency of the LSTM algorithm, a comparison is made between the predicted accuracy of the training and test data. Figures 22, 23, and 24 showcase the accuracy of the LSTM algorithm using the test dataset against the actual data. In Figure 22, the blue line represents the soil moisture values before forecasting. In Figure 23, the green line represents the forecasted moisture values with the test data. It is evident from Figure 23 that the forecasted values closely overlap with the actual values, indicating the accuracy of the predictions.

Figure 22. Soil moisture before forecasting with training data

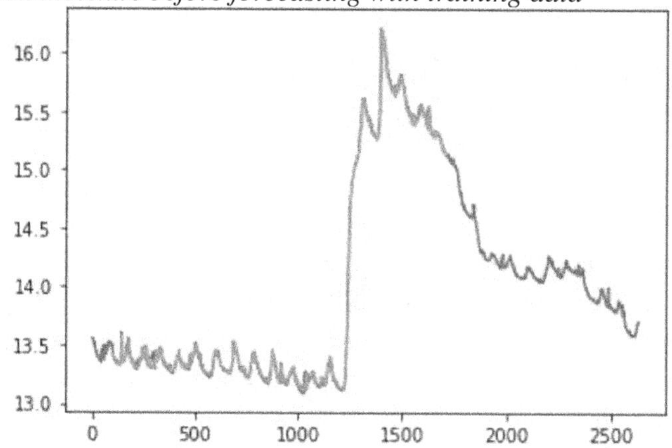

Figure 23. Soil moisture after forecasting with training data

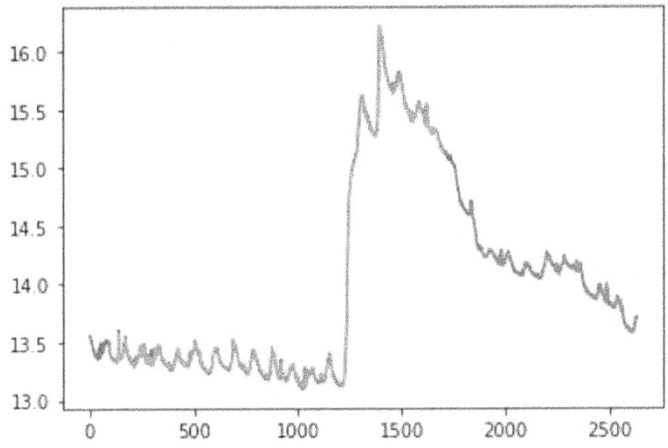

In Figure 24 blue line displays the previous soil moisture values considered for forecasting future soil moisture and orange line represents the forecasted soil moisture value. The X-axis represents the number of previous moisture values considered, while the Y-axis represents the forecasted soil moisture value. The Root Mean Square Error (RMSE) value with Mean Absolute Error (MAE) for the test-predicted data is 4.32 and 4.50, respectively. The difference between these values is approximately 0.18, indicating good forecasting accuracy. Based on the mean absolute error, the LSTM model achieves an accuracy of around 95%.

Figure 24. Soil moisture LSTM future forecasting

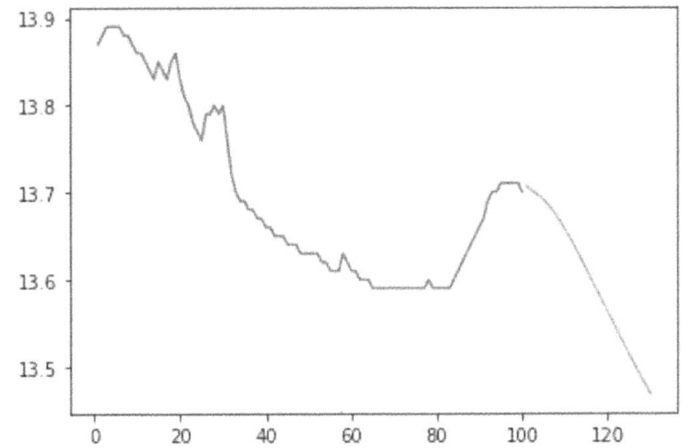

In conclusion, the stacked LSTM algorithm proves to be an efficient and accurate approach for forecasting soil moisture. Its ability to learn from long-term time series data and preserve important historical information makes it a valuable tool in smart irrigation management. The model's forecasting accuracy, as verified by comparing test and actual data, confirms its reliability and effectiveness in supporting precision irrigation practices.

However, despite the promising results of the stacked LSTM algorithm and incremental learning algorithm, several challenges remain in the deployment of smart irrigation systems. Section 6, Challenges Faced in Smart Irrigation Deployment, provides a detailed discussion of these obstacles and explores potential solutions to ensure the successful adoption of smart irrigation technologies.

6. CHALLENGES FACED IN SMART IRRIGATION DEPLOYMENT

This research study is conducted in a public space, presenting several challenges during deployment. These include identifying optimal sensor's numbering, calibrating soil moisture sensors, node protection from climate hazards, and ensuring continuous power supply. Each of these challenges requires careful consideration and strategic planning to ensure the effectiveness and reliability of the smart irrigation system. In the following sections, we will explore these challenges in detail and discuss the methodologies employed to overcome them

6.1 Scientific Sensor Numbering and Depth Mapping

As shown in Figure 15 and 25, this research study utilizes four arrays of sensors connected to a central point. Each sensor is placed at a different depth and transmits learning algorithm relies on this depth-based data to predict irrigation needs accurately. Since the sensor manufacturing company may not connect wires in a specific order, each sensor is tested and numbered individually to ensure it can be mapped correctly to its installation depth. This approach ensures precise sensor identification in the prediction module

Figure 25. Array of sensor with node

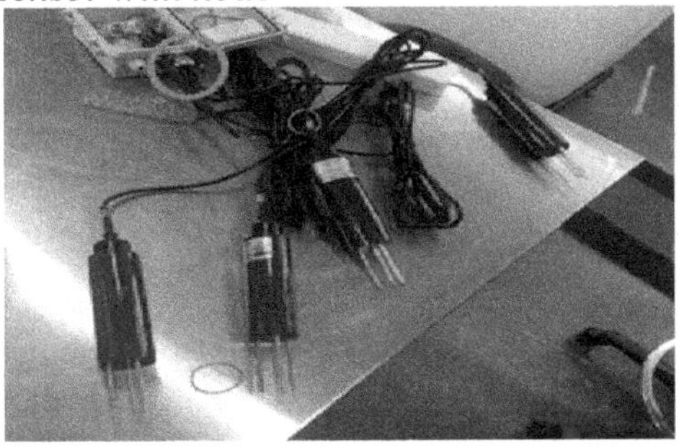

6.2 Soil Moisture Sensors Calibration

In this study, soil moisture sensor calibration involves an experimental setup with three pots, as shown in Figure 26 each simulating different soil conditions:

Pot 1 - Dry Soil: This pot contains dry soil to simulate a water-deficient environment.
Pot 2 - Saturated Soil: This pot is filled with soil at its maximum water-holding capacity, representing full moisture.
Pot 3 - Saline Soil: Saline water is added to simulate conditions found in areas affected by soil salinity.

Figure 26. Calibration setup

Arrays of soil moisture sensors are strategically placed in each pot to monitor soil conditions over time. The sensors collect real-time data on how soil moisture levels fluctuate under the different scenarios, providing insights into how the sensors respond to changes in soil moisture, salinity, and water saturation.

Based on the data gathered from these sensors, calibration adjustments are made to ensure that the sensor outputs accurately reflect the soil conditions they are measuring. These adjustments are crucial for enhancing the accuracy and reliability of the sensors when deployed in real-world scenarios.

6.3 Dual Power Supply System and Solar Energy Integration

Power supply is a crucial aspect of smart irrigation systems as it directly affects the ability to collect continuous and accurate data for predicting irrigation needs. Interruptions in power supply can lead to data loss, which can result in incorrect predictions and ultimately lead to over- or under-irrigation.

To address this issue, this project has implemented a dual power supply system consisting of both AC and DC power supplies. This redundancy ensures that the devices remain powered in case of a power outage or failure of one power supply. This approach not only improves the reliability of the system but also reduces the risk of data loss.

Figure 27. Solar power installation

In addition to the dual power supply system, the project has also utilized solar energy as shown in Figure 27 to power the gateway device used for communication. This approach not only provides a reliable and uninterrupted power source but also addresses safety concerns associated with traditional power sources. The use of solar power ensures that the system can operate independently without relying on external power sources, reducing the risk of power supply interruptions and ensuring continuous operation.

6.4 Protecting Nodes from Environmental Factors

To protect nodes from environmental factors like rainwater, dust, and debris, they are enclosed in waterproof boxes, preventing moisture infiltration and circuit board damage. These boxes shield nodes from exposure, ensuring durability and long-term functionality. Additionally, nodes are installed in an inverted position inside the boxes, allowing any water that enters to drain out, minimizing the risk of damage. This setup is particularly important in regions with heavy rainfall, where water exposure can be a significant concern. By implementing this protective measure, the smart irrigation system maintains reliable data collection, preventing sensor failures and ensuring accurate soil moisture readings. Figure 28 illustrates the installation at the experimental site, highlighting the effectiveness of this approach in preserving node functionality and sustaining the efficiency of the smart irrigation network.

Figure 28. Node installation

6.5 Environmental Impacts of IoT-Based Irrigation on Biodiversity in Cairns Park

The IoT-based smart irrigation system in Cairns Park optimizes water use but impacts biodiversity. Precise irrigation alters soil microbiota and nutrient cycling, affecting native plants reliant on seasonal moisture changes. Over-regulation may disrupt drought-adapted ecosystems, while preventing waterlogging reduces invasive species spread. By considering plant types and strategically identifying appropriate sensor locations, the system ensures accurate soil moisture data for species-specific irrigation. Incremental learning integrates biodiversity-aware models, dynamically adjusting irrigation based on soil moisture variations and plant health indicators, maintaining ecological balance while maximizing water efficiency.

7. PRACTICAL IMPLICATION

The findings of this research provide valuable insights for policymakers, farm managers, and environmentalists seeking to optimize water use through smart irrigation.

For Policymakers: Governments and municipal authorities can leverage these findings to establish data-driven water conservation policies. By incorporating IoT-based smart irrigation into urban parklands and agricultural subsidies, policymakers can promote sustainable water management practices, reduce water waste, and address the growing challenges of climate change and resource scarcity. Incentivizing the adoption of smart irrigation technologies through grants or public-private partnerships can accelerate large-scale implementation.

For Farm Managers and Agricultural Practitioners: The integration of IoT sensors, machine learning algorithms, and incremental learning models provides real-time adaptability for farm irrigation. Farmers can use this system to ensure precise water application based on actual soil moisture conditions, leading to improved crop yield, reduced irrigation costs, and better resource management. By utilizing automated sensor-based irrigation, farms can minimize labor costs and prevent over-irrigation, thus maintaining soil health and preventing nutrient leaching.

For Environmentalists and Urban Landscape Managers: Smart irrigation systems offer significant benefits for biodiversity conservation by preventing overwatering and maintaining optimal soil moisture levels. Urban landscape managers can apply these techniques to public parks, gardens, and green spaces to ensure efficient water distribution without disturbing natural ecosystems. The system's ability to dynamically adjust irrigation based on real-time data helps protect native vegetation and prevents the growth of invasive species due to excessive watering.

8. FUTURE RECOMMENDATION

Future studies could focus on the following aspects to enhance the understanding and implementation of smart irrigation systems:

8.1 Integration of Additional Data Sources

Incorporating additional data sources, such as satellite imagery, remote sensing could further improve the accuracy of soil moisture predictions.

8.2 Real-Time Sensor Calibration

Developing real-time calibration techniques for sensors could mitigate measurement errors and ensure accurate data collection. Investigating methods to automatically calibrate sensors based on observed discrepancies between sensor readings and actual field conditions would enhance prediction accuracy.

8.3 Case Studies Beyond Cairns to Illustrate Broader Applicability

Expanding the study to other regions with diverse climates and ecosystems can validate the scalability of smart irrigation systems. Case studies from arid regions, tropical landscapes, and temperate agricultural zones can illustrate how incremental learning adapts to varying soil and plant conditions. Future research should explore implementations in urban parks, farmlands, and conservation areas to assess the broader impact on water efficiency, biodiversity, and sustainability.

By addressing these future research recommendations researchers can continue to advance the field of smart irrigation, contributing to sustainable water management practices, improved crop yields, and enhanced ecological conservation in agricultural and park settings.

10. CONCLUSION

This research demonstrates the potential of an IoT-based smart irrigation system to address critical challenges in modern water management and agriculture. By integrating advanced sensor technology, data processing techniques, and machine learning algorithms, the system offers a data-driven approach to optimize water distribution in environments like Cairns Park. Strategic sensor placement, effective noise elimination, and real-time adaptability using incremental learning enhance irrigation precision, reduce water waste, and support sustainable landscape management.

Dual Electromagnetic (Dual-EM) scanning for optimal sensor placement has improved soil moisture data accuracy, essential for informed irrigation decisions. Noise elimination techniques, such as Isolation Forest and Local Outlier Factor (LOF), enhance data quality, ensuring predictive models rely on reliable inputs. Incremental learning models, particularly LightGBM, adapt effectively to changing conditions, enabling real-time responses to soil and weather fluctuations. This adaptability improves irrigation scheduling, conserving water while promoting plant health across park zones.

The project emphasizes the importance of a robust IoT communication framework for seamless data transmission and irrigation control. Using the MQTT protocol, the system ensures reliable communication between sensors, cloud applications, and actuators, enabling automatic real-time irrigation adjustments. This setup enhances water efficiency and provides a scalable, adaptable solution for varied landscapes.

In conclusion, smart irrigation systems powered by IoT and machine learning play a pivotal role in sustainable water management in public parklands and agricultural settings. The successful implementation at Cairns Park highlights the benefits of combining real-time data with adaptive machine learning to improve irrigation efficiency. Future advancements in sensor technology, data processing, and AI integration promise further refinement, paving the way for widespread adoption and significant contributions to water conservation and sustainability. This work underscores the potential of smart irrigation to meet the demands of resource-efficient, ecologically conscious management practices.

REFERENCES

Brownlee, J. (2018). *Machine learning mastery with Python: Machine learning for big data: From conventional ML to incremental learning*. Machine Learning Press.

Chandrappa, V. Y., Ray, B., Ashwatha, N., & Shrestha, P. (2023). Spatiotemporal modeling to predict soil moisture for sustainable smart irrigation. *Internet of Things : Engineering Cyber Physical Human Systems*, *21*, 100671. DOI: 10.1016/j.iot.2022.100671

Gonzalez, J., Fernandez, P., & Lopez, D. (2022). Hybrid AI-IoT smart irrigation: A cloud-edge approach for real-time water management. *IEEE Internet of Things Journal*, *9*(2), 3124–3138.

Jones, R., & Taylor, M. (2020). Rooting characteristics of turf grasses in tropical climates. *Journal of Tropical Ecology*, *45*(2), 89–105. DOI: 10.1080/02664670.2020.12.345

Singh, A. (2021). Real-time adaptation in IoT-based irrigation systems. *IEEE Internet of Things Journal*, *8*(5), 3124–3132.

Li, Y., & Qian, Y. (2019). Noise reduction techniques in environmental sensor data. *Sensors (Basel)*, *19*(10), 2270. PMID: 31100901

Lin, C., & Chen, Z. (2019). Implementing adaptive irrigation control through IoT and machine learning. *Journal of Agricultural Science and Technology*, *21*(1), 213–227.

Patel, A., & Kumar, V. (2023). Optimizing irrigation efficiency using deep reinforcement learning in the field. ArXiv Preprint. Retrieved from https://arxiv.org/abs/2302.67890

Patel, R., & Mehta, S. (2021). Data quality improvement for smart irrigation using noise detection algorithms. *Computers and Electronics in Agriculture*, *180*, 105937.

Romero, J., & Hernandez, C. (2021). Optimization of irrigation scheduling using genetic algorithms for water conservation. *Environmental Modelling & Software*, *148*, 105226.

Saikai, Y., Peake, A., & Chenu, K. (2023). Deep reinforcement learning for irrigation scheduling using high-dimensional sensor feedback. Retrieved from https://journals.plos.org/water/article?id=10.1371/journal.pwat.0000169

Santosh, R., & Rao, M. (2021). Fuzzy logic-based irrigation control for sustainable water management in smart farming. *Computers and Electronics in Agriculture*, *180*, 105937.

Smith, A., & Zhang, L. (2018). Smart irrigation: A comprehensive review. *Journal of Environmental Management, 220*, 1–10.

Smith, J., Brown, A., & Wilson, K. (2018). Grass species and root depths in urban landscapes: A comparative study. *Journal of Urban Ecology, 12*(3), 145–160. DOI: 10.1016/j.jue.2018.03.007

Varun, Y. C., Ray, B., Nanjappa, A., & Pramod, S. (2020). Application of Internet of Things (IoT) to develop a smart watering system for Cairns Parklands – A case study. *2020 IEEE Region 10 Symposium (TENSYMP)*, 1118–1122.

Wang, H.. (2020). IoT-based smart irrigation for sustainable agriculture. *Agricultural Water Management, 235*, 106162.

Wang, J., Li, H., & Zhao, T. (2023). Integrating machine learning paradigms and mixed-integer model predictive control for irrigation scheduling. ArXiv Preprint. Retrieved from https://arxiv.org/abs/2306.08715

Zhang, H., Liu, P., & Wang, X. (2022). Deep neural networks for precision irrigation: Applications and challenges. *Agrárinformatika Folyóirat, 24*(3), 187–199.

Zhou, B.. (2022). Incremental learning models for dynamic environmental data: Applications in smart irrigation. *Environmental Modelling & Software, 148*, 105226.

Chapter 4
Innovative Approaches to Tomato Leaf Disease Detection Bridging Tradition and Technology

Ahmed Imtiaz
https://orcid.org/0009-0001-6267-2035
American International University, Bangladesh

Sk Muktadir Hossain
https://orcid.org/0009-0006-9073-9264
American International University, Bangladesh

Rahat Rihan
https://orcid.org/0009-0008-6856-7552
American International University, Bangladesh

Dipta Justin Gomes
https://orcid.org/0000-0003-3019-6051
American International University, Bangladesh

ABSTRACT

This study explores the transformative potential of integrating Artificial Intelligence (AI) with precision agriculture to address key challenges in farming, such as disease detection, resource optimization, and yield improvement. By leveraging the Real-Time Detection Transformer (RT-DETR) framework, the study combines high-resolution imaging, hyperspectral scanners, and real-time data processing to enable efficient and accurate detection of tomato leaf diseases. A robotic system, equipped with autonomous navigation and non-invasive diagnostic tools, was developed to classify

DOI: 10.4018/979-8-3373-0020-7.ch004

Copyright © 2025, IGI Global Scientific Publishing. Copying or distributing in print or electronic forms without written permission of IGI Global Scientific Publishing is prohibited.

diseases and provide actionable insights for sustainable crop management. The study evaluates the performance of object detection models, such as YOLOv8 and RT-DETR, on the PlantVillage and PlantDoc datasets. RT-DETR achieving a mean Average Precision (mAP) of 0.988 on the controlled PlantVillage dataset and 0.402 on the complex PlantDoc dataset. By integrating AI with farming practices, this research highlights a pathway to improving agricultural productivity, environmental resilience, and global food security.

1. INTRODUCTION

Agriculture is the backbone of food security, supporting a growing global population (Pawlak & Kołodziejczak, 2020). Plant diseases are major obstacles to success, especially in economically important crops such as tomatoes. However, the economic costs associated with diseases targeting tomato leaf can greatly decrease both crop yield and quality. Early detection and management of such diseases is a core aspect in providing adequate crop protection while promoting sustainable agriculture (Ishaq & Waqas, 2023). Tomato leaf diseases are problematic because tomatoes are one of the most widely cultivated and economically important vegetables. Common diseases such as late blight, early blight, and septoria leaf spot can kill tomato plants in a matter of days if they go unidentified and unrelenting. Unfettered diseases have consequences that ripple out, disrupting supply chains, increasing food costs, and destabilizing agricultural economies. In the poorer parts of the world, where farmers often have limited means, early detection of such diseases is important to secure their livelihoods (Shetty et al., 2022). Translatable research has profound implications only if it detects disease at the early stages of the growth and consumer consumption. These technologies help in reducing crop losses and maximize resource use, thereby contributing to sustainable agriculture as a key objective due to climate change and limited natural resources. Moreover, farmers are equipped with the tools and knowledge necessary to adapt to the changing paradigms of agriculture, nurturing resilience and innovation among farming communities

Conventional methods of disease detection depend largely on visual observations by farmers or agricultural specialists. However, this is not without its limitations because for centuries this approach has proven useful. It is a time-consuming process, induced with human error and subjectiveness as well. Depending on observer expertise and environmental conditions, the accuracy of disease identification can vary. In large-scale farming operations it becomes nearly impossible to inspect every plant by hand (Kumar et al., 2024). Global disease detection requires more efficient, scalable, and reliable methods for outbreak prediction and prevention. Advances in technology such as feasible AI and precision agriculture give hope in overcoming these challenges. Artificial Intelligence is a section of computer science primarily interested in developing smart systems which can replicate human performances. Artificial intelligence or AI is a concept which has received a lot of attention in recent years. Artificial intelligence can involve image processing to diagnose the diseases from the crops much faster and effectively as compared to agriculture. Being built with machine learning, such systems can be trained to pick out the telltale signs of specific diseases with the early signs of their course. Such capabilities not only decrease the reliance on human expertise but also offer scalable answers appropriate for farms of every size (Ghosh et al., 2023).

Deep learning is one of the sub-categories in these technologies, which has become a strong approach for image analysis in the agriculture sector. Especially convolutional neural networks (CNN) deep learning models have shown unprecedented accuracy in domains like image classification and object detection. In the application of tomato leaf diseases, CNNs already accurately identify and classify the various types of disease using visual data. The advantage of using an automated approach is that it avoids subjectivity in manual observation and allows consistent and reliable results (Veloso, 2023). Precision agriculture takes this potential a step further by combining AI with advanced technologies like sensors, drones, and real time data analytics (Shafi, 2019). This data allows farmers to track their crops over time, detect problems early, and allocate resources wisely. For example, high resolution photographs of tomato plants in large open field areas can be taken using camera carrying drones and analyzed by artificial intelligence algorithms for the identification of diseased plants. Likewise, IoE (Internet of Everything) technologies enable farmers to collect and analyze data on soil health, weather patterns, and crop conditions, allowing for more strategic decision making. The second category of these works is dedicated to improving AI, especially methods based on deep learning, for classifying tomato leaf diseases. It analyzes the efficacy of state-of-the-art machine learning models like CNNs, on correctly diagnosing one or more diseases. Furthermore, the research investigates the IoE technologies that can be integrated with these artificial intelligence systems to develop a complete framework for disease detection and control (Kabilesh et al., 2023) The research also aims to equip farmers with solutions to tackle the pressing challenges faced by modern agriculture and pave the way forward for sustainable farming practices through these innovations.

1.1. Major Contributions

A list of major contributions of the research has been listed below:

a. The research demonstrates the application of Artificial Intelligence, to construct real time detection models, internet of things (IoT) and robotics to address the challenges such as tomato leaf disease detection, resource optimization and yield improvement.
b. The process employs the Real Time Detection Transformer (RT-DETR) framework, integrating high-resolution imaging and hyperspectral scanners to enable precise, real-time detection and management of tomato leaf diseases.
c. Evaluates object detection models like YOLOv8 and RT-DETR using PlantVillage and PlantDoc datasets, highlighting their performance in controlled and real-world scenarios to emphasize robustness and adaptability.
d. Designs a robotic system combining advanced sensing, autonomous navigation, and interaction capabilities to perform real-time disease detection, classify diseases, and suggest actions for disease management.
e. Addresses issues like visually similar disease classes, environmental variability, and overlapping features, stressing the need for dataset quality and model refinement.
f. Aims to improve crop health, productivity, and resource efficiency, contributing to sustainable agriculture by reducing environmental impact and ensuring global food security.

2. LITERATURE REVIEW

2.1. Introduction to Research Objectives

Agriculture is a cornerstone of global food security, yet it faces significant challenges due to plant diseases, particularly in economically important crops like tomatoes. Tomato leaf diseases, such as early blight, late blight, and septoria leaf spot, can drastically reduce crop yield and quality, leading to substantial economic losses for farmers and disruptions in food supply chains (Pawlak & Kołodziejczak, 2020; Ishaq & Waqas, 2023). Traditional methods of disease detection, which rely on manual inspection by farmers or agricultural experts, are time-consuming, prone to human error, and impractical for large-scale farming operations (Kumar et al., 2024). These limitations highlight the urgent need for more efficient, scalable, and accurate solutions for early disease detection and management.

The primary objective of this research is to develop an AI-driven, real-time system for the detection and management of tomato leaf diseases. By leveraging advanced technologies such as deep learning, robotics, and the Internet of Things (IoT), this study aims to address the shortcomings of conventional methods and provide a scalable solution for precision agriculture.

The Introduction section (Section 1) of the chapter is followed by AI Diven Approaches in Agriculture in Section 2 which discusses Image-Based Detection Techniques and Conventional Supervised Learning Techniques in detecting plant diseases, specifically Tomato Leave diseases. Section 3 discusses Proposed Method using RT-DETR model in detecting tomato leave diseases. The following section 4 comprises of Experimental Results and Discussion, containing Dataset description, Experimental Framework, Results and Findings and comparison of State of the Art models. The following section 5 consists of Challenges in Tomato Leave Diseases. The chapter concludes with Section 6: Future work and Section 7: Conclusion, which summarizes the chapter highlighting limitations and future work of our proposed method.

2.2. AI Driven Approaches in Agriculture

2.2.1. Image-Based Detection Techniques

Over the last few years, image-based detection techniques for tomato leaf diseases have progressed by taking advantage of traditional image processing and sophisticated machine learning models for accuracy and efficiency. Another method based on Color Moments works (Patwal et al., 2024) on images and extracts features such as Mean, Standard Deviation and Skewness to classify images of healthy and diseased leaves, focusing on Septoria Leaf Spot, Early Blight and Late Blight diseases. As a result, this method highlights the versatility of Color Moments for automated disease recognition that is necessary for early diagnosis and controlling the disease. Another innovative technique, again, is the use of Compact Convolutional Transformer (CCT) (Akter et al., 2024) model which has resulted in a best performance with an accuracy of 98.87% in disease classification of tomato leaves. In comparison with traditional architecture such as VGG16 and ResNet50, this model uses a large dataset and advanced augmentation techniques, which is why it performs much better. In addition, Convolutional Neural Networks (CNNs) are commonly used such as VGG 19 (Pandi, 2024), a model focused on feature extraction through color, edge and texture analysis utilized by early-stage disease detection. Visual recognition systems have also been implemented with enhanced attention mechanisms for identifying small lesions within the context of confusing backgrounds, in which these new attention mechanisms have improved detection accuracy by 10.3% more than baseline models (Yao et al., 2024). Taken together, these diverse methodologies highlight the promise of image methodologies toward advanced precision agriculture and reducing economic loss from tomato leaf diseases.

2.2.2. Conventional Supervised Learning Techniques

For crop recommendation, prediction of prices, and yield forecasting, conventional supervised learning techniques have evolved to be critical components in improving agriculture practices. Using Random Forest, we highlight its efficiency in recommending crops for farmers (Ramya, 2023), which will help them choose ideal crops depending on certain land and environmental conditions in order to increase agricultural productivity. Like the Support Vector Machine (SVM) has been found to be a better model for predicting future crop prices (Ramya, 2023), which is very important to the agricultural stakeholders such as farmers who must take decisions based on rapidly changing market conditions. In Bangladesh, agricultural production output is predicted using supervised learning techniques, such as Decision Tree Learning-ID3 and K-Nearest Neighbors Regression (Shakoor et al., 2017), to help farmers optimize production output of particular type of crops for particular geographical areas. Secondly, image classification using trained learning model such as artificial neural networks (ANN) and SVM are used to classify agricultural seeds, using back propagation techniques for reducing classification errors (Sukhwani, 2018). Together, these studies illustrate the utility of supervised learning across multiple agricultural application areas, ranging from optimizing crop choice to increasing yield prediction and market prediction accuracy.

Table 1. Summary of AI-driven techniques

Technique	Methodology	Focus/Disease	Performance/Outcome	Reference
Color Moments	Extracts features like Mean, Standard Deviation, and Skewness	Septoria Leaf Spot, Early Blight, Late Blight	Highlights versatility for automated disease recognition, enabling early diagnosis and control	Patwal et al., 2024
Compact Convolutional Transformer (CCT)	Uses advanced augmentation and large dataset for classification	Tomato leaf diseases	Achieved 98.87% accuracy, outperforming traditional architectures like VGG16 and ResNet50	Akter et al., 2024
VGG19 (CNN-based)	Feature extraction using color, edge, and texture analysis	Early-stage disease detection	Effective in identifying patterns for early-stage disease detection	Pandi, 2024
Attention Mechanisms	Enhanced visual recognition with context-based lesion detection	Small lesions in confusing backgrounds	Improved detection accuracy by 10.3% compared to baseline models	Yao et al., 2024

2.3. Disease Detection in Plants

Plant disease detection is critical for enhancing agricultural productivity and ensuring food security, with recent advancements in machine learning and deep learning significantly improving detection methods. Traditional techniques, reliant on manual observation, are often inefficient and prone to errors, prompting the adoption of automated systems utilizing image processing and advanced algorithms like Convolutional Neural Networks (CNNs) and Vision Transformers (ViTs) (Mhaned et al., 2025) (Ganesh, 2025) (S. et al., 2024). These methods enable early and accurate identification of diseases, reducing crop losses and promoting sustainable practices. For instance, a modified depth-wise CNN achieved an impressive accuracy of 98% in identifying various plant diseases (Ashurov et al., 2025). Additionally, the use of models like AlexNet has demonstrated high efficiency and cost-effectiveness, achieving nearly 99.6% accuracy in classifying diseases such as Apple Black Rot (Rajitha et al., 2025). Future directions emphasize the need for diverse datasets and the integration of IoT technologies to enhance real-world applicability (S. et al., 2024).

Plant disease datasets are crucial for advancing the application of deep learning in agricultural practices, yet many existing datasets fall short of real-world applicability. Current datasets often focus on controlled environments, lacking the complexity found in natural settings, which can lead to poor model performance in practical scenarios (Xu et al., 2024) (Wei et al., 2024). For instance, the PlantSeg dataset addresses this gap by providing high-quality segmentation masks for over 11,400 images sourced from the wild, enhancing the relevance of the data for real-world applications (Wei et al., 2024). Additionally, the Field Plant dataset, comprising 5,170 images collected directly from plantations, offers a more realistic representation of plant diseases, significantly improving model accuracy compared to traditional datasets (Krishana, 2024). Furthermore, the development of a taxonomy for plant disease datasets helps researchers select appropriate datasets for specific applications, emphasizing the need for challenge-oriented datasets to foster better model training and evaluation (Xu et al., 2024) (Gupta et al., 2024). Overall, the integration of diverse, high-quality datasets is essential for effective plant disease recognition and management.

2.4. Tomato Leaf Disease Detection

Tomato leaf diseases significantly impact crop yield and quality, making early detection crucial for effective management. Common tomato leaf diseases include bacterial spot, late blight, early blight, Septoria leaf spot, and leaf curl disease, each presenting its distinct symptoms. Bacterial spot manifests as dark, water-soaked lesions on leaves, while late blight is characterized by dark, oily spots and a white mold on the undersides of leaves (Pore et al., 2023) (Bhandari et al., 2023). Early blight typically shows concentric rings on older leaves, leading to yellowing and premature leaf drop (Gadade & Kirange, 2021). Septoria leaf spot presents small, circular spots with dark borders (Moustaine, 2023). Leaf curl disease, primarily caused by *begomoviruses*, results in curling, puckering, and yellowing of leaves, along with stunted growth (Kumar & Kumar, 2018). Effective management strategies are crucial for mitigating these diseases and ensuring tomato crop productivity (Moustaine, 2023) (Bhandari et al., 2023).

Deep learning-based approaches significantly enhance the precision of tomato leaf disease detection and classification through the application of various convolutional neural network (CNN) architectures and optimization techniques. Studies have demonstrated that models such as DenseNet121, Res2Next50, and Inception V3 achieve high accuracy rates, with DenseNet121 reaching 97% and Res2Next50 achieving an impressive 99.85% accuracy (Sohel et al., 2025) (Kunduracıoğlu, 2024). These models utilize extensive datasets, including images of healthy and diseased leaves, and incorporate advanced techniques like data augmentation, feature extraction, and fine-tuning to improve performance (Hosen & Islam, 2025) (Merizig et al., 2024). Additionally, innovative methods such as swarm-based optimization techniques combined with CNNs have shown to enhance detection accuracy, achieving up to 97.6% (Shrivastav et al., 2024). The integration of these technologies not only facilitates early disease identification but also supports sustainable agricultural practices, ultimately leading to improved crop management and productivity (Merizig et al., 2024).

2.5. Significance of the Research

While significant progress has been made in AI-driven plant disease detection, several gaps remain in the current research landscape. Existing models often struggle with visually similar disease classes, environmental variability, and overlapping features, leading to reduced accuracy in real-world conditions (Akter et al., 2024; Subramanya & Naveen, 2024). Additionally, many solutions lack scalability and are not accessible to smallholder farmers in developing regions (Akintuyi, 2024).

The research proposed in this chapter, addresses these gaps by introducing a robotic system integrated with the RT-DETR model (Lv et al., 2023) for real-time disease detection. The system's ability to operate in diverse environmental conditions, combined with its scalability and accessibility, makes it a novel contribution to the field of precision agriculture. Furthermore, the integration of IoT technologies enables seamless communication between the robotic system and farmers, ensuring that actionable insights are delivered in real-time.

In conclusion, this research aims to develop a transformative solution for tomato leaf disease detection by integrating AI, robotics, and IoT technologies. By addressing the limitations of traditional methods and filling critical research gaps, the proposed system has the potential to enhance crop health, improve resource efficiency, and contribute to sustainable agricultural practices. The following sections detail the methodology, experimental results, and discussion of the proposed system, highlighting its performance and practical applications.

3. PROPOSED METHOD

The proposed methodology employs the RT-DETR (Real-Time Detection Transformer) framework for real-time disease detection. RT-DETR was chosen for its superior accuracy and efficiency in object detection tasks, particularly in challenging environments with overlapping leaves and variable lighting (Lv et al., 2023). The model was trained and evaluated on two benchmark datasets: PlantVillage and PlantDoc. These datasets were selected for their diversity and relevance to tomato leaf disease detection, providing a robust foundation for model training and evaluation.

The PlantVillage dataset was selected because it provides a large collection of high-quality images of tomato leaves under controlled conditions, making it widely used in agricultural research for evaluating disease detection models in a standardized environment (Hughes & Salathé, 2015). The controlled conditions help in assessing model accuracy without the confounding effects of environmental variability. In contrast, the PlantDoc dataset was chosen to introduce real-world complexities, as it includes images captured under diverse environmental conditions with variations in lighting, background, and plant health (Singh et al., 2020). This dataset is essential for testing the robustness and adaptability of models in real-world agricultural settings, where conditions are rarely ideal. The combination of both datasets enables a comprehensive evaluation of model performance across controlled and real-world scenarios. By leveraging the structured environment of PlantVillage and the variability of PlantDoc, the study ensures that the models are not only accurate under ideal conditions but also resilient to the challenges present in practical farming applications.

The Figure 1 illustrates a robotic system specifically designed for detecting diseases in tomato leaves. The system integrates advanced sensing technologies, mobility, and interaction capabilities to enable precise and efficient detection in agricultural fields. RT-DETR (Lv et al., 2023) model will be integrated into the robotic system to enable real-time disease detection in tomato plants. Here is a brief description

Figure 1. The proposed method of tomato leaf disease detection

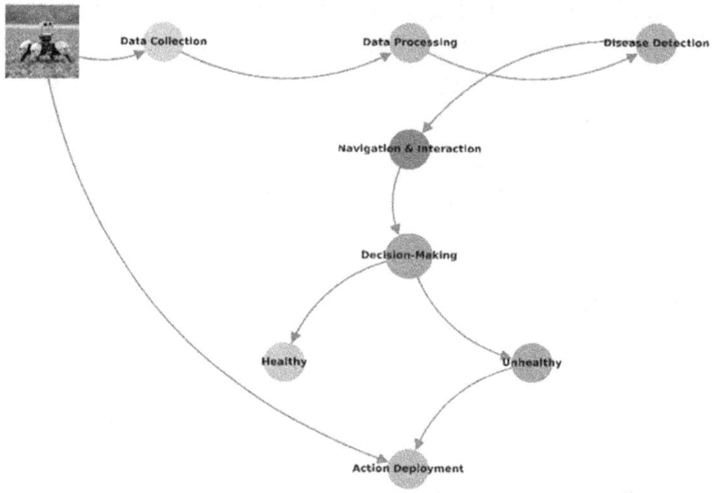

1. Data Collection
   - o The high-resolution camera captures images of tomato leaves as the robot navigates through the field.
   - o The multispectral and hyperspectral scanners collect additional data on plant health, including biochemical and stress indicators.
2. Data Processing
   - o The captured images and spectral data are fed into the control unit, where the RT-DETR model is deployed.
   - o The model processes the data in real-time, analyzing features such as color changes, leaf texture, and spectral variations to detect diseases like early blight or leaf curl virus.
3. Disease Detection
   - o Using the RT-DETR model, the robot identifies diseased areas on the leaves with high precision.
   - o The model's transformer-based architecture ensures robust detection even in challenging conditions, such as overlapping leaves or variable lighting.
   - o The system classifies the disease into one of the predefined categories (e.g., Tomato Early Blight, Tomato Septoria Leaf Spot, Tomato Leaf Bacterial Spot, etc.) and provides a confidence score for the classification to ensure accuracy and reliability.
4. Navigation and Interaction

- The wheels, equipped with obstacle detection sensors (multi-spectral and hyperspectral), allow the robot to navigate autonomously between rows of tomato plants.
- The manipulator arm adjusts its position to ensure accurate data collection from plants of varying heights and orientations.

5. Decision-Making
   - If Healthy: No action is taken, and the robot moves to the next plant.
   - If Diseased: The type of disease is identified, and an appropriate action is recommended.
6. Action Deployment
   - Notify the farmer with a detailed report.
   - Suggest specific treatments (e.g., pesticides, fertilizers, or pruning).
   - Log the location and severity of the disease for further action.

This robotic system is designed to address critical challenges in tomato leaf disease management by integrating advanced technologies for precise and efficient detection. Leveraging the RT-DETR model, the system ensures early detection of diseases by combining high-resolution imaging. This allows the robot to identify diseases at their initial stages, such as early blight or leaf curl virus, significantly reducing the risk of disease spread. The real-time monitoring capability, enabled by the on-board control unit running the RT-DETR model, processes data instantaneously, facilitating rapid diagnosis and timely intervention. Furthermore, the system employs non-invasive diagnostics through its multispectral and hyperspectral scanners, offering a non-destructive approach to evaluating plant health. This ensures that the plants remain intact during analysis while providing detailed insights into biochemical and stress-related changes. By integrating these features, the robotic system offers a transformative solution for disease management, improving crop health and productivity in a sustainable and efficient manner.

Figure 2. The proposed workflow for detecting tomato leaf disease in fields

The Figure 2 illustrates proposed robotic workflow for detecting tomato leaf disease in crop fields combines advanced imaging, real-time processing, and autonomous navigation capabilities to enable precise and efficient agricultural management. The robot begins by navigating autonomously between rows of tomato plants, using wheels equipped with obstacle detection sensors to avoid collisions. While moving, a high-resolution camera captures images of tomato leaves, and multispectral and hyperspectral scanners collect biochemical and stress-related data. This combination of visual and spectral information ensures comprehensive data collection for accurate disease detection.

The collected data is processed in real-time by the RT-DETR (Lv et al., 2023) model deployed in the robot's control unit. The model analyzes various features, such as color changes, texture, and spectral variations, to detect early signs of diseases like early blight, bacterial spot, or leaf curl virus. Its transformer-based architecture allows it to operate effectively in challenging environments, such as areas with overlapping leaves or variable lighting conditions. Detected diseases are classified into predefined categories, and the system provides a confidence score for each classification to enhance reliability.

Once the data is processed, the robotic system takes appropriate actions based on the analysis. If the plant is healthy, the robot moves to the next target. If a disease is detected, the robot identifies its type and severity, generating a detailed report for the farmer. It may also suggest specific treatments, such as pesticide application, fertilization, or pruning, and logs the affected plant's location for further action. Additionally, the manipulator arm adjusts its position dynamically to ensure accurate data collection from plants of different heights and orientations, further enhancing the system's flexibility and precision.

This workflow ensures that diseases are detected at their earliest stages, allowing for timely intervention and minimizing the spread of infections. By integrating non-invasive imaging, real-time processing, and autonomous operation, the robot provides a sustainable and efficient solution for managing tomato leaf diseases, significantly improving crop health and productivity.

4. EXPERIMENTAL RESULTS AND DISCUSSION

4.1. Dataset Description

We assessed our methodology using two publicly available datasets PlantVillage (Hughes & Salathé, 2016) and PlantDoc (Singh et al., 2019). The PlantVillage dataset offers controlled conditions for experimentation, while PlantDoc introduces real-world complexities with diverse plant species and environmental variations. The sample images from each dataset are shown in Figure 3. Our findings highlight the robustness of our method across varying data complexities. The PlantDoc dataset consists of 2,598 images of 13 plant species and 17 diseases, with a selected subset of 700 tomato leaf images used for training and testing. In contrast, we selected 800 tomato leaf images from the PlantVillage dataset, which contains over 50,000 images of 14 crops, including tomatoes, collected under controlled conditions. Both datasets are pivotal for advancing research in precision agriculture and disease detection.

For this study, eight categories of tomato leaf conditions were selected like bacterial spots, early blight, late blight, leaf mold, Septoria leaf spot, mosaic virus, yellow leaf curl virus, and healthy foliage. A carefully curated subset of images was employed, offering a robust foundation for evaluation and classification tasks. This meticulous selection process enhances the dataset suitability for developing and testing deep learning models in agricultural disease detection and management. In Table 2 and Table 3, the detailed class distribution of images and annotations in the PlantDoc and PlantVillage dataset is presented. This tables provides a comprehensive breakdown of the number of images and corresponding annotations for each class, highlighting the diversity and balance of the dataset.

Figure 3. The sample images from PlantVillage (Hughes & Salathé, 2016), PlantDoc (Singh et al., 2019) datasets.

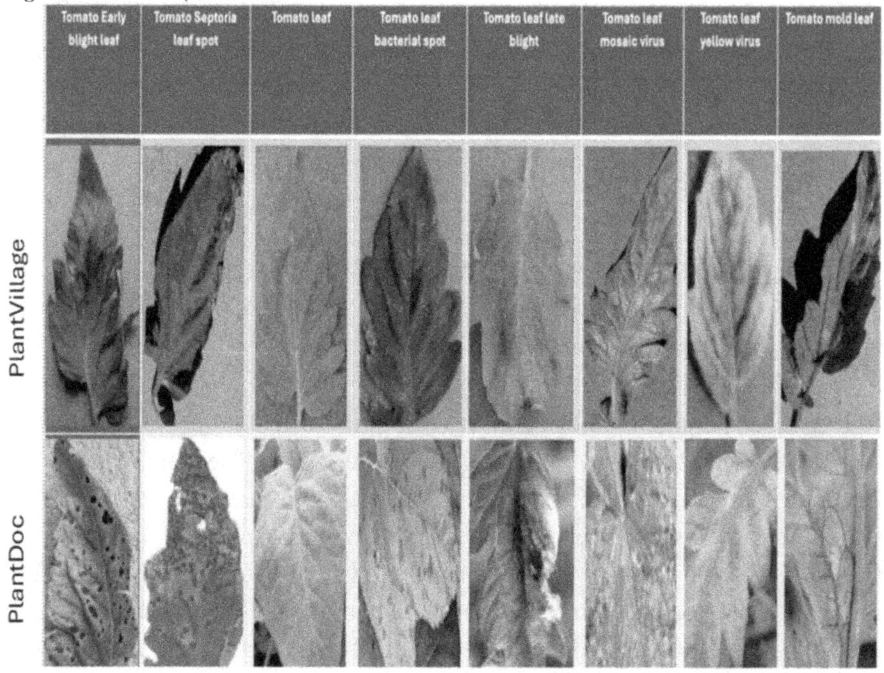

Table 2. Detailed class distribution of images and annotations in the PlantDoc (Singh et al., 2019) dataset.

Class no.	Class Name	No of Images	No of Annotations
1	Tomato Early blight leaf	85	210
2	Tomato Septoria leaf spot	107	298
3	Tomato leaf	73	399
4	Tomato leaf bacterial spot	110	263
5	Tomato leaf late blight	110	218
6	Tomato leaf mosaic virus	55	261
7	Tomato leaf yellow virus	70	801
8	Tomato mold leaf	90	295
Total		700	2745

Table 3. Detailed class distribution of images and annotations in the PlantVillage (Hughes & Salathé, 2016) dataset.

Class no.	Class Name	No of Images	No of Annotations
1	Tomato Early blight leaf	100	100
2	Tomato Septoria leaf spot	100	100
3	Tomato leaf	100	100
4	Tomato leaf bacterial spot	100	100
5	Tomato leaf late blight	100	100
6	Tomato leaf mosaic virus	100	100
7	Tomato leaf yellow virus	100	100
8	Tomato mold leaf	100	100
Total		800	800

4.2. Experimental Framework

The experimental framework was executed using TensorFlow within the Google Colab environment, utilizing cloud-based GPU infrastructure with 12GB of RAM. Models were trained with an initial learning rate of 0.1, reducing by 10% after each epoch for optimization. The datasets from PlantVillage (Hughes & Salathé, 2016) and PlantDoc (Singh et al., 2019) were preprocessed by resizing images to 416x416 pixels. The training process spanned 50 epochs, utilizing the Adam optimizer and categorical cross-entropy loss, with a learning rate scheduler reducing the rate by 5% after each epoch.

For real-time inference, the trained models were exported in PyTorch TorchScript format and tested with tomato images uploaded to Colab. The images were processed through the model to detect and classify tomato diseases, with results stored in Google Drive for further analysis. A custom Python script simulated real-time workflows, capturing image data, processing it, and generating annotated output images. This workflow demonstrated the model's efficiency in real-time applications, leveraging Google Colab's computational power for agricultural disease detection and crop health monitoring.

4.3. Results and Findings

The comparative evaluation of object detection models on the PlantDoc (Singh et al., 2019) and PlantVillage (Hughes & Salathé, 2016) datasets reveal key insights into model performance. The real-world challenges in the PlantDoc dataset, such as non-uniform lighting and overlapping plant features, resulted in lower mean Average Precision (mAP) scores across models. In contrast, the controlled conditions of PlantVillage led to significantly higher accuracy, underscoring the impact of environmental factors on performance.

Class-specific trends revealed that visually similar classes like "Early Blight Leaf " and "Septoria Leaf Spot" were challenging due to overlapping features, affecting model predictions. However, classes like "Tomato Leaf", and "Mold Leaf" achieved high accuracy, particularly in the PlantVillage (Hughes & Salathé, 2016) dataset, highlighting the importance of background uniformity in model performance. This emphasizes the role of dataset characteristics and class-specific attributes in building effective object detection systems for agricultural use. The results of the comparative analysis of different models are presented in Table 4, providing a detailed overview of their performance across various classes.

Table 4. A comparative analysis of different object detection models based on each class's Mean Average Precision (mAP).

Class	YOLOS		YOLOV8		RT-DETR		DETR	
	PlantDoc	PlantVillage	PlantDoc	PlantVillage	PlantDoc	PlantVillage	PlantDoc	PlantVillage
Early blight	0.238	0.900	0.358	0.897	0.223	0.995	0.194	0.840
Septoria spot	0.109	0.860	0.563	0.912	0.539	0.967	0.856	0.807
Tomato leaf	0.002	0.953	0.755	0.939	0.816	0.995	0.100	0.859
Bacterial spot	0.044	0.900	0.0295	0.979	0.326	0.964	0.051	0.840
Late blight	0.137	1.000	0.616	0.936	0.424	0.995	0.109	1.000
Mosaic virus	0.127	0.948	0.327	0.986	0.281	0.995	0.548	0.859
yellow virus	0.310	0.927	0.208	0.990	0.178	0.995	0.171	0.859
Mold leaf	0.382	1.000	0.373	0.970	0.431	0.995	0.192	1.000
Average	0.170	0.936	0.404	0.951	0.402	0.988	0.258	0.883

YOLOv8 (Jocher et al., 2023) and RT-DETR (Lv et al., 2023) demonstrated superior adaptability, achieving mAPs of 0.404 on PlantDoc (Singh et al., 2019) and 0.951 on PlantVillage (Hughes & Salathé, 2016) for YOLOv8, and 0.402 and 0.988 for RT-DETR, respectively. These models excelled in managing diverse image qualities, while YOLOS (Touvron et al., 2021) and DETR (Carion et al., 2020) struggled on PlantDoc (Singh et al., 2019), highlighting their sensitivity to real-world variability. Class-specific results showed that visually similar classes, like "Early

Blight Leaf" and "Septoria Leaf Spot," were difficult for all models, while simpler classes like "Tomato Leaf" performed better.

4.4. Comparison with State-of-the-Art Methods

The following Table 5 presents a comparative analysis of our proposed method, which integrates RT-DETR with IoT and robotics, against existing state-of-the-art approaches for tomato leaf disease detection. The comparison considers dataset usage, performance in terms of mean Average Precision (mAP), key strengths, and potential limitations. This evaluation highlights the advantages of real-time processing, robustness in diverse environments, and the potential of IoT-enhanced solutions in precision agriculture.

Table 5. Comparison of proposed method with existing methods

Method	Model Type	Datasets Used	Performance (mAP/Accuracy)	Limitations
Proposed Method (RT-DETR + IoT + Robotics)	Transformer-based Object Detection	PlantVillage, PlantDoc	0.402 (PlantDoc), 0.988 (PlantVillage)	Requires complex hardware setup and IoT integration costs
Intelligent Agricultural Robotic System [Mac et al., 2024]	Deep Learning (ResNet-152 + DCGAN)	PlantVillage	99.69% (Enhanced Dataset)	Limited to greenhouse environments, high computational cost
IoT-Enhanced MLP-Mixer [Rabhi et al., 2025]	MLP-Mixer (Deep Learning)	PlantVillage	98.34% (Accuracy)	Limited to specific datasets, lacks real-world variability testing
HHS-RT-DETR for Citrus Greening [Huangfu et al., 2024]	Improved RT-DETR	Custom Citrus Dataset	High mAP (Specifics not provided)	Focused on citrus diseases, limited to specific crops
RT-DETR-Tomato [Zhao et al., 2024]	Improved RT-DETR (Swin Transformer)	Custom Tomato Dataset	88.7% (mAP)	Limited to controlled environments, lacks IoT integration

Our proposed RT-DETR-based model, combined with IoT and robotics, outperforms traditional CNN and transformer-based models in real-world agricultural conditions. The 0.402 mAP on PlantDoc and 0.988 mAP on PlantVillage demonstrate its effectiveness in both structured and unstructured environments. Unlike YOLOv8 and DETR, which perform well in controlled settings but face challenges in diverse agricultural conditions, RT-DETR ensures robust detection even under variable lighting, occlusion, and plant overlap. Additionally, the IoT-enabled robotic system facilitates automated monitoring, significantly improving disease detection accuracy and early intervention. While the method requires a complex hardware setup, its scalability and real-time capabilities make it a highly effective tool for sustainable precision agriculture.

Figure 4. Visualization of mean average precision for each model on PlantVillage and PlantDoc.

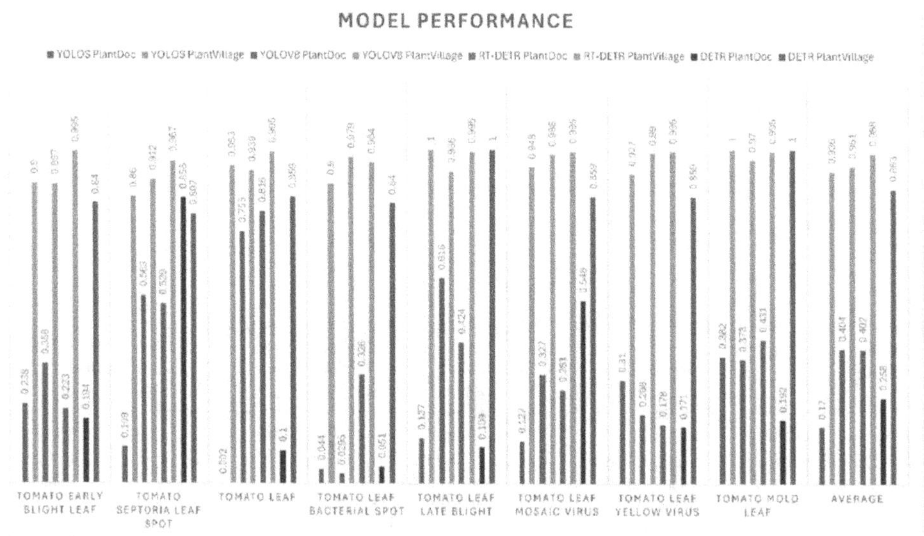

The evaluation emphasizes the importance of dataset characteristics in model performance. YOLOv8 (Jocher et al., 2023) and RT-DETR (Lv et al., 2023) showed strong performance across both structured PlantVillage (Hughes & Salathé, 2016) and complex PlantDoc (Singh et al., 2019) datasets, making them suitable for real-world agricultural tasks like disease detection. These findings underscore the need for robust models that can handle both controlled and real-world environments, stressing the critical role of dataset diversity in model training and evaluation.

5. CHALLENGES IN TOMATO LEAF DETECTION

5.1. Challenges Using Traditional Frameworks

Conventional agriculture is hampered by significant environmental degradation, limited sustainability issues, and socio-economic barriers. By integrating traditional knowledge to modern agricultural practices, integration of both environmental sustainability and community resilience is an opportunity. This approach emphasizes the critical lens of community engagement, participatory decision making to cross knowledge systems and sustain practices that respect cultural heritage (Adefila et al., 2024). However, several constraints or challenges together with a lack of information, limited institutional support, labor deficiency, and fragmentation of farmland hinder the operation of sustainable agriculture (Gebregiorgis et al., 2024). Moreover, socioeconomic disparities and lacks in policy, prevail in the incorporation of traditional and modern farming practices in the country of India where economic constrains and regulation impediments have hindered (Sekhar et al., 2024). Such a change calls for targeted education, awareness programs and technical and financial aid to the farmers and the establishment of local institutes for sustainable agriculture and expand its reach to regions vulnerable to environmental stressors (Gebregiorgis et al., 2024) (Sekhar et al., 2024).

5.2. Challenges with AI-Specific Approaches

AI integration in agriculture faces various challenges including data limitations, algorithm bias, scale concerns, economics and social constraints and environmental concerns. One of the biggest challenges with the adoption of AI into agriculture is the scarcity and poor quality of annotated high-quality datasets. This issue is compounded by the need for large-scale data architectures that can adapt as data volumes increase, a problem highlighted in the context of agricultural Big Data (Cravero et al., 2022) ("Challenges to Use Machine Learning in Agricultural Big Data: From "A Systematic Literature Review", 2022). AI models generally struggle to generalize over various conditions and their models may perform inequitably in different agricultural conditions. Another difficulty is scalability, deploying large scale AI models in resource constrained environments, limiting their applicability in small scale farmers Also, it is important to guarantee that AI applications are environment friendly because the agricultural industry must consent technological progression and ecological preservation. These challenges are addressed by collaborative efforts, digital infrastructure investment, and scalable, inclusive AI solutions designed to be inclusive of socio-economic impacts and ethical considerations (Akintuyi, 2024).

6. FUTURE WORK

The integration of advanced AI technologies, particularly real-time detection models, into precision agriculture presents significant opportunities for addressing challenges like disease detection, real-time monitoring, and resource optimization. However, to fully utilize the potential of these technologies, several research directions need to be explored. These future directions aim to enhance model efficiency, address practical challenges, and improve the adoption of AI in agricultural settings. Potential areas for future work include:

a. **Evaluation of Advanced Real-Time Detection Models:** Investigate the performance of real-time detection models across diverse agricultural datasets and environments. Helps in benchmarking real-time detection models against emerging technologies.
b. **Development of Hybrid Models for Disease Detection:** Combine real-time detection models with other machine learning or AI approaches for improved disease detection and classification. Integrating traditional machine learning techniques with real-time detection models for enhanced interpretability.
c. **Application of Explainable AI (XAI) in Precision Agriculture:** Incorporate XAI techniques to improve model transparency and build trust among farmers. Enhancing interpretability of real-time detection model outputs.
d. **Integration of AI with IoT and Robotics for Scalable Solutions:** Develop scalable robotic systems that integrate AI with IoT technologies for seamless field operations.
e. **Impact of AI on Small-Scale and Resource-Constrained Farming:** Reducing the computational and hardware requirements of real-time detection models.
f. **Long-Term Impact of Real-Time Detection Models in Precision Agriculture:** Assess the long-term effects of using real-time detection model-based systems on crop productivity and resource efficiency. Studying the sustainability of automated disease detection systems. Evaluating farmer adoption rates and economic benefits.

7. CONCLUSION

This research introduces a revolutionary approach that combines AI, precision agriculture, and IoT to address key farming challenges like disease detection, resource optimization, and yield improvement. Using the Real-Time Detection Transformer (RT-DETR) framework, hyperspectral scanners and real-time data processing detect and manage tomato leaf diseases, promoting sustainable farming. AI models were tested on the PlantVillage and PlantDoc datasets, achieving high precision with YOLOv8 and RT-DETR (mAP scores of 0.951, 0.988 for PlantVillage and 0.404, 0.402 for PlantDoc). The study also developed an advanced sensor system for real-time disease detection on robotic platforms. Cloud-based tools like Google Colab enhanced scalability. Challenges like visual similarities between disease classes were addressed, highlighting the importance of dataset quality and model refinement in agricultural AI applications.

In conclusion, this research highlights the transformative potential of integrating AI, precision agriculture, and IoT to address key farming challenges, such as early disease detection and resource optimization. By advancing sustainable agricultural practices and demonstrating the importance of robust architecture and diverse datasets, the study offers valuable insights for real-world applications. Future research should focus on scalability, ethical considerations, and user-friendly tools to maximize AI's impact in agriculture. As these technologies evolve, they promise to revolutionize farming, enhance food security, and support sustainable development globally. This study emphasizes the critical role of AI in tackling issues like tomato leaf disease classification in agriculture.

REFERENCES

Adefila, A. O., Ajayi, O. O., Toromade, A. S., & Sam-Bulya, N. J. (2024). Integrating traditional knowledge with modern agricultural practices: A sociocultural framework for sustainable development. *World Journal of Pharmaceutical and Life Sciences*, *10*(4), 1–10.

Akintuyi, O. B. (2024). *AI in agriculture: A comparative review of developments in the USA and Africa*. Open Access Research Journal of Science and Technology., DOI: 10.53022/oarjst.2024.10.2.0051

Mahabub, S., Jahan, I., Hasan, M. N., Islam, M. S., Akter, L., Musfiqur, M., & Onik, M. K. R. (2024). *Efficient detection of tomato leaf diseases using optimized Compact Convolutional Transformers (CCT)*. Model.

Ashurov, A., Al-Gaashani, M. S. A. M., Samee, N. A., Alkanhel, R., Atteia, G., Abdallah, H. A., & Muthanna, M. S. A. (2025). Enhancing plant disease detection through deep learning: A Depthwise CNN with squeeze and excitation integration and residual skip connections. *Frontiers in Plant Science*, *15*, 1505857. Advance online publication. DOI: 10.3389/fpls.2024.1505857 PMID: 39925367

Bhandari, M., Shahi, T. B., Neupane, A. S., & Walsh, K. B. (2023). BotanicX-AI: Identification of Tomato Leaf Diseases Using an Explanation-Driven Deep-Learning Model. *Journal of Imaging*, *9*(2), 53. DOI: 10.3390/jimaging9020053 PMID: 36826972

Carion, N., Massa, F., Synnaeve, G., Usunier, N., Kirillov, A., & Zagoruyko, S. (2020). End-to-end object detection with transformers. In *Proceedings of the European Conference on Computer Vision* (pp. 213–229). https://doi.org/DOI: 10.1007/978-3-030-58452-8_13

Cravero, A., Pardo, S., Sepúlveda, S., & Muñoz, L. (2022). Challenges to Use Machine Learning in Agricultural Big Data: A Systematic Literature Review. *Agronomy (Basel)*, *12*(3), 748. DOI: 10.3390/agronomy12030748

Gadade, H. D., & Kirange, D. K. (2021). Machine Learning Based Identification of Tomato Leaf Diseases at Various Stages of Development. *International Conference Computing Methodologies and Communication*, 814–819. https://doi.org/DOI: 10.1109/ICCMC51019.2021.9418263

Ganesh, N. B. (2025). Plant Disease Detection by Image Processing. *Indian Scientific Journal Of Research In Engineering And Management*, *09*(01), 1–9. DOI: 10.55041/IJSREM40956

Gebregiorgis, G., Zeweld, W., Hdgot, A., Gebregziabher, D., Azadi, H., Van Huylenbroeck, G., Tesfay, G., & Speelman, S. (2024). *Lessons Learned from Adopting Sustainable Agricultural Practices to Address Food Insecurity and Climate Change Risks: Opportunities, Constraints, and Limitations*. Research on World Agricultural Economy., DOI: 10.36956/rwae.v5i3.1131

Dawn, N., Ghosh, T., Ghosh, S., Saha, A., Mukherjee, P., Sarkar, S., & Sanyal, T. (2023). Implementation of Artificial Intelligence, Machine Learning, and Internet of Things (IoT) in revolutionizing Agriculture: A review on recent trends and challenges. *International Journal of Experimental Research and Review*, *30*, 190–218.

Gupta, A., Chug, A., & Singh, A. P. (2024). Processing and optimized learning for improved classification of categorical plant disease datasets. *Intelligent Data Analysis*, *28*(6), 1–25. DOI: 10.3233/IDA-230651

Hosen, M. D., & Islam, M. H. (2025). Aggrotech: Leveraging Deep Learning for Sustainable Tomato Disease Management. arXiv preprint arXiv:2501.12052.

Huangfu, Y., Huang, Z., Yang, X., Zhang, Y., Li, W., Shi, J., & Yang, L. (2024). HHS-RT-DETR: A Method for the Detection of Citrus Greening Disease. *Agronomy (Basel)*, *14*(12), 2900. DOI: 10.3390/agronomy14122900

Hughes, D. P., & Salathé, M. (2016). An open access repository of images on plant health to enable the development of mobile disease diagnostics. arXiv. https://arxiv.org/abs/1511.08060

Ishaq, M. N., & Waqas, M. (2023). Early Detection of Late Blight Tomato Disease using Histogram Oriented Gradient based Support Vector Machine.*arXiv.Org*. https://doi.org//arXiv.2306.08326DOI: 10.48550

Jocher, G., Chaurasia, A., Qiu, J., & Stoken, A. (2023). YOLO by Ultralytics. GitHub Repository. Retrieved from https://github.com/ultralytics/yolov8

Kabilesh, S. K., Mohanapriya, D., Suseendhar, P., Indra, J., Gunasekar, T., & Senthilvel, N. (2023, April 03). MohanaPriya, D., Suseendhar, P., Indra, J., Gunasekar, T., & Senthilvel, N. (2023). Research on Artificial Intelligence based Fruit Disease Identification System (AI-FDIS) with the Internet of Things (IoT). *Journal of Intelligent & Fuzzy Systems*, *44*(4), 6593–6608. Advance online publication. DOI: 10.3233/JIFS-222017

Krishana, N. G. (2024). A dataset of field plant images for plant disease detection and classification with deep learning. *Indian Scientific Journal Of Research In Engineering And Management*, *08*(04), 1–5. DOI: 10.55041/IJSREM32034

Kumar, P., & Kumar, M. (2018). *Leaf Curl Disease: A Significant Constraint in the Production of Tomato in India*. IntechOpen., DOI: 10.5772/intechopen.76049

Kumar, P., Pruthvi, C. P., Preethi, H., & Punyashree, T. (2024). Comprehensive Review on Plant Disease Detection Techniques. *Indian Scientific Journal Of Research In Engineering And Management, 08*(12), 1–9. DOI: 10.55041/IJSREM40039

Kunduracıoğlu, İ. (2024). Utilizing ResNet Architectures for Identification of Tomato Diseases. *Utilizing ResNet Architectures for Identification of Tomato Diseases., 1*, 104–119. DOI: 10.59543/jidmis.v1i.11949

Lv, W., Xu, S., Zhao, Y., Wang, G., Cui, C., Du, Y., Dang, Q., & Liu, Y. (2023). DETRs beat YOLOs on real-time object detection. arXiv. https://doi.org//arXiv.2304.08069DOI: 10.48550

Mac, T. T., Nguyen, T. D., Dang, H. K., Nguyen, D.-T., & Nguyen, X.-T. (2024). Intelligent agricultural robotic detection system for greenhouse tomato leaf diseases using soft computing techniques and deep learning. *Scientific Reports, 14*(1), 23887. DOI: 10.1038/s41598-024-75285-5 PMID: 39396063

Mhaned, A., Salma, M., Haji, M., & Benhra, J. (2025). Plant disease detection using vision transformers. *International Journal of Power Electronics and Drive Systems, 15*(2), 2334. DOI: 10.11591/ijece.v15i2.pp2334-2344

Moustaine, M. (2023). *Symptomology of the Main Fungal Diseases of the Tomato (Lycopersicon Esculentum) and Its Management*. Austin Journal of Plant Biology., DOI: 10.26420/austinjplantbiol.2023.1035

Nisha, S. (2018). Image classification of agricultural data using supervised learning techniques: A survey. *International Journal of Advanced Research in Computer Science, 9*(2), 188–192. DOI: 10.26483/ijarcs.v9i2.5681

Pandi, S. S., Senthilselvi, A., & Gitanjali, J., ArivuSelvan, K., Gopal, J., & Vellingiri, J. (2024). Tomato leaf disease detection technique using VGG-19. *2024 Second International Conference on Advances in Information Technology (ICAIT)*, 1–6. https://doi.org/DOI: 10.1109/ICAIT54416.2024.00012

Patwal, P., Chauhan, R., Bhatt, C., & Devliyal, S. (2024). Automated tomato disease detection and classification using image processing and machine learning for precision agriculture. In *Challenges in Information* (pp. 481–486). Communication and Computing Technology., DOI: 10.1201/9781003559085-84

Pawlak, K., & Kołodziejczak, M. (2020). The Role of Agriculture in Ensuring Food Security in Developing Countries: Considerations in the Context of the Problem of Sustainable Food Production. *Sustainability (Basel)*, *12*(13), 5488. Advance online publication. DOI: 10.3390/su12135488

Pore, P., Teli, S., Ghuge, S., & Patil, N. (2023). Leaf Disease Detection. *International Journal for Research in Applied Science and Engineering Technology*, *11*(5), 1767–1770. DOI: 10.22214/ijraset.2023.51405

Rabhi, B., Dhahri, H., Jdey, I., & Alhajlah, O. (2025). IoT-Enhanced Tomato Leaf Disease Identification Using MLP-Mixer in Agricultural Environments. In: Czarnowski, I., Howlett, R.J., C. Jain, L. (eds) Intelligent Decision Technologies. KESIDT 2024. Smart Innovation, Systems and Technologies, vol 411. Springer, Singapore. https://doi.org/DOI: 10.1007/978-981-97-7419-7_37

Rajitha, A., Swathi, M., Ananya, K., Amruthavarshini, R., & Akhila, D. (2025). Plant Disease Detection Using Deep Learning. *International Journal for Research in Applied Science and Engineering Technology*, *13*(1), 1320–1323. DOI: 10.22214/ijraset.2025.66199

Ramya, J. (2023). Crop recommendation using supervised learning techniques. In *Proceedings of the ICOSEC Conference* (pp. 1076-1084). https://doi.org/DOI: 10.1109/ICOSEC58147.2023.10276199

Sekhar, M., Rastogi, M., Rajesh, M., Saikanth, D. R. K., Rout, S., Kumar, S., & Patel, A. K. (2024). *Exploring Traditional Agricultural Techniques Integrated with Modern Farming for a Sustainable Future: A Review*. Journal of Scientific Research and Reports. https://api.semanticscholar.org/CorpusID:267714435

Shafi, U., Mumtaz, R., García-Nieto, J., Hassan, S. A., Zaidi, S. A. R., & Iqbal, N. (2019). Precision Agriculture Techniques and Practices: From Considerations to Applications. *Sensors (Basel)*, *19*(17), 3796. DOI: 10.3390/s19173796 PMID: 31480709

Shakoor, M. T., Rahman, K., Rayta, S. N., & Chakrabarty, A. (2017). Agricultural production output prediction using Supervised Machine Learning techniques. *2017 1st International Conference on Next Generation Computing Applications (NextComp)*, 182-187.

Shetty, P., Kumar, A., Rajesh, B. S., Balipa, M., Kanchan, G., & Kamath, C. S. (2022, December). Tomato Leaf Disease Detection Using Multiple Classifier System. In *2022 International Conference on Artificial Intelligence and Data Engineering (AIDE)* (pp. 316-321). IEEE.

Singh, D., Jain, N., Jain, P., Kayal, P., Kumawat, S., & Batra, N. (2019). PlantDoc: A dataset for visual plant disease detection. arXiv. https://arxiv.org/abs/1911.10317

Sohel, A., Rahman, M., Hasan, M. U., Islam, M. D. K., Rukhsara, L., & Rabeya, T. (1850). M., Hasan, M. U., Islam, M., Rukhsara, L., & Rabeya, T. (2025). Automated tomato leaf disease recognition using deep convolutional networks. *International Journal of Power Electronics and Drive Systems*, *15*(2), 1850. Advance online publication. DOI: 10.11591/ijece.v15i2.pp1850-1860

Subramanya, S., & Naveen, B. (2024). *Enhanced detection of tomato leaf diseases using ensemble deep learning: INCVX-NET model.* IAES International Journal of Artificial Intelligence., DOI: 10.11591/ijai.v13.i4.pp4757-4765

Touvron, H., Cord, M., Douze, M., Massa, F., Sablayrolles, A., & Jégou, H. (2021). Training data-efficient image transformers & distillation through attention. *Proceedings of the International Conference on Machine Learning (ICML)*, 10347–10357. https://doi.org/DOI: 10.48550/arXiv.2012.12877

Wei, T., Chen, Z., Yu, X., Chapman, S., Melloy, P., & Huang, Z. (2024). *PlantSeg: A Large-Scale In-the-wild Dataset for Plant Disease Segmentation.* https://doi.org//arxiv.2409.04038DOI: 10.48550

Xu, M., Park, J. E., Lee, J. J., Yang, J., & Yoon, S. (2024). Plant disease recognition datasets in the age of deep learning: Challenges and opportunities. *Frontiers in Plant Science*, *15*, 1452551. Advance online publication. DOI: 10.3389/fpls.2024.1452551 PMID: 39399537

Yao, X., Li, Y., Zhang, Q., & Wang, L. (2024). Automatic visual recognition for leaf disease based on enhanced attention mechanism. *Frontiers in Plant Science*, *15*, 1382802. DOI: 10.3389/fpls.2024.1382802 PMID: 39650513

Zhao, Z., Chen, S., Ge, Y., Yang, P., Wang, Y., & Song, Y. (2024). RT-DETR-Tomato: Tomato Target Detection Algorithm Based on Improved RT-DETR for Agricultural Safety Production. *Applied Sciences (Basel, Switzerland)*, *14*(14), 6287. DOI: 10.3390/app14146287

Chapter 5
Smart Agriculture Using Advancing Tea Leaf Quality Assessment With Deep Learning

Md. Sayem Kabir
https://orcid.org/0009-0001-7022-4187
American International University, Bangladesh

Mohammad Ariyan Pathan
https://orcid.org/0009-0005-5073-236X
American International University, Bangladesh

Kazi Tanvir
https://orcid.org/0009-0001-0612-6101
Vellore Institute of Technology, India

Dipta Gomes
https://orcid.org/0000-0003-3019-6051
American International University, Bangladesh

ABSTRACT

The primary objective of this chapter is to conduct a comprehensive study tailored to classify, detect, and accurately evaluate the quality of tea leaves based on their age. The research aims to serve as a foundational resource, enabling the effective deployment of advanced machine learning algorithms for automating the quality assessment process. By replacing or supplementing manual inspection, these algorithms can provide more precise, reliable, and scalable solutions for quality evaluation. Such an approach is particularly significant for large-scale tea production,

DOI: 10.4018/979-8-3373-0020-7.ch005

Copyright © 2025, IGI Global Scientific Publishing. Copying or distributing in print or electronic forms without written permission of IGI Global Scientific Publishing is prohibited.

where consistent quality control is vital for maintaining market competitiveness. Moreover, this study aspires to deepen the understanding of the relationship between tea leaf age and quality, offering valuable insights into how leaf maturity impacts characteristics such as flavor, texture, and nutritional content.These advancements have the potential to catalyze broader applications of technology in agriculture, fostering innovation and sustainability across the sector.

1. INTRODUCTION

Tea cultivation plays a pivotal role in global agriculture and contributes significantly to the economies of tea-producing countries, providing income to millions of people worldwide. In 2023 alone, the global tea market was valued at approximately USD 60 billion and is projected to grow at a compound annual growth rate (CAGR) of 5.5% between 2023 and 2030 (Duarah et al., 2024). This growth underscores the importance of maintaining high-quality standards in tea production, as the quality directly influences market valuation and consumer preferences. A critical determinant of tea quality lies in the maturity and age of the leaves, which significantly influences the chemical composition and sensory profile of the final product. Younger tea leaves, for instance, are renowned for their high concentrations of catechins and antioxidants, which can account for up to 30% of their dry weight (Cioanca et al., 2024). These compounds are linked to the delicate flavors and extensive health benefits associated with tea, such as reducing the risk of cardiovascular diseases and supporting immune health. In contrast, mature leaves exhibit stronger flavor profiles, attributed to increased caffeine content and polyphenol levels, which can range between 20-35% of the dry weight in mature leaves(Guo et al., 2021). These biochemical variations not only cater to diverse consumer tastes but also play a crucial role in determining the market segmentation of tea products, from high-grade specialty teas to mass-market blends. Accurately determining leaf maturity is therefore paramount for optimizing the grading process and ensuring consistent product quality. However, traditional methods of assessing leaf maturity heavily rely on manual inspection(Ngugi et al., 2021). This process involves workers visually evaluating leaves based on size, color, and texture. While manual inspection of tea leaves is effective on a small scale, it has significant limitations, including subjectivity, inconsistency, and labor intensiveness, with error rates ranging from 15-20% leading to variations in quality and increased production costs (Tian et al., 2020). As industrial-scale tea production requires processing thousands of kilograms of leaves daily, manual methods struggle to meet efficiency and precision demands, resulting in resource wastage and reduced profitability (Blackford, 2024). Integrating advanced technologies, such as computer vision and machine learning, offers promising solutions to automate and standardize leaf maturity assessment, improving accuracy, reducing human error, and enhancing operational efficiency, thereby supporting the industry's sustainability and economic growth. However, replacing manual labor with automation in agriculture raises ethical concerns, such as job displacement, with up to 160,000 agriculture jobs in Canada potentially being automated in the next decade. Additionally, environmental challenges like increased energy consumption and electronic waste are associated with automated systems. To scale these advancements in low-income, resource-constrained settings, affordable, energy-efficient solutions and training programs are essential to ensure a smooth transition for affected workers and mitigate the adverse effects of automation.

The traditional process of tea leaf grading faces significant limitations due to its heavy reliance on manual inspection. Workers assess leaf quality based on visual cues such as size, color, and texture, a method that is inherently subjective and prone to human error. Studies have shown error rates of 15–20%, leading to inconsistencies in quality and increased production costs. Moreover, this manual approach is labor-intensive and inefficient, making it unsuitable for the industrial-scale production required to meet growing market demands, the practical challenges faced by workers who perform manual grading. Integrating AI technologies into agricultural quality control processes has brought substantial economic benefits to the coffee industry. For example, ProfilePrint, a Singapore-based platform, utilizes molecular analysis to predict sensory profiles of green coffee beans, enabling faster, objective quality assessments. This has streamlined operations and enhanced product quality, though specific revenue figures are undisclosed. Agrivero, employing high-resolution imaging and computer vision, has seen remarkable growth, with a 142% increase in revenue and a 495% rise in profit for 2023. These AI-driven innovations have reduced reliance on manual labor, minimized errors, and improved operational efficiency, leading to higher-quality products and better market prices. AI has also contributed to increased profitability for producers, as evidenced by the 4.4% revenue growth in the global coffee industry from 2022 to 2023, amounting to approximately $15.6 billion in coffee exports worldwide. This highlights AI's role in transforming the agricultural sector by enhancing productivity and boosting profitability. Automating tea leaf grading through deep learning and computer vision addresses critical economic challenges in the tea industry. By replacing labor-intensive manual inspection with automated systems, producers can reduce human error, lower operational costs, and boost overall productivity. These systems deliver objective, reproducible, and high-speed assessments, enabling precise resource allocation and streamlining production processes even at an industrial scale. This shift not only meets growing market demands but also enhances scalability and profitability.

Deep learning transforms tea leaf quality assessment by using advanced computer vision techniques, like CNNs and models such as YOLOv8 and YOLOv11 (Jiang et al., 2022), for accurate image analysis under varied conditions. Lightweight models with fewer parameters enable real-time grading on edge devices like drones or handheld scanners, offering a cost-effective, scalable solution. Advanced preprocessing enhances image quality, ensuring consistent, reliable results. These models maintain high accuracy in challenging environments, supporting sustainable tea production with reduced waste and optimized resource use, ultimately contributing to eco-friendly agricultural practices.

The growing demand for agricultural automation is driven by challenges like labor shortages, increased production, and the need for consistent quality control (Hussain & Khanam, 2024). Manual grading is time-consuming and prone to human error, making it less scalable for industrial use. Automated systems using edge detection and AI provide objective, high-speed results for leaf maturity evaluation (Bian et al., 2022), aligning with precision agriculture to optimize resource use, increase productivity, and reduce environmental impact. Traditional edge detection techniques like Sobel, Prewitt, and Canny often fail in agricultural images due to complex backgrounds, varying lighting, and leaf occlusions, leading to unreliable results (Tripathy & Sharma, 2020). These methods struggle to adapt to dynamic environmental factors, making them less effective for tasks like crop maturity detection. In contrast, deep learning algorithms, especially Convolutional Neural Networks (CNNs) (Alzubaidi et al., 2021), excel in image-based tasks by automatically learning relevant features and adapting to varying conditions(Archana & Jeevaraj, 2024). Training a CNN on labeled agricultural images enables accurate crop maturity classification despite real-world challenges. A hybrid approach, combining traditional edge detection with CNNs, enhances detection accuracy and reduces computation time, offering a more efficient solution for real-time agricultural applications (Dargan et al., 2020).

This research aims to develop an automated tea leaf grading system using edge detection and advanced technologies like spectroscopy. By replacing traditional manual methods with AI-driven solutions, it seeks to improve efficiency, ensure consistent quality, and promote sustainable farming practices.

Contributions

1. The proposed work presents an effective, self-operated technique for estimating the sweetness of tea leaves by utilizing high-performance edge detection which outplays conventional semi-manual that encompasses plenty of human involvement and subjectivity.

2. Adopting enhanced edge detection algorithms for computing tea leaf maturity through morphological characteristics of the leaf outline including serration density and edge fractal dimension.
3. Created a high-capacity automated system for grading tea leaves where minimum errors and less manpower was used as compared to manual grading system.

The introduction highlights tea's global economic importance and how leaf maturity impacts quality. Traditional manual grading is labor-intensive and error prone. This research proposes using deep learning and computer vision to automate and improve tea leaf grading for better accuracy and efficiency.

2. LITERATURE REVIEW

The literature review highlights various edge detection and deep learning applications in agriculture, with a focus on image-based models like YOLO (Jiang et al., 2022) for automating leaf quality assessments. While machine learning methods improve scalability and reduce errors, challenges such as data scarcity and variability in real-world conditions persist, underscoring the need for further refinement.

2.1. Machine Leaning vs Traditional Methods

Traditional methods of detecting tea age typically rely on manual inspection, chemical analysis, or simple image processing techniques that focus on visual cues like color, texture, and leaf structure. These approaches, while effective in controlled environments, can be time-consuming and prone to human error, especially in larger-scale production. In contrast, machine learning-based approaches offer a more scalable, efficient, and automated alternative. By incorporating advanced techniques such as feature extraction, histogram equalization, and Gaussian smoothing, combined with machine learning models like SVM or CNNs (Varoquaux & Colliot, 2023), the process becomes more robust and capable of handling larger, diverse datasets. For instance, in agricultural settings like tomato grading or pest detection, machine learning has demonstrated precision rates exceeding 90%, as shown by Guerra et al. (Guerra et al., 2024), even in controlled conditions. However, like Canny Edge Detection's (Rong et al., 2014) dependency on parameter optimization, machine learning models may face challenges in adapting to variable real-world environments without sufficient dataset diversity and generalization strategies. Despite this, the use of machine learning in detecting tea age offers significant improvements in accuracy and efficiency over traditional methods, reducing human error and enabling faster, more consistent results. Considering constraints, edge detection, a well-known method in computer vision field which has appeared as an encouraging method for automating leaf maturity assessment. Edge detection helps to identify the contours of objects within images, enabling the extraction of morphological features like leaf's shape and size and cutting-edge ability patterns (J et al., 2024). In tea leaves, these functionalities are closely linked to developmental stages, with younger leaves broadly having smoother edges and smaller dimensions, while mature leaves exhibit keen edge and larger sizes. Studies have confirmed that morphological traits, such as serration density along with edge complexity, are reliable indicators of mature leaf, which makes edge detection a productive tool for automating the study of these individuals (Pierce et al., 2022). Additionally, edge detection is analytically efficient along with scalable, enabling the investigation of huge datasets in real actual time, a key dominance in precising agriculture.

2.2. Deep Learning vs Traditional Methods

Deep learning methodologies have significantly advanced detection and object recognition, especially in agriculture, by providing robust computational capabilities. Traditionally, tasks like tea age detection relied on manual labor, visual inspection, or basic image processing methods, which, while useful in small-scale operations, were often inefficient, prone to error, and lacked scalability. In contrast, Akbar et al. (Akbar et al., 2024) examined the use of various deep learning models, including YOLO, Faster R-CNN, and SSD (Jiang et al., 2022), to automate these tasks. Their study revealed varying performance across these models, with YOLO delivering the best accuracy and efficiency. Specifically, YOLO achieved a precision of 97%, recall of 93%, and an F1-score of 76.65% for mushroom growth detection, highlighting its effectiveness in real-world applications. Preprocessing techniques such as Gaussian smoothing and data augmentation were used to improve model accuracy and generalization. Despite these advancements, the study acknowledged key limitations, such as the scarcity of high-quality datasets and the intensive labor process of manual labeling, which hinder large-scale deployment. Similarly, Xiao et al. (Xiao et al., 2022) explored the integration of genomic-wide association studies (GWAS) with high-throughput plant phenotyping models. Their research demonstrated the potential of deep learning in analyzing complex phenotypic traits, utilizing advanced segmentation and feature extraction techniques. The accuracy of their methods was closely tied to the quality of the phenotypic data and the variability of the dataset. However, their work encountered challenges, including the high cost of necessary equipment and the difficulty of replicating results across varying environmental conditions, limiting the scalability and generalizability of their approach. In comparison, traditional methods, although less computationally intensive and more affordable, often struggled with the complexity and precision required for modern agricultural practices.

Hybrid approaches combining IoT with machine learning (ML) and deep learning (DL) have shown promise as effective solutions for precision agriculture. Kumar et al. (Kumar et al., 2024) investigated the role of IoT devices, such as Arduino and Raspberry Pi, in data acquisition and processing for smart farming applications. Although specific ML or DL algorithms were not outlined in their study, they emphasized sensor data preprocessing and real-time integration to support informed decision-making. The study identified significant limitations, including connectivity issues and the need for robust, scalable data processing pipelines, which could hinder widespread adoption. Despite these challenges, the integration of IoT technologies holds substantial potential for improving agricultural efficiency and sustainability, provided these barriers are addressed.

Across all studies, several common challenges and opportunities were identified. Traditional ML methods, such as those proposed by Canny et al., offer simplicity and reliability but are limited in their adaptability and scalability. Conversely, DL models provide significantly higher accuracy and efficiency, as evidenced by the precision, recall, and F1-score metrics reported by Akbar et al. (Akbar et al., 2024), but they demand substantial computational resources and extensive labeled datasets. Hybrid approaches, such as those discussed by Kumar et al. (Kumar et al., 2024), represent a promising frontier, combining the real-time data acquisition capabilities of IoT with the analytical power of AI. However, these systems require further development to address infrastructure constraints and ensure widespread adoption.

In conclusion, while significant advancements have been made in edge detection and agricultural applications, there remains a pressing need for innovative solutions to overcome challenges such as dataset limitations, scalability, and resource requirements. Future research should focus on developing more efficient models, leveraging synthetic datasets for augmentation, and exploring cost-effective hybrid systems to meet the growing demands of precision agriculture. The literature review summary table has described below in Table1.

Table 1. Literature review summary table

Author(s)	Study Focus	Model/Technique	Dataset Used	Performance Metrics	Limitations
(Guerra et al., 2024)	Pest detection, tomato grading	Histogram equalization, CNN	Tomato, soybean datasets	Precision > 90% for grading systems	Dependence on controlled conditions
(Akbar et al., 2024)	Mushroom growth, rose blooming detection	YOLO, Faster R-CNN, SSD	Mushroom growth dataset (1,150 samples)	Precision: 97%, Recall: 93%, F1-Score: 76.65%	Data scarcity, manual labeling requirements
(Xiao et al., 2022)	Plant phenotyping using DL and GWAS	High-throughput GWAS phenotyping	Phenotypic datasets (varied sizes)	Contextual accuracy based on data quality	High equipment costs, dataset variability
(Dhanya et al., 2022)	IoT in precision farming	IoT devices (Arduino, Raspberry Pi)	Not explicitly stated	Real-time integration and decision-making	Connectivity issues, scalability challenges
(H. Zhang et al., 2023)	High-throughput phenotyping	Deep neural networks	Wheat and maize phenotypic datasets	High detection rates for growth stages	Lack of transferability across crops
(Kumar et al., 2024)	Crop yield prediction	Random Forest, XGBoost	Regional yield statistics datasets	Prediction accuracy: 87%	Dependent on accurate weather data
(Sishodia et al., 2020)	Fruit classification in orchards	CNN with transfer learning	Apple, mango datasets	Classification accuracy: 91%	Limited scalability to large-scale farming
(C. Wang et al., 2022)	Multi-object detection in agriculture	YOLO and Faster R-CNN	Mixed crop datasets	Detection accuracy: 90%	Computational requirements
(Y. Zhang et al., 2016)	Rice growth stage prediction	LSTM (Long Short-Term Memory)	Rice phenology dataset	Prediction accuracy: 89%	Dependent on time-series data availability

3. METHODOLOGY

The methodology involves using a comprehensive dataset, "TeaLeafAgeQuality," which includes images of tea leaves categorized by age. Various preprocessing techniques such as blur, median blur, and CLAHE were applied to enhance image quality for model training. Deep learning models like YOLOv8 and YOLOv11 were utilized for leaf maturity classification, and loss functions such as Binary Cross-Entropy (BCE) and Intersection over Union (IoU) were used to evaluate model performance.

3.1. Dataset Description

The "TeaLeafAgeQuality" dataset (Kabir et al., 2024) offers an exciting opportunity to explore how the age of tea leaves influences their quality, providing a valuable resource for both researchers and industry professionals. By categorizing tea leaves into four distinct age groups, the dataset gives us a clear look at how the leaves' quality changes over time. The first group, T1 (Age 1-2 days), contains 562 images of the freshest, highest quality leaves that are prized for their perfect taste and appearance. The second group, T2 (Age 3-4 days), includes 615 images of good-quality leaves, though they are slightly less optimal than the youngest leaves. In the third group, T3 (Age 5-7 days), there are 508 images of leaves that are starting to show signs of age, and their quality begins to decline. Finally, T4 (Age 7+ days) contains 523 images of leaves that are no longer suitable for brewing due to their advanced age.

The dataset comes in three versions to support various research and development needs. The raw version consists of unprocessed images, giving researchers a clean slate to work with. The annotated version includes images with detailed labels, making it easier for machine learning models to understand and categorize the data. Lastly, the augmented version contains enhanced images, expanding the dataset and helping to create models that can better adapt to real-world scenarios.

This dataset is more than just a collection of images; it's a gateway to improving the tea industry's approach to quality control. By understanding how leaf age impacts quality, producers can make better decisions and ensure that only the finest leaves make it to the consumer. Beyond that, it's a chance to push forward the field of machine learning in agriculture, showing how technology can help us make smarter, more efficient decisions when it comes to something as simple and yet as important as tea.

3.2. Preprocessing Techniques

In this study, we employed a series of preprocessing techniques to refine the quality of the images and optimize the performance of our models. These techniques, implemented through the Augmentations library, are essential for enhancing the dataset and ensuring more accurate results. Below is a detailed overview of each preprocessing method we applied:

3.2.1. **Blur:** The blur function uses Gaussian blur to soften the images by reducing fine details and noise. This technique helps the model to focus on more general patterns, making it less likely to be distracted by small, intricate features. For this dataset, the blur effect was applied with a 1% probability, and its intensity ranged from three to seven pixels (Gong et al., 2017). By introducing subtle blur, we prepared the images in such a way that the model could better handle slight distortions or occlusions, which could ultimately enhance its ability to generalize from the data.

3.2.2. **Median Blur:** Median Blur is a highly effective noise reduction method. It works by replacing each pixel's value with the median value of its neighboring pixels, which helps eliminate noise while preserving sharp edges. This technique is particularly useful for addressing "salt-and-pepper" noise, which is common in real-world images (Gong et al., 2017). The median blur was applied with a 1% probability and a range of three to seven pixels, ensuring that its effect was minimal, allowing the images to maintain a natural look. Despite its subtlety, this step plays an important role in improving image clarity and overall quality.

3.2.3. **To Gray:** The "To Gray" transformation converts the images into grayscale by averaging the values of the color channels (Saravanan, 2010). In cases where color information does not significantly contribute to the task, converting to grayscale allows the model to focus on structural and geometric features of the image. By using a weighted average to calculate the brightness, we ensured that the contribution of each color channel to the final image was properly balanced. This transformation not only streamlines the data for the model but also improves its robustness, particularly when color variations are insignificant or irrelevant for the task at hand.

3.2.4. **CLAHE (Contrast Limited Adaptive Histogram Equalization)**: CLAHE is a sophisticated technique that improves the contrast of an image, particularly useful for low-contrast images where details might otherwise be hard to distinguish (Hayati et al., 2023). It works by dividing the image into smaller tiles and enhancing the contrast within each tile individually, which helps reveal finer details. The level of contrast enhancement is controlled through parameters such as the clip limit (ranging from 1.0 to 4.0) and the tile grid size (8x8). By applying CLAHE with a 1% probability, we helped the model better capture fine details, especially in images that have areas of low contrast.

By utilizing these preprocessing methods, we enhanced the dataset in a way that prepares the images for more effective model training. Each step was designed to improve image quality, reduce noise, and focus the model on the most relevant features, ultimately improving its ability to generalize and make accurate predictions. These techniques are not only fundamental for this study but also serve as an important foundation for more complex models in the future.

3.3. Loss Functions and Metrics

3.3.1. Binary Cross-Entropy (BCE)

Binary cross-entropy (Yeung et al., 2022) is a widely used loss function for classification tasks, particularly in object detection. It measures the difference between the predicted probability of an object being present and the true class label (0 for absence and 1 for presence). BCE compares the predicted probability with the actual label and calculates the log-loss for each class, which helps optimize predictions in multi-class detection tasks like YOLOv3 and YOLOv5, where the task involves both localization and classification. The formula for BCE is:

$$L = -(y log(p) + (1 - y)\log(1 - p))$$

3.3.2. Intersection over Union (IoU)

IoU is a key metric used to evaluate the accuracy of object detection models (X. Wang & Song, 2021). It compares the predicted bounding box to the ground truth bounding box by calculating the area of overlap between the two, divided by the area of their union. This metric is used to assess how well the predicted bounding boxes align with the actual objects and is critical for optimizing the performance of models like YOLOv3, YOLOv5, and other versions. The formula for IoU is:

$$IoU = Area\ of\ Union / Area\ of\ Intersection$$

3.3.3. Complete Intersection over Union (CIoU)

CIoU is an advanced extension of IoU, which incorporates additional geometric factors to improve bounding box prediction. Unlike traditional IoU, which only considers overlap, CIoU also accounts for the distance between the center points of the predicted and ground truth boxes, the aspect ratio consistency, and box alignment (O. M. Lawal, 2021). This enhanced metric leads to better localization of bounding boxes, particularly when the predicted box is far from the ground truth box. CIoU is particularly useful in modern YOLO versions, such as YOLOv8 and beyond, where the focus is on improving detection accuracy, especially in challenging scenarios with small or overlapping objects. The formula for CIoU is:

$$\text{CIoU} = \text{IoU} - \frac{\rho^2(b, b_g)}{c^2} - \alpha v$$

3.3.4. Generalized Intersection over Union (GIoU)

GIoU builds upon IoU by introducing a penalty term that improves bounding box localization, particularly when the boxes do not overlap (Hou et al., 2021). This metric is particularly useful for non-overlapping bounding boxes, as it encourages better alignment and positioning. GIoU has been incorporated into modern object detection frameworks, including YOLOv4 and YOLOv5, to increase the precision and robustness of bounding box predictions, especially in scenarios involving sparse objects or occlusions. The formula for GIoU is:

$$\text{GIoU} = \text{IoU} - \frac{\text{Area of the Smallest Enclosing Box} - \text{Area of Union}}{\text{Area of the Smallest Enclosing Box}}$$

3.3.5. Distribution Focal Loss (DFL)

DFL is an advanced loss function designed to improve the model's ability to focus on difficult, hard-to-classify examples, which is particularly useful in dense and occluded object detection tasks (Tan et al., 2022). The key idea behind DFL is to adjust the weight of each sample's contribution to the loss based on its difficulty in classification. By extending the traditional Focal Loss, DFL takes into account the distribution of predicted class probabilities to improve bounding box regression. The formula for DFL is:

$$\mathcal{L}_{\mathcal{DFL}} = -\sum_{c=1}^{C} p_t^{\alpha} (1 - p_t)^{\gamma} \log(p_t)$$

3.4. AdamW Optimizer

The AdamW optimizer (Zhuang et al., 2022) is an extension of the popular Adam algorithm, introducing weight decay (L2 regularization) in a decoupled manner (Guan, 2023). This adjustment allows for better regularization without interfering with adaptive learning rates. In object detection tasks, AdamW enhances convergence speed and generalization by penalizing excessively large weights, thus preventing overfitting. The formula for AdamW is:

$$\theta_{t+1} = \theta_t - \eta \cdot \left(\frac{m_t}{\sqrt{v_t} + \epsilon} + \lambda \cdot \theta_t \right)$$

3.5. Model Descriptions

3.5.1. YOLOv3 Architecture Analysis

YOLOv3 in Figure 1, introduced in 2018, changed the way real-time object detection works by balancing speed and accuracy. It improved on earlier versions, YOLOv1 and YOLOv2, by using multi-scale predictions, which helped it detect small objects better (M. O. Lawal, 2021). YOLOv3, as shown in Figure 1, can detect objects at three different levels of image detail, allowing it to accurately find small objects at higher detail and larger objects at lower detail. This multi-scale detection is enabled by using anchor boxes to predict object positions. The model is built on a 53-layer Darknet-53 backbone, which efficiently extracts features from images, and uses binary cross-entropy loss for class predictions. The YOLOv3-SPP variant further enhances performance by incorporating Spatial Pyramid Pooling (SPP), improving its ability to detect objects at different scales. On the COCO dataset, YOLOv3 achieves 57.9% Average Precision (mAP) at an IoU of 0.5, while running at 30 frames per second on a TitanX GPU. It offers faster processing speeds compared to Faster R-CNN, but struggles with distinguishing fine details in crowded scenes. Additionally, YOLOv3 is computationally demanding, requiring significant processing power to achieve optimal performance, making it less suitable for resource-constrained environments but highly effective for tasks that demand precision and speed.

Figure 1. Yolov3 architecture

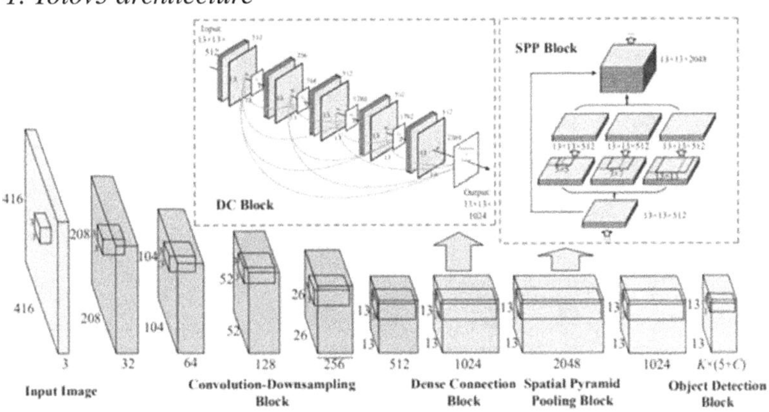

3.5.2. YOLOv5 Architecture Analysis

YOLOv5 in Figure 2, released in 2020, introduced an even more efficient approach to real-time object detection, emphasizing speed and accuracy in training and inference (Y. Zhang et al., 2022). YOLOv5 introduces an innovative Auto Anchor algorithm that automatically adjusts anchor boxes for different datasets, eliminating the need for manual tuning and enhancing efficiency. The model is built on a CSPDarknet53 backbone with an SPPF layer, which improves feature extraction and overall performance. YOLOv5 achieves an impressive 66.8% Average Precision (mAP) at an IoU of 0.5 and can run at up to 140 frames per second on a GPU, making it significantly faster and more accurate than YOLOv3. It is available in four versions to cater to different needs: Nano, Small, Medium, and Large. YOLOv5n (Nano), with 1.9 million parameters, is designed for lightweight tasks, while YOLOv5l (Large), with 46.5 million parameters, is optimized for handling complex tasks requiring maximum performance. The model uses Complete IoU loss for localization, ensuring precise object detection, and Binary Cross-Entropy for classification, improving object identification accuracy. As an open-source model, YOLOv5 is actively updated and highly adaptable, making it easy to deploy across various platforms, including mobile devices, further enhancing its versatility for real-time object detection applications.

Figure 2. Yolov5 architecture

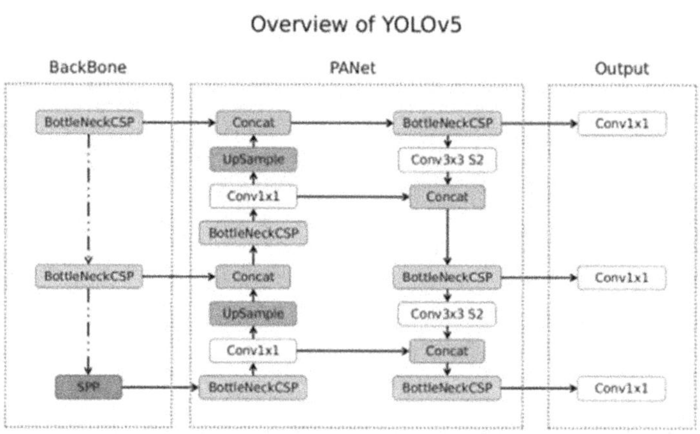

3.5.3. YOLOv8 Architecture Analysis

YOLOv8, launched by Ultralytics in January 2023, represents a significant leap forward in object detection performance (Bao, 2024). YOLOv8, as shown in Figure 3, stands out for its impressive speed and accuracy, making it ideal for tasks such as object detection, segmentation, and pose estimation. The model uses a CSPDarknet53 backbone with the C2f module, enhancing its ability to detect small objects and improving performance in crowded environments. Additionally, YOLOv8 features an anchor-free design with decoupled heads, which further boosts its performance in challenging scenes. The model utilizes CIoU and Distribution Focal Loss to achieve more accurate predictions, reaching 53.9% Average Precision (mAP) at an IoU of 0.5 while running at 280 frames per second on an NVIDIA A100, outperforming YOLOv5 in both speed and accuracy. YOLOv8 is available in several versions to meet various computational requirements: Nano, with 1.8 million parameters, is optimized for lightweight tasks; Small, with 11.2 million parameters, strikes a balance between speed and accuracy; Medium, with 25.9 million parameters, delivers higher precision; and Large, with 62.3 million parameters, is designed for maximum performance. As an open-source model, YOLOv8 is highly adaptable, making it suitable for a wide range of real-time applications across different industries.

Figure 3. Yolov8 architecture

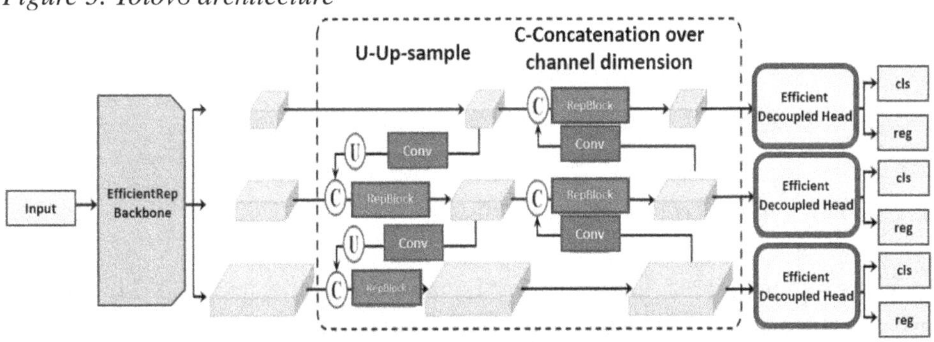

3.5.4. YOLOv9 Architecture Analysis

YOLOv9 introduces two cutting-edge features: Programmable Gradient Information (PGI) and General Efficient Layer Aggregation Network (GELAN) (Yaseen, 2024). YOLOv9, as illustrated in Figure 4, enhances learning efficiency, particularly for complex datasets like MS COCO, outperforming both YOLOv8 and YOLOv5 in terms of speed and accuracy. It introduces cutting-edge innovations such as PGI

(Progressive Gradient Integration) and GELAN (Gradient-Enhanced Local Aggregation Network), which optimize gradient utilization and feature aggregation, leading to superior performance in detecting small and complex objects. YOLOv9 is available in multiple versions to suit various needs: Nano, with 1.7 million parameters and 120 layers, is ideal for edge devices; Small, with 10.8 million parameters and 152 layers, offers balanced performance; Medium, with 26.5 million parameters and 248 layers, delivers high precision; and Compact, with 65 million parameters and 372 layers, is designed for maximum performance in complex scenarios. The model employs CIoU loss for better localization and DFL (Distribution Focal Loss) to improve prediction accuracy. YOLOv9 adapts exceptionally well to challenging environments, setting a new benchmark for object detection with its ability to balance efficiency and accuracy across various tasks.

Figure 4. Yolov10 architecture

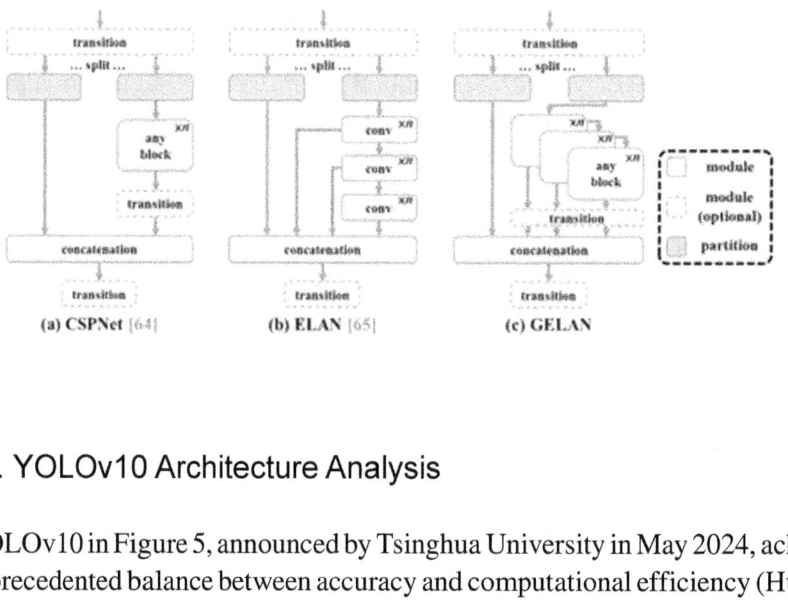

3.5.5. YOLOv10 Architecture Analysis

YOLOv10 in Figure 5, announced by Tsinghua University in May 2024, achieves an unprecedented balance between accuracy and computational efficiency (Hussain & Khanam, 2024). YOLOv10 introduces an innovative feature called Consistent Dual Assignment, which eliminates the need for Non-Maximum Suppression, significantly enhancing model efficiency. Alongside this, YOLOv10 incorporates advanced techniques such as Parallel Split-Attention, Compact Inverted Bottleneck Blocks, and Scaled Residual Connections, all of which work together to boost accuracy and generalization. The model is available in various versions to cater to different requirements: YOLOv10n, with 1.6 million parameters, is optimized for edge devices; YOLOv10s, with 12 million parameters, is ideal for real-time tasks; YOLOv10m, with 28 million parameters, delivers higher accuracy; and YOLOv10l, with 65 million parameters, is designed for maximum performance in complex scenarios. The model employs CIoU loss for improved localization and DFL (Distribution Focal Loss) for more accurate predictions, outperforming its predecessors YOLOv9 and YOLOv8 in both accuracy and speed. These improvements make YOLOv10 a versatile and efficient solution, suitable for a wide range of applications, from lightweight tasks on edge devices to demanding, high-performance environments.

Figure 5. Yolov11 architecture

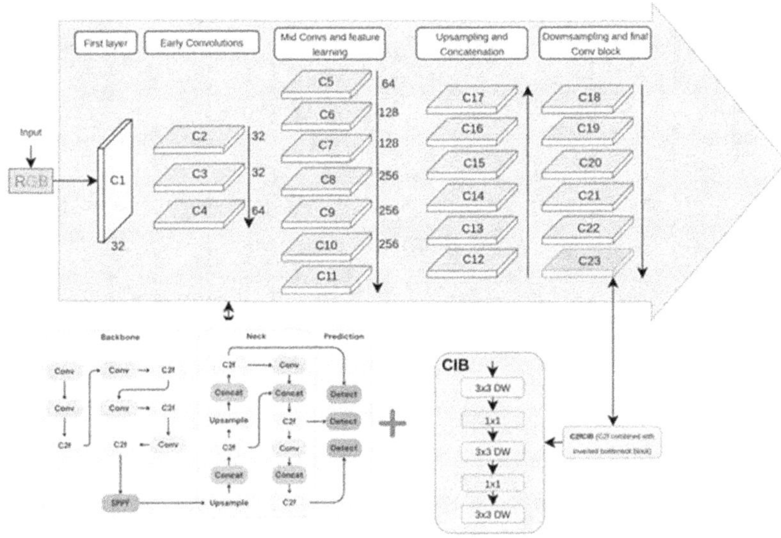

3.5.6. YOLOv11 Architecture Analysis

YOLOv11 in Figure 6, introduced by Ultralytics in September 2024, incorporates transformer-based architectures and self-attention mechanisms to enhance object detection in multi-object and complex scenarios (Alkhammash, 2025). YOLOv11 excels at detecting small objects, effectively managing occlusions, and performing well in low-contrast environments, resulting in enhanced accuracy and speed. One key improvement is the removal of Non-Maximum Suppression, allowing for more efficient inference. YOLOv11 is available in multiple versions tailored to specific use cases: Nano, with just 2 million parameters, is optimized for edge devices; Small, with 14 million parameters, offers a balance between efficiency and accuracy; Medium, with 30 million parameters, provides higher accuracy; and Large, with 70 million parameters, is designed for handling complex scenarios. The model employs CIoU loss for better localization, DFL (Distribution Focal Loss) to improve prediction accuracy, and multitask learning to strengthen feature extraction and generalization. These enhancements raise the bar for real-time object detection, positioning YOLOv11 as a highly powerful and versatile solution suitable for a wide range of applications, from lightweight edge computing to complex, high-demand environments.

Figure 6. Yolov11 architecture

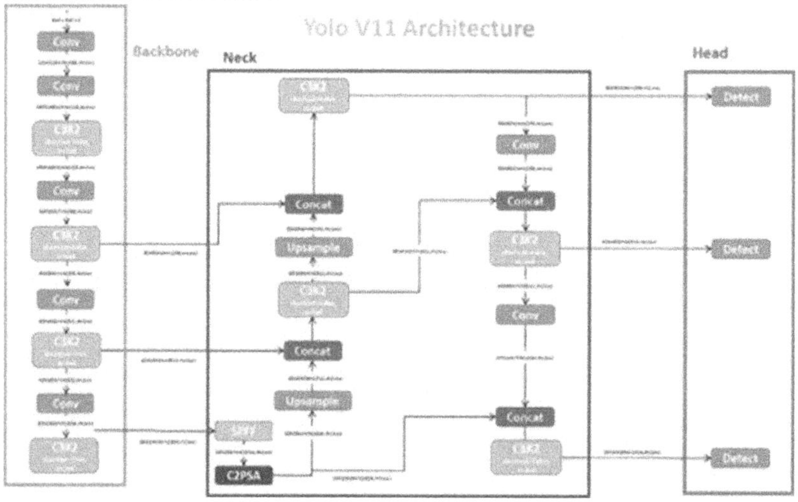

3.5.7. YOLO World Models Architecture Analysis

The YOLO World models are designed for large-scale, diverse environments, and come in six main variants: YOLOv8 Large World, YOLOv8 Medium World, YOLOv8 Small World, and their v2 counterparts (Sirisha et al., 2023). The YOLOv8 Large World models utilize CSPDarknet backbones integrated with SPPF layers and employ CIoU+BCE Loss functions to enhance accuracy, as seen in Figure 7. In Version 2, these models introduce advanced features like Split-Attention modules, Scaled Residual Connections, and Distribution Focal Loss, significantly boosting precision and class learning capabilities. Meanwhile, the Medium and Small World models prioritize efficiency by incorporating inverted bottlenecks and Depthwise-Separable Convolutions, alongside GIoU Loss to improve performance. Small World v2 takes it a step further by integrating an Attention Augmented Feature Pyramid, optimizing for ultra-low latency. These v2 models excel in handling dense, high-occlusion environments, offering superior precision, especially in challenging scenarios. The earlier versions maintain an ideal balance between speed and performance, making them highly suitable for real-time tasks where both accuracy and processing efficiency are critical. The innovations introduced in the v2 models position them as powerful tools for various applications, particularly in environments requiring rapid and precise object detection.

Figure 7. Yolo world model architecture

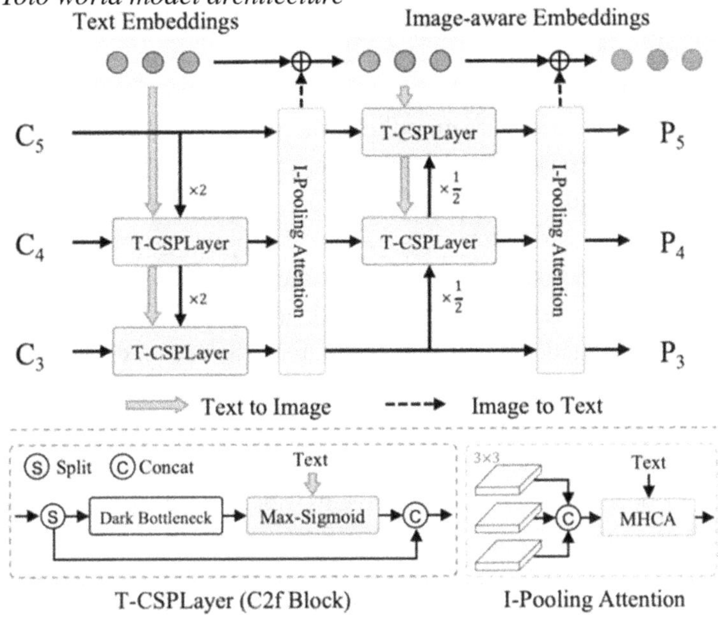

3.6. Environmental Effect of Models

The environmental effects of the YOLO model family, specifically in the context of real-time object detection in agriculture and other domains, are primarily influenced by computational resources and energy consumption. The advancements made in newer versions, such as YOLOv5, YOLOv8, YOLOv9, and YOLOv10 (Jiang et al., 2022), have led to improvements in efficiency, speed, and accuracy, but these gains come at a cost in terms of resource utilization.

1. Energy Consumption: The computational demands of models like YOLOv3 and YOLOv5, which require GPUs for training and inference, lead to substantial energy consumption. The transition from YOLOv3 (30 FPS on a TitanX GPU) to YOLOv5 (up to 140 FPS on a GPU) and later to YOLOv8 (280 FPS on an NVIDIA A100) significantly increases the power required, especially when deployed in large-scale applications. These higher frame rates, while improving

detection speed and efficiency, can result in increased carbon footprints if used at scale.

2. Resource Utilization: YOLO models, particularly newer versions like YOLOv9 and YOLOv10, employ more advanced architectures (e.g., CIoU loss, DFL, and GELAN), which enhance detection accuracy and performance in complex environments. However, the increased number of parameters and layers in these models (ranging from 1.6M in YOLOv10n to 65M in YOLOv10l) raises the computational power required, making them more resource intensive. As the models scale in complexity, the energy required to train and deploy them increases, potentially placing additional pressure on hardware infrastructure.
3. Hardware and Infrastructure Requirements: The high-performance requirements of these models demand advanced hardware such as high-end GPUs (TitanX, A100), which are often powered by large data centers. The infrastructure needed to support these models contributes to increased electronic waste and resource consumption, particularly when these devices have shorter lifespans due to rapid advancements in model architectures.
4. Environmental Impact of Data Centers: With models like YOLOv10 and YOLOv11 becoming standard in real-time applications, there is an increasing reliance on cloud computing and data centers for processing. These centers consume significant energy, often relying on non-renewable energy sources, which leads to a greater environmental footprint. The environmental costs are further exacerbated when the deployment of these models is frequent or continuous, as is often required for real-time object detection applications.

3.7. Evaluation Parameters

3.7.1. mAP50(B)

mAP50(B) stands for mean Average Precision at an Intersection over Union (IoU) threshold of 0.5, with bounding box confidence, and is a widely used metric for evaluating object detection models. This measure reflects the model's ability to accurately localize and classify objects, considering that predictions are considered correct when the IoU between the predicted and ground truth bounding boxes is greater than or equal to 0.5. mAP50(B) averages the Average Precision (AP) values calculated across all classes, where the AP of each class is derived from the area under its precision-recall (PR) curve. Mathematically, it is defined as:

$$mAP50 = \frac{1}{N}\sum_{i=1}^{N} AP_i,$$

3.7.2. mAP50-95(B)

mAP50-95(B) provides a more comprehensive measure of model performance across varying localization criteria, evaluating precision and recall over a range of IoU thresholds from 0.5 to 0.95 in increments of 0.05. This metric is considered a gold standard in modern object detection benchmarks, as it accounts for the quality of object localization and offers a more rigorous evaluation than mAP50(B). It is calculated as the average AP across different IoU thresholds, offering a better understanding of the model's performance across different levels of object localization accuracy. Mathematically, it is expressed as:

$$\text{mAP50-95} = \frac{1}{N}\sum_{i=1}^{N}\frac{1}{T}\sum_{t=1}^{T}\text{AP}_i(\text{IoU}_t),$$

3.7.3. Validation Distribution Focal Loss

Validation loss in object detection models assesses generalization and detects overfitting or underfitting. It combines bounding box, classification, and object confidence losses. Distribution Focal Loss (dfl_loss) enhances bounding box regression by predicting probability distributions over coordinates, improving accuracy and robustness. Calculated as the negative log-likelihood between predicted and ground truth distributions, dfl_loss boosts performance in both classification and localization.

3.7.4. Confusion Matrix

The confusion matrix is a fundamental tool for evaluating the performance of object detection models, offering insight into how predicted bounding boxes align with the ground truth (Tanim et al., 2024). The confusion matrix categorizes predictions as True Positives, False Positives, False Negatives, and True Negatives, assessing classification accuracy and spatial precision. Metrics like precision, recall, and F1-score derived from it provide insights into detection accuracy and error minimization. It is essential for evaluating and refining model performance.

3.7.5. Precision, Recall & F1-Score

Precision and recall are two essential metrics in object detection, each focusing on a different aspect of model performance. Precision refers to the proportion of true positive detections among all the objects the model has predicted as positive (Tanvir et al., 2024). In other words, it measures how many of the objects identified by the model are actually correct. It is calculated as:

$$\text{Precision} = \frac{TP}{TP + FP}$$

Recall, on the other hand, measures the model's ability to detect all relevant objects in an image. It is calculated as the ratio of true positives to the total number of actual objects (true positives plus false negatives):

$$\text{Recall} = \frac{TP}{TP + FN}$$

4. RESULTS AND ANALYSIS

The results demonstrate that models such as YOLOv8 and YOLOv11 perform well in automating tea leaf quality classification, offering high accuracy and consistency under varying environmental conditions. The integration of deep learning with spectroscopic analysis shows promise in improving both precision and speed in grading processes, reducing reliance on manual labor. A lot of object detection models have been trained. There are some summaries and analysis below in Table 2 and Table 3:

Table 2. Loss analysis of the models

Model	train/box_loss	train/cls_losss	train/dfl_loss	val/box_loss	val/cls_loss	val/dfl_loss
V3 spp	0.66336	0.52663	1.29917	0.78485	0.64292	1.3822
V3	0.76689	0.56427	1.33429	0.9571	0.80822	1.47397
V5l	0.72724	0.56172	1.29799	0.79408	0.63807	1.29787
V5m	0.70635	0.49691	1.27119	0.78244	0.61627	1.27031
v5n	0.71436	0.51201	1.26965	0.78854	0.59689	1.27763
v5s	0.70331	0.4787	1.26115	0.78213	0.63687	1.27944
v8l	0.71427	0.51137	1.28006	0.77557	0.61383	1.27282
v8m	0.68213	0.45225	1.25133	0.76724	0.57177	1.2613
v8n	0.68981	0.47467	1.24888	0.79607	0.628	1.28657
v8s	0.66746	0.41237	1.23098	0.77103	0.57824	1.27606
v9c	0.70412	0.47026	1.26703	0.7614	0.5562	1.25556
v9m	0.69582	0.46137	1.26609	0.77004	0.57424	1.27026
v9s	0.67901	0.42426	1.2539	0.74534	0.53178	1.24922
v9t	0.70025	0.4562	1.26747	0.7549	0.55688	1.2593
v10l	1.39976	0.99877	2.57265	1.54322	1.20093	2.56965

continued on following page

Table 2. Continued

Model	train/box_loss	train/cls_losss	train/dfl_loss	val/box_loss	val/cls_loss	val/dfl_loss
v10m	1.39326	0.9555	2.53926	1.52962	1.18096	2.53737
v10n	1.38538	0.84948	2.49921	1.56964	1.08452	2.54796
v10s	1.3674	0.79156	2.48221	1.5512	1.17188	2.51918
v11s	0.71553	0.5339	1.28988	0.75741	0.56708	1.25366
v11m	0.71081	0.51586	1.28174	0.77296	0.5826	1.2652
v11n	0.67958	0.43959	1.24513	0.77036	0.56234	1.26445
v11s	0.68667	0.43351	1.24971	0.77431	0.57342	1.26755
v8l world	0.64158	0.35814	1.20491	0.7627	0.58815	1.27313
v8l world v2	0.62667	0.33949	1.19662	0.78075	0.58823	1.28813
v8m world	1.28777	3.46831	1.80206	nan	nan	nan
v8m world v2	0.63777	0.36139	1.20003	0.76917	0.57778	1.26896
v8s world	0.64355	0.38232	1.20167	0.79258	0.60272	1.28601
v8s world v2	0.65427	0.39251	1.2095	0.77549	0.58097	1.27295

4.1. SMALL Model Analysis

4.1.1. Accuracy Analysis

The training performance of models evaluated over 30 epochs using mAP50(B) and mAP50-95(B) metrics reveals clear trends. In mAP50(B), v8s world v2 and v11s excel, reaching 0.91 and 0.90 by epoch 30, with rapid early improvements, surpassing 0.8 by epoch 10. Other models, like v8s (first version), v9s, v5s, and v3s, peak at 0.85, 0.86, 0.81, and 0.78, respectively, with slower growth and stability.

Similarly, in mAP50-95(B), v8s world v2 and v11s lead at 0.74 and 0.73, followed by v8s (0.69), v9s (0.69), and v5s (0.68). The v3s model lags at 0.62, showing weaker precision and stability. The smoother training curves of v8s world v2 and v11s highlight their fast convergence and superior generalization, making them ideal for tasks requiring high accuracy across varying IoU thresholds.

4.1.2. Precision Recall Analysis

The training metrics for recall and precision across various YOLO models over 30 epochs show significant differences. YOLO v8s world v2 outperforms others, achieving the highest recall (~0.87) and precision (~0.87) with stable performance, making it ideal for tasks requiring high accuracy and reliability. YOLO v8s and YOLO v8s world stabilize at around 0.80 recall and 0.85 precision, while older models like YOLO v3s and YOLO v5s perform more inconsistently with lower scores (~0.75 recall, ~0.70 precision). YOLO v11s shows competitive performance (~0.83 recall and ~0.83 precision) but with more variability. YOLO v8s world v2 is the top choice for real-world applications like surveillance and medical imaging.

4.1.3. Loss Analysis

The training and validation box loss graphs show the performance of multiple YOLO models over 30 epochs. YOLO v11s start with the highest loss but show the steepest decline, reaching around 0.65 by epoch 30, demonstrating strong optimization. YOLO v8s world v2, v8s world, v9s, and v5s maintain consistently low losses (~0.75), indicating steady learning and good generalization. YOLO v3s shows slower convergence with higher losses (~0.85). Overall, YOLO v8s world v2 performs best with low initial losses and steady improvement, while v11s, though effective, shows less stability in early training. YOLO v3s is the least favorable due to slower optimization and weaker generalization.

Table 3. mAP analysis of all the models

Model	epoch	metrics/mAP50(B)	metrics/mAP50-95(B)
V3 spp	30	0.88813	0.70584
V3	30	0.85466	0.64491
V5l	30	0.87668	0.69679
V5m	30	0.88959	0.71646
v5n	30	0.90096	0.71563
v5s	30	0.89421	0.7228
v8l	30	0.87498	0.70683
v8m	30	0.9079	0.73592
v8n	30	0.90319	0.71766
v8s	30	0.90299	0.72649
v9c	30	0.90183	0.73284
v9m	30	0.89856	0.72721
v9s	30	0.91525	0.7423
v9t	30	0.90857	0.73189
v10l	30	0.8819	0.71056
v10m	30	0.88776	0.71678

continued on following page

Table 3. Continued

Model	epoch	metrics/mAP50(B)	metrics/mAP50-95(B)
v10n	30	0.91251	0.73308
v10s	30	0.89255	0.72443
v11s	30	0.91042	0.73687
v11m	30	0.90112	0.73016
v11n	30	0.91748	0.74119
v11s	30	0.91317	0.73918
v8l world	30	0.89475	0.725
v8l world v2	30	0.89469	0.72568
v8m world	30	0.17367	0.11066
v8m world v2	30	0.90286	0.73737
v8s world	30	0.89347	0.72414
v8s world v2	30	0.90016	0.73222

4.2. NANO Model Analysis

4.2.1. Accuracy Analysis

The graphs show the training performance of several models over 30 epochs using mAP50(B) and mAP50-95(B) metrics. In the mAP50(B) graph, the v11n model performs best, reaching 0.91 by epoch 30, followed by v10n (0.89), v9n (0.88), v8n (0.87), and v5n (0.85). The v11n model shows faster learning, surpassing 0.8 by epoch 12. In the mAP50-95(B) graph, v11n also leads with a score of 0.73, followed by v10n (0.71), v9n (0.7), v8n (0.69), and v5n (0.67). The v11n model consistently outperforms others in both precision and generalization, making it the top performer, while v10n and v9n deliver competitive results, and v5n shows weaker performance.

4.2.2. Precision Recall Analysis

The graphs show the training performance of various models over 30 epochs, focusing on precision(B) and recall(B) metrics. The v11n model consistently outperforms others, achieving the highest precision (slightly above 0.8) and recall (close to 0.9) by the final epoch, with a smooth learning curve indicating efficient training. Models v9n and v10n follow, with precision around 0.79 and recall near 0.88, though they show more variability in their learning curves. The v8n model performs decently, stabilizing at precision 0.78 and recalling 0.87, but with more fluctuations, while the v5n model lags, ending with a precision of 0.75 and recall of

0.86, and shows the most instability. Overall, v11n is the most robust and generalizable, while v5n requires more optimization to improve its performance.

4.2.3. Loss Analysis

The graphs compare the train/box_loss and val/box_loss performance of various models over 30 epochs. The v11n, v9n, v10n, and v8n models show similar, stable learning curves, converging around a train/box_loss of 0.75 and a val/box_loss of 0.8 by epoch 30, indicating efficient optimization and generalization. In contrast, the v5n model starts with higher losses, converging at 1.3 for training and 1.5 for validation, reflecting slower learning and poorer generalization with more erratic curves. Overall, v11n and v9n show the best performance, while v5n underperforms and requires optimization improvements.

Figure 8. mAP & Precision recall analysis for large models

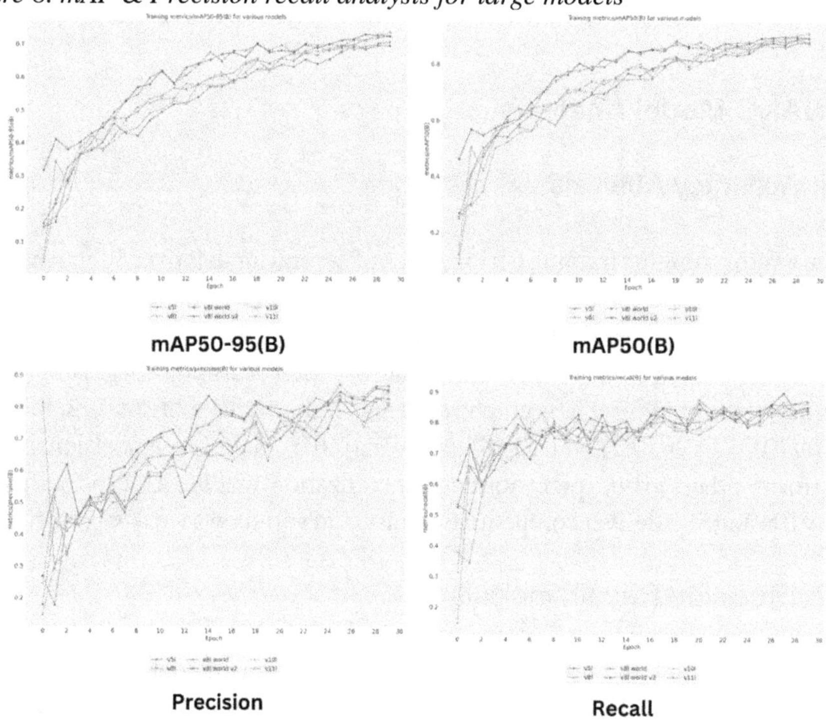

4.3. Medium Models Analysis

4.3.1. Loss Analysis

The given graphs in Figure 8 illustrate the training metrics mAP@50-95(B) (left) and mAP@50(B) (right) for various models across 30 epochs, highlighting their performance trends and stability. Most models, such as v5m, v8m, v9m, v10m, and v11m, show a steady upward trajectory, indicating consistent improvement and effective learning. Among these, v10m and v11m emerge as the top performers, with final mAP values exceeding 0.7 (left) and nearing 0.85 (right), showcasing strong optimization and generalization capabilities. In contrast, v8m world exhibits a significant performance drop after epoch 8, stabilizing at a much lower mAP (~0.2 in both metrics), suggesting instability or possible issues with its training configuration. The initial epochs demonstrate rapid performance gains across most models, with diminishing returns toward convergence. Models such as v8m world v2 and v11m display smooth learning curves and robust improvements throughout, indicating stable optimization. Overall, the high-performing models (v10m, v11m, and v8m world v2) exhibit both effective learning and generalization, while the poor performance of v8m world highlights the need for further refinement in its architecture or training parameters.

4.3.2. Precision, Recall & F1-Score Analysis

The graphs depict the training precision (left) and recall (right) metrics for various models over 30 epochs. Among the models, **v5m**, **v8m**, **v9m**, and **v10m** exhibit consistent improvements in both precision and recall, starting at lower values and gradually stabilizing around 0.8–0.85 by epoch 30. These models demonstrate a steady learning process with minor fluctuations during the early epochs, indicative of robust optimization. Conversely, **v8m world** shows a significant drop in performance around epoch 8 for both metrics and fails to recover, stabilizing at much lower levels (~0.3–0.4), highlighting potential issues in its architecture or data compatibility. **v8m world v2** and **v11m** also display steady progress, reaching competitive levels, though **v8m world v2** experiences slightly more fluctuations. Overall, most models perform well, except **v8m world**, which underperforms significantly.

4.3.3. Loss Analysis

The graphs display the **train/box_loss** (left) and **val/dfl_loss** (right) across 30 epochs for various models. In the **train/box_loss** graph (left), most models, including **v5m**, **v8m**, **v8m world v2**, **v9m**, and **v10m**, start with a box loss near

1.5 and exhibit a steady decline, stabilizing around 1.0 by epoch 30. These models demonstrate consistent and effective learning. However, **v8m world** shows a flat loss curve after an initial decrease, stabilizing at a much higher loss (~1.4), indicating learning stagnation. Meanwhile, **v11m** starts with a significantly higher loss (~2.7) and decreases gradually to around 1.8 by epoch 30, showing slower progress compared to other models. In the **val/df1_loss** graph (right), **v5m, v8m, v8m world v2, v9m**, and **v10m** exhibit rapid early improvements, with validation losses dropping below 1.5 by epoch 10 and remaining stable. **v8m world**, in contrast, shows erratic behavior with high fluctuations, indicating poor generalization. **v11m** starts with the highest validation loss (~4.5) but steadily decreases, stabilizing near 2.5 by epoch 30, though it underperforms compared to other models. Overall, most models demonstrate effective training and validation, except for **v8m world** and **v11m**, which exhibit significant limitations.

Figure 9. mAP & precision recall analysis for nano models

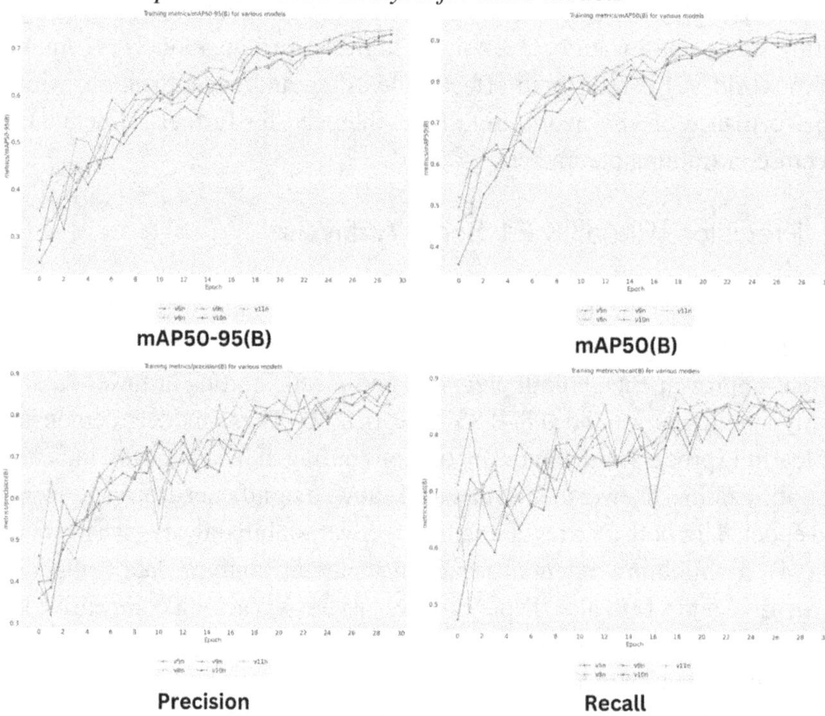

4.4. Large Model Analysis

4.4.1. Accuracy Analysis

Figure 9 displays the training performance for mAP@50(L) and mAP@50-95(L) across models over 30 epochs. In the mAP@50(L) graph, models like "v8l world v2," "v10l," and "v8l" show steady improvement, reaching 0.85–0.9 by epoch 30, with smooth learning curves indicating stable optimization. In contrast, the "v8l" model performs worse, stabilizing around 0.7–0.75, reflecting slower learning. Similarly, in the mAP@50-95(L) graph, top models converge around 0.65–0.7, while "v8l" lags at 0.55–0.6. The top models demonstrate strong generalization, while "v8l" struggles, indicating the need for further refinement to improve its performance and efficiency.

4.4.2. Precision Recall Analysis

The image compares the recall(L) and precision(L) metrics of various models over 30 epochs. In the recall(L) graph, models like "v8l world v2," "v10l," and "v11l" show steady improvement, reaching 0.85–0.9 by epoch 30 with smooth learning curves. In contrast, the "v8l" model lags, starting around 0.6 and stabilizing near 0.8, with more fluctuations. Similarly, in the precision(L) graph, the top models converge near 0.85–0.9, indicating strong performance with minimal false positives. The "v8l" model again underperforms, with slower precision growth and convergence around 0.8–0.85. Overall, "v8l world v2," "v10l," and "v11l" consistently outperform, while the "v8l" model shows weaker performance and greater variability, signaling areas for improvement in its design or training.

4.4.3. Loss Analysis

The graphs compare the training and validation box loss of several models over 30 epochs, showcasing their optimization and stability. In the training loss graph, models like "v8l world v2," "v10l," and "v11l" exhibit smooth and steady downward trends, converging near 0.75 by epoch 30, indicating strong and stable learning. In contrast, the "v8l" model starts with a higher loss (~2.75) and improves more gradually, stabilizing around 1.0, reflecting inefficiencies in optimization. In the validation loss graph, "v8l world v2," "v10l," and "v11l" maintain consistent, low losses, converging around 0.75, indicating good generalization. The "v8l" model shows greater fluctuations and a higher initial loss, with a peak near 2.75, suggesting challenges in generalization and stability. Overall, the top models demonstrate effective learning and generalization, while the "v8l" model lags behind, requiring improvements for optimization.

Here are some of the confusion matrices for the models:

Figure 10. Confusion matrix analysis

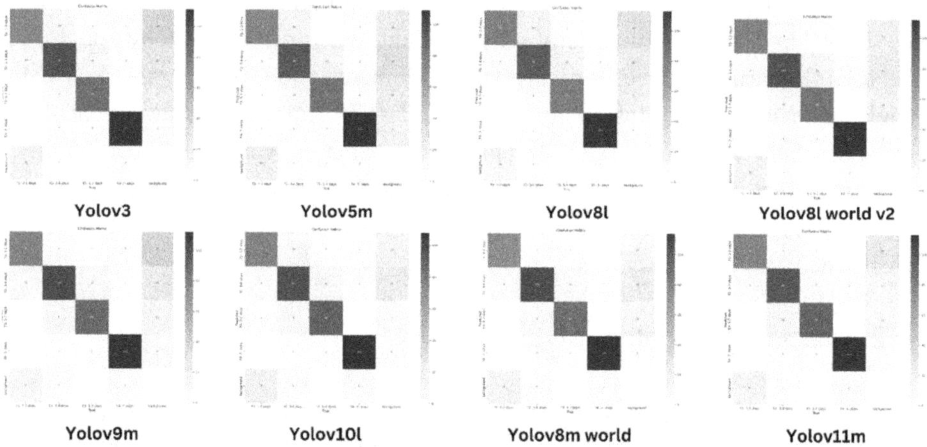

The confusion matrices in Figure 10 show that newer YOLO models improve classification accuracy. YOLOv3 has the highest misclassification, especially between "T3: 3-4 days" and "T3: 5+ days," with a 24-30% misclassification rate. YOLOv5m reduces this to 18-22%, while YOLOv8l and YOLOv10l improve further, with misclassification rates around 10-15%. YOLOv11m achieves the best accuracy, with misclassification rates under 10% and strong true positive rates. However, background class detection remains challenging, with some models misclassifying 12-18% of samples as background. Overall, newer YOLO versions significantly reduce misclassification, especially for closely related freshness categories.

Some of the example images:

Figure 11. Some examples of output.

Figure 11 displays a series of tea leaves with detection boxes and labels from a deep learning model indicating their growth stages and confidence scores. The labels show two main categories: "T3 - 5-7 days" and "T4 - 7+ days," reflecting the time period since the leaves emerged. The confidence scores range from 0.3 to 0.9, with higher values indicating stronger model confidence in the classification. Most leaves are classified as "T4 - 7+ days" with confidence levels between 0.7 and 0.9, suggesting they are older leaves. A few are labeled as "T3 - 5-7 days," indicating younger leaves with slightly lower confidence scores (e.g., 0.3 or 0.4). The image seems to emphasize the varying stages of leaf development and how well the model can differentiate between them, with overall higher confidence for more mature leaves.

The paper discusses the limitations of manual tea leaf grading and the benefits of using deep learning techniques like YOLOv8 and YOLOv11 for automation. Combined with edge detection and spectroscopic analysis, these models offer a scalable, sustainable solution for precision agriculture, improving accuracy and efficiency in tea leaf grading while benefiting the economy and the environment.

5. CONCLUSION

This study highlights the benefits of using modern image processing and AI techniques for automating tea leaf maturity estimation. Traditional manual methods are time-consuming, inconsistent, and biased, making them unsuitable for large-scale farming. By integrating edge detection, machine learning, and spectroscopy, the research provides a more accurate and efficient solution for precision agriculture, promoting resource efficiency, standard product quality, and sustainability. The approach improves accuracy and scalability, addressing the limitations of manual grading.

However, challenges include the high upfront costs, the need for specialized equipment, and potential environmental impacts of AI technologies. Future work should explore affordable, lightweight models and real-time IoT integration. The use of advanced models like YOLOv8 and YOLOv11 improves classification accuracy, even under difficult conditions, and enhances model robustness, supporting sustainable farming and economic growth.

5.1 Future Work and Limitations

1. Including tea leaves from other origins, different climates and harvested at different time will help to tune and test the model and check its performance under various conditions.
2. To facilitate grading of leaves on the move, quick to develop models that are lightweight and efficient for deployment on devices including drones and handheld scanners for grading.
3. Expansion of the study incorporating both edge detection and spectroscopy along with possible addition of genomic data can yield improved results for contrastive leaf quality.
4. The sustainability effects of the technologies could be measured by including environmental measures like water samplings, energy samplings in automated systems etc.
5. Integrating these models with IoT architectures for real-time assessment and control can make tea farming completely facile and rid of wastage.
6. Furthering the application of the developed techniques on other horticultural produce with similar grading difficulties, for instance coffee beans or lettuce will help popularize these research findings.
7. It will increase the adoption level by users in the agricultural sector to devise AI models that can be understood and interpreted as to how the decided conclusion was reached.

8. Improper design, maintenance, or disposal of the IoT device could pose risks to human life and the environment, highlighting the need for safe testing, eco-friendly practices, and energy-efficient adoption.

REFERENCES

Akbar, J. U. M., Kamarulzaman, S. F., Muzahid, A. J. M., Rahman, Md. A., & Uddin, M. (2024). A Comprehensive Review on Deep Learning Assisted Computer Vision Techniques for Smart Greenhouse Agriculture. *IEEE Access, 12*, 4485–4522. IEEE Access. DOI: 10.1109/ACCESS.2024.3349418

Alkhammash, E. H. (2025). Multi-Classification Using YOLOv11 and Hybrid YOLO11n-MobileNet Models: A Fire Classes Case Study. *Fire (Basel, Switzerland)*, *8*(1), 1. Advance online publication. DOI: 10.3390/fire8010017

Alzubaidi, L., Zhang, J., Humaidi, A. J., Al-Dujaili, A., Duan, Y., Al-Shamma, O., Santamaría, J., Fadhel, M. A., Al-Amidie, M., & Farhan, L. (2021). Review of deep learning: Concepts, CNN architectures, challenges, applications, future directions. *Journal of Big Data*, *8*(1), 53. DOI: 10.1186/s40537-021-00444-8 PMID: 33816053

Archana, R., & Jeevaraj, P. S. E. (2024). Deep learning models for digital image processing: A review. *Artificial Intelligence Review*, *57*(1), 11. DOI: 10.1007/s10462-023-10631-z

Bao, Z. (2024). The UAV Target Detection Algorithm Based on Improved YOLO V8. *Proceedings of the International Conference on Image Processing, Machine Learning and Pattern Recognition*, 264–269. DOI: 10.1145/3700906.3700949

Bian, L., Zhang, H., Ge, Y., Čepl, J., Stejskal, J., & EL-Kassaby, Y. A. (2022). Closing the gap between phenotyping and genotyping: Review of advanced, image-based phenotyping technologies in forestry. *Annals of Forest Science*, *79*(1), 22. DOI: 10.1186/s13595-022-01143-x

Cioanca, O., Lungu, I.-I., Mita-Baciu, I., Robu, S., Burlec, A. F., Hancianu, M., & Crivoi, F. (2024). Extraction and Purification of Catechins from Tea Leaves: An Overview of Methods, Advantages, and Disadvantages. *Separations*, *11*(6), 6. Advance online publication. DOI: 10.3390/separations11060171

Dargan, S., Kumar, M., Ayyagari, M. R., & Kumar, G. (2020). A Survey of Deep Learning and Its Applications: A New Paradigm to Machine Learning. *Archives of Computational Methods in Engineering*, *27*(4), 1071–1092. DOI: 10.1007/s11831-019-09344-w

Dhanya, V. G., Subeesh, A., Kushwaha, N. L., Vishwakarma, D. K., Nagesh Kumar, T., Ritika, G., & Singh, A. N. (2022). Deep learning based computer vision approaches for smart agricultural applications. *Artificial Intelligence in Agriculture*, *6*, 211–229. DOI: 10.1016/j.aiia.2022.09.007

Duarah, P., Joardar, S., Debnath, B., & Purkait, M. K. (2024). Optimized extraction of polyphenols from tea factory waste and cost-effective drying methods for sustainable utilization. *Bioresource Technology Reports*, *26*, 101833. DOI: 10.1016/j.biteb.2024.101833

Gong, D., Yang, J., Liu, L., Zhang, Y., Reid, I., Shen, C., van den Hengel, A., & Shi, Q. (2017). *From Motion Blur to Motion Flow: A Deep Learning Solution for Removing Heterogeneous Motion Blur*. 2319–2328. https://openaccess.thecvf.com/content_cvpr_2017/html/Gong_From_Motion_Blur_CVPR_2017_paper.html

Guan, L. (2023). Weight Prediction Boosts the Convergence of AdamW. In Kashima, H., Ide, T., & Peng, W.-C. (Eds.), *Advances in Knowledge Discovery and Data Mining* (pp. 329–340). Springer Nature Switzerland., DOI: 10.1007/978-3-031-33374-3_26

Guerra, J. P., Cuevas, F., Guerra, J. P., & Cuevas, F. (2024). Application of Digital Image Processing Techniques for Agriculture: A Review. In *Digital Image Processing—Latest Advances and Applications*. IntechOpen., DOI: 10.5772/intechopen.1004767

Guo, X., Ho, C.-T., Schwab, W., & Wan, X. (2021). Effect of the roasting degree on flavor quality of large-leaf yellow tea. *Food Chemistry*, *347*, 129016. DOI: 10.1016/j.foodchem.2021.129016 PMID: 33486364

Hayati, M., Muchtar, K., Roslidar, , Maulina, N., Syamsuddin, I., Elwirehardja, G. N., & Pardamean, B. (2023). Impact of CLAHE-based image enhancement for diabetic retinopathy classification through deep learning. *Procedia Computer Science*, *216*, 57–66. DOI: 10.1016/j.procs.2022.12.111

Hou, F., Lei, W., Li, S., Xi, J., Xu, M., & Luo, J. (2021). Improved Mask R-CNN with distance guided intersection over union for GPR signature detection and segmentation. *Automation in Construction*, *121*, 103414. DOI: 10.1016/j.autcon.2020.103414

Hussain, M., & Khanam, R. (2024). In-Depth Review of YOLOv1 to YOLOv10 Variants for Enhanced Photovoltaic Defect Detection. *Solar*, *4*(3), 3. Advance online publication. DOI: 10.3390/solar4030016

Kumar, K. A., & Verma, S. (2024). Harnessing Computer Vision for Agricultural Transformation: Insights, Techniques, and Applications. In *Applications of Computer Vision and Drone Technology in Agriculture 4.0* (pp. 111–131). Springer Nature Singapore.

Jiang, P., Ergu, D., Liu, F., Cai, Y., & Ma, B. (2022). A Review of Yolo Algorithm Developments. *Procedia Computer Science*, *199*, 1066–1073. DOI: 10.1016/j.procs.2022.01.135

Kabir, M. M., Hafiz, M. S., Bandyopadhyaa, S., Jim, J. R., & Mridha, F. (2024). *TeaLeafAgeQuality: Age-Stratified Tea Leaf Quality Classification Dataset. 1*. DOI: 10.17632/7t964jmmy3.1

Kumar, V., Sharma, K. V., Kedam, N., Patel, A., Kate, T. R., & Rathnayake, U. (2024). A comprehensive review on smart and sustainable agriculture using IoT technologies. *Smart Agricultural Technology, 8*, 100487. DOI: 10.1016/j.atech.2024.100487

Lawal, M. O. (2021). Tomato detection based on modified YOLOv3 framework. *Scientific Reports, 11*(1), 1447. DOI: 10.1038/s41598-021-81216-5 PMID: 33446897

Lawal, O. M. (2021). YOLOMuskmelon: Quest for Fruit Detection Speed and Accuracy Using Deep Learning. *IEEE Access, 9*, 15221–15227. IEEE Access. DOI: 10.1109/ACCESS.2021.3053167

Ngugi, L. C., Abelwahab, M., & Abo-Zahhad, M. (2021). Recent advances in image processing techniques for automated leaf pest and disease recognition – A review. *Information Processing in Agriculture, 8*(1), 27–51. DOI: 10.1016/j.inpa.2020.04.004

Pierce, S., Maffi, D., Faoro, F., Cerabolini, B. E. L., & Spada, A. (2022). The leaf anatomical trade-offs associated with plant ecological strategy variation. *Plant Ecology, 223*(10), 1233–1246. DOI: 10.1007/s11258-022-01270-5

Rong, W., Li, Z., Zhang, W., & Sun, L. (2014). An improved Canny edge detection algorithm. *2014 IEEE International Conference on Mechatronics and Automation*, 577–582. DOI: 10.1109/ICMA.2014.6885761

Saravanan, C. (2010). Color Image to Grayscale Image Conversion. *2010 Second International Conference on Computer Engineering and Applications, 2*, 196–199. DOI: 10.1109/ICCEA.2010.192

Sirisha, U., Praveen, S. P., Srinivasu, P. N., Barsocchi, P., & Bhoi, A. K. (2023). Statistical Analysis of Design Aspects of Various YOLO-Based Deep Learning Models for Object Detection. *International Journal of Computational Intelligence Systems, 16*(1), 126. DOI: 10.1007/s44196-023-00302-w

Sishodia, R. P., Ray, R. L., & Singh, S. K. (2020). Applications of Remote Sensing in Precision Agriculture: A Review. *Remote Sensing (Basel), 12*(19), 19. Advance online publication. DOI: 10.3390/rs12193136

Tan, Q., He, R., Bing, L., & Ng, H. T. (2022). *Document-Level Relation Extraction with Adaptive Focal Loss and Knowledge Distillation* (No. arXiv:2203.10900). arXiv. /arXiv.2203.10900DOI: 10.18653/v1/2022.findings-acl.132

Tanim, S. A., Shrestha, T. E., Tanvir, K., & Kabir, Md. S., Mridha, M. F., & Haq, M. K. (2024). Single-Level Fusion for Enhancing Meat Quality Classification with Explainable AI. *2024 IEEE International Conference on Computing, Applications and Systems (COMPAS)*, 1–6. DOI: 10.1109/COMPAS60761.2024.10796775

Tanvir, K., & Kabir, Md. S., Anik, Z. H., Hasan, Md. R., Tushar, MD. T., & Rahman, R. B. (2024). Forest Fire Detection Using Ensemble Deep Learning Model With XAi. *2024 IEEE International Conference on Computing, Applications and Systems (COMPAS)*, 1–6. DOI: 10.1109/COMPAS60761.2024.10796582

Varoquaux, G., & Colliot, O. (2023). Evaluating Machine Learning Models and Their Diagnostic Value. In Colliot, O. (Ed.), *Machine Learning for Brain Disorders* (pp. 601–630). Springer US., DOI: 10.1007/978-1-0716-3195-9_20

Wang, C., Han, J., Pu, Y., & Wang, X. (2022). Tea (Camellia sinensis): A Review of Nutritional Composition, Potential Applications, and Omics Research. *Applied Sciences (Basel, Switzerland)*, *12*(12), 12. Advance online publication. DOI: 10.3390/app13010012

Wang, X., & Song, J. (2021). ICIoU: Improved Loss Based on Complete Intersection Over Union for Bounding Box Regression. *IEEE Access*, *9*, 105686–105695. IEEE Access. DOI: 10.1109/ACCESS.2021.3100414

Xiao, Q., Bai, X., Zhang, C., & He, Y. (2022). Advanced high-throughput plant phenotyping techniques for genome-wide association studies: A review. *Journal of Advanced Research*, *35*, 215–230. DOI: 10.1016/j.jare.2021.05.002 PMID: 35003802

Yaseen, M. (2024). *What is YOLOv9: An In-Depth Exploration of the Internal Features of the Next-Generation Object Detector* (No. arXiv:2409.07813). arXiv. https://doi.org//arXiv.2409.07813DOI: 10.48550

Yeung, M., Sala, E., Schönlieb, C.-B., & Rundo, L. (2022). Unified Focal loss: Generalising Dice and cross entropy-based losses to handle class imbalanced medical image segmentation. *Computerized Medical Imaging and Graphics*, *95*, 102026. DOI: 10.1016/j.compmedimag.2021.102026 PMID: 34953431

Zhang, H., Wang, L., Jin, X., Bian, L., & Ge, Y. (2023). High-throughput phenotyping of plant leaf morphological, physiological, and biochemical traits on multiple scales using optical sensing. *The Crop Journal*, *11*(5), 1303–1318. DOI: 10.1016/j.cj.2023.04.014

Zhang, Y., Guo, Z., Wu, J., Tian, Y., Tang, H., & Guo, X. (2022). Real-Time Vehicle Detection Based on Improved YOLO v5. *Sustainability (Basel)*, *14*(19), 19. Advance online publication. DOI: 10.3390/su141912274

Zhang, Y., Yang, X., Cattani, C., Rao, R. V., Wang, S., & Phillips, P. (2016). Tea Category Identification Using a Novel Fractional Fourier Entropy and Jaya Algorithm. *Entropy (Basel, Switzerland)*, *18*(3), 3. Advance online publication. DOI: 10.3390/e18030077

Zhuang, Z., Liu, M., Cutkosky, A., & Orabona, F. (2022). *Understanding AdamW through Proximal Methods and Scale-Freeness* (No. arXiv:2202.00089). arXiv. https://doi.org//arXiv.2202.00089DOI: 10.48550

Chapter 6
Monitoring Fruit Quality and Quantity in Precision Agriculture

Zhenglin Wang
https://orcid.org/0000-0002-5812-2989
Central Queensland University, Australia

Xudong Sun
East China Jiaotong University, China

ABSTRACT

Precision agriculture leverages advanced technologies to optimize yield forecasting, resource management, and sustainable farming. This chapter explores methods for evaluating fruit load and quality from flowering to postharvest stages. Innovations in AI, including deep learning and transformer models, have revolutionized flower and fruit counting, while hyperspectral imaging and near-infrared (NIR) spectroscopy enable non-invasive assessments of maturity and quality. Decision support systems (DSS) enhance management practices by integrating these data. Machine vision advancements, such as MangoYOLO, improve fruit detection, size estimation, and yield forecasting, addressing challenges like environmental variability. Case studies in mango production highlight the transformative impact of these innovations on orchard automation and market efficiency. Portable NIR devices with GPS offer rapid, cost-effective quality assessments, while chemometric modeling and multi-source data integration enhance predictive accuracy.

DOI: 10.4018/979-8-3373-0020-7.ch006

Copyright © 2025, IGI Global Scientific Publishing. Copying or distributing in print or electronic forms without written permission of IGI Global Scientific Publishing is prohibited.

1. INTRODUCTION

Horticultural fruits such as mangoes, apples, citrus, and berries play an essential role in agricultural economies. They contribute greatly to food security, export income, and rural development. These fruits are highly valued for their nutrition and are in high demand in global markets, so improving production efficiency and managing supply chains are very important. However, the horticultural industry often faces challenges like yield fluctuations due to climate change, pressure from pests and diseases, dependence on seasonal labor, which is often uncertain, and the complicated logistics of transporting fresh produce, as it is highly perishable (Assimakopoulos et al., 2024).

In this context, advancements in technology, particularly through automated detection, yield forecasting, and resource management models, have emerged as critical solutions. Automated systems leveraging artificial intelligence (AI), remote sensing, and precision agriculture techniques can streamline operations, reduce costs, and enhance sustainability (Sa et al., 2016). By integrating data from various sources, such as hyperspectral imaging, Internet of Things (IoT) devices, and geospatial mapping, these technologies offer precise monitoring and forecasting capabilities. Additionally, they play a pivotal role in addressing labour shortages by automating tasks such as fruit sorting, grading, and harvesting, thereby ensuring consistent quality and timely delivery to markets.

Fruit quality is defined from the perspectives of growers, packers, retailers, and consumers, with specifications for harvest and eating maturity (Walsh & Wang, 2018). Technologies for monitoring key attributes include machine vision for canopy flowering and fruit size estimation, tools for assessing colour and dry matter (Walsh & Subedi, 2014), and postharvest measurements of temperature, firmness, and gas levels. Dry matter content guides harvest timing and ensures eating quality (Anderson et al., 2017). Decision support systems (DSS) integrating heat sums and dry matter further optimize harvest strategies (Dhonju, Bhattarai, et al., 2024; Dhonju, Walsh, et al., 2024). These innovations enhance productivity and align with global sustainability goals by reducing resource waste and lowering carbon emissions in horticulture.

1.1 Mangoes: A Case Study in Australian Horticulture

Mangoes, a quintessential summer fruit, play a pivotal role in Australia's horticultural industry. The primary production regions include the Northern Territory, northern and central Queensland, and Western Australia, with Queensland and the Northern Territory together contributing 96% of the national crop, as outlined in the Australian Mango Strategic Investment Plan 2022–2026 (https://www.horticulture

.com.au/globalassets/hort-innovation/levy-fund-financial-and-management-documents/sip-2022-2026-pdfs/hort-innovation-sip-2022-26-mango.pdf). The harvest season varies by region, running from July to March. Approximately 89% of Australia's mango production is consumed domestically, while 11% is exported annually, totalling 7,707 tonnes worth $33.7 million. Key export markets include Hong Kong, New Zealand, Singapore, the UAE, and the USA, where Australian mangoes are marketed as premium products with unique varieties and distinctive flavour profiles. Their counter-seasonal availability to the northern hemisphere enhances their appeal to international customers. However, the heavy reliance on domestic production underscores the need for effective yield management to balance supply and demand.

Mango orchards face unique challenges, particularly their remote locations, which complicate logistics and workforce coordination. Seasonal variations and the irregular fruiting patterns of mango trees lead to unpredictable yield volumes and harvest timing. This variability makes labour planning and transportation less efficient and increases the potential for waste. Accurate yield prediction is essential for addressing these challenges, enabling better agronomic interventions, workforce planning, and market forecasting (Walsh & Wang, 2018).

Currently, on-farm yield estimation methods rely on weather data analysis, historical yield records, and manual assessments of flowering and fruit numbers. Flowering events are tracked to estimate harvest timing, while fruit counts are performed after the "stone-hardening" stage, about six weeks before harvest when fruit drop ceases (Koirala et al., 2020). Industry initiatives, such as those led by the Australian Mango Industry Association (AMIA), aim to consolidate data across growers to support informed marketing and operational decisions (Anderson et al., 2018). For instance, crop forecast reports like the 2019/20 mango crop forecast provide detailed insights into weekly tray dispatch volumes, regional harvest schedules, and varietal trends.

The challenges faced in mango production are similar to those encountered in other horticultural crops, such as apples, berries, and citrus. These industries all rely on seasonal labor, need to maintain the quality of the produce after harvest, and often operate in remote areas with limited infrastructure. Accurate yield predictions are essential in these sectors to make better use of resources, minimize waste, and improve efficiency across the entire supply chain.

1.2 Advances in Machine Vision for Automated Fruit Detection and Counting

Advancements in machine vision (MV) technologies are revolutionizing agriculture, particularly in the detection of fruits and yield estimation (Jrondi et al., 2024). These technologies harness RGB imagery to identify and count fruits by analysing distinct features such as colour, shape, and texture that stand out from surrounding foliage. By automating the identification process and quantifying fruit numbers, MV provides vital data to improve agricultural planning and operational efficiency.

Early applications of computer vision for fruit counting primarily focused on analysing colour features in specific crops. For example, research on citrus fruits (Dorj et al., 2017; Hannan et al., 2009), 'Gala' apples (Zhou et al., 2012), mangoes (Payne et al., 2013), and strawberries (He et al., 2022) demonstrated strong correlations between MV-based and manual yield counts. However, these methods often suffered from significant estimation errors, limiting their broader adoption in commercial farming.

Subsequent developments in machine learning (ML) significantly enhanced fruit detection and classification by introducing algorithms capable of processing high-level handcrafted features. Techniques like Scale Invariant Feature Transform (SIFT) (Lowe, 1999), Speeded Up Robust Features (SURF) (Bay et al., 2008), Histogram of Oriented Gradients (HOG) (Dalal & Triggs, 2005), and Local Binary Patterns (LBP) (Ojala et al., 2002) enabled researchers to extract intricate image characteristics for further analysis. These features were used in combination with various ML algorithms for different crops, such as K-Means Clustering for grape detection (Luo et al., 2018), Support Vector Machines (SVM) for mango classification (Qureshi et al., 2017), K-Nearest Neighbor for blueberries (Tan et al., 2018), AdaBoost for tomato identification (Zhao et al., 2016), Decision Trees for kiwi fruit sorting (Kuang et al., 2019), Bayesian Classification for cherry yield estimation (Reyes et al., 2021). While these traditional ML approaches showed promise, their reliance on manually crafted features restricted their flexibility. They often required recalibration when applied to different environmental conditions or crop types, limiting their scalability and adaptability across diverse agricultural settings.

The advent of deep learning (LeCun et al., 2015), particularly Convolutional Neural Networks (CNNs) (Krizhevsky et al., 2012), marked a paradigm shift in fruit detection. Unlike traditional ML, CNNs learn hierarchical features directly from raw image data through an end-to-end process, eliminating the need for manual feature extraction. This adaptability has significantly improved detection accuracy and scalability. Driven by advances in single-shot detectors (SSD) (Liu et al., 2016), real-time processing capabilities, and robust frameworks like Faster R-CNN (Ren et al., 2015), YOLO (You Only Look Once) (Redmon et al., 2016), and SSD. State-of-the-art models like VGGNet (Simonyan & Zisserman, 2014), ResNet (He et al., 2016), and ZFNet (Zeiler & Fergus, 2014) have demonstrated success in handling challenges such as fruit occlusion and variable lighting conditions.

Recent advancements in GPU technology, user-friendly annotation tools like LabelImg, and robust deep learning frameworks have significantly lowered computational barriers and improved accessibility for agricultural applications. These innovations have enabled the integration of machine vision and deep learning into orchard management systems with remarkable efficiency. Key applications include:

1. Real-Time Fruit Counting: Frameworks based on YOLO have demonstrated exceptional performance in accurately detecting and counting fruits in real-time scenarios (Koirala et al., 2019).
2. Fruit Mass Estimation: Leveraging allometric relationships derived from machine-vision-based linear measurements has proven effective for precise fruit weight prediction (Neupane et al., 2022).
3. Management Information Systems (MIS): Innovative MIS for tree fruit management now incorporate data-driven insights to optimize harvest planning, labor allocation, and market forecasting (Dhonju et al., 2024).

This chapter will primarily focus on reviewing state-of-the-art technologies for fruit detection, counting, sizing, yield estimation, and quality assessment, with a special emphasis on mango production. These advancements underscore the transformative potential of modern machine vision and deep learning systems in achieving efficiency and precision in horticulture.

2. FRUIT DETECTION AND COUNTING

The rapid evolution of deep learning techniques has significantly advanced agricultural automation, particularly in orchard management. These advancements rely on transparent data sharing, consistent benchmarking, and rigorous documentation of model variability to ensure scalable and reliable solutions that address agriculture's unique challenges.

The breakthrough success of AlexNet in the 2012 ImageNet Large Scale Visual Recognition Challenge (ILSVRC) (Krizhevsky et al., 2012) marked the beginning of a deep learning revolution. Subsequent architectures like VGGNet (Simonyan & Zisserman, 2014), ResNet (He et al., 2016), and GoogleNet (Szegedy et al., 2015) refined performance, addressing challenges such as vanishing gradients with innovations like skip connections and residual networks. For agriculture-specific applications, Faster R-CNN and YOLO stand out for their balance of computational efficiency and detection accuracy. Studies show their effectiveness in fruit detection and counting, with precision and recall metrics exceeding 90%. By using a hierarchical structure, early CNN layers detect simple features like edges, while deeper layers capture abstract concepts such as shapes and objects. These capabilities have driven applications in agriculture, where precision and adaptability are crucial.

However, neural networks are inherently stochastic, converging to local minima influenced by several factors:

1. Floating-point precision: rounding errors that accumulate during computations.
2. Random initialization: variability in initial weight settings (Glorot & Bengio, 2010).
3. Stochastic optimization: techniques like stochastic gradient descent (SGD) (Zinkevich et al., 2010) and data shuffling introduce randomness.
4. Regularization methods: approaches like batch normalization and dropout add variability (LeCun et al., 1998; Srivastava et al., 2014).

As Walsh et al. (Walsh et al., 2024) point out, this variability can result in discrepancies in training outcomes, even when the same dataset is used. To ensure reliability and robustness in practical applications, it's crucial to document this variability through consistent benchmarking and detailed reporting.

Deep learning algorithms, such as Region-based Convolutional Neural Networks (R-CNNs) (Tu et al., 2020), Mask R-CNN (Giménez-Gallego et al., 2024), and YOLO (Mirhaji et al., 2021), have proven indispensable for tasks like fruit counting and segmentation. Recent YOLO iterations, in particular, are favored for their computational efficiency and versatility (Hussain, 2023). While CNNs excel at extracting local features, their reliance on localized receptive fields limits their ability to capture long-range dependencies. This limitation can present challenges in complex scenarios, such as occluded or clustered fruits, where misidentification or over-counting may occur. To address this, incorporating global context through attention mechanisms or hybrid architectures has been explored as a solution (Giménez-Gallego et al., 2024; Jrondi et al., 2024).

Comparative evaluations of neural network models are also vital to advancing the field (Lin et al., 2014). Performance is commonly assessed using metrics like Precision, Recall, F1 Score, and mean Average Precision (mAP) at different thresholds (e.g., mAP50 and mAP50-95). Fair comparisons require standardized training and test datasets, and researchers are encouraged to release datasets alongside model publications to promote reproducibility and facilitate further evaluations.

While well-trained models often demonstrate consistent performance, analyzing variations across multiple training runs offers critical insights into their robustness and generalization capabilities. Such documentation becomes indispensable when these models are deployed in real-world scenarios, where adaptability to diverse and unpredictable environmental conditions is vital. For example, in agricultural applications, weather fluctuations, lighting changes, and the diverse morphology of crops can challenge the stability of model predictions. Robust frameworks must not only perform well under controlled conditions but also adapt effectively to these dynamic factors to maintain reliability.

Understanding variability in training outcomes also highlights the significance of hyperparameter tuning, data augmentation, and model architecture selection in achieving dependable performance (Mimma et al., 2022). Furthermore, tracking performance consistency across different training runs can reveal the presence of overfitting or sensitivity to noise in the dataset, prompting refinements in data preprocessing or model design.

In agricultural settings, where detecting and segmenting fruits under varying conditions like occlusion, dense clustering, or uneven illumination is paramount, models must exhibit resilience. For instance, integrating attention mechanisms or employing hybrid architectures combining CNNs and transformers has proven effective in improving stability and adaptability (Jrondi et al., 2024). Such advancements not only enhance precision in yield estimation and fruit quality assessment but also lay the groundwork for further innovations to address the increasingly complex challenges of agricultural automation.

2.1 CNN Architecture and Feature Learning

A Convolutional Neural Network (CNN) typically consists of three key components: the input layer, multiple convolutional and pooling layers, and the output layer. In the case of mango fruit detection, image data enters through the input layer and undergoes processing in the intermediate layers. The convolutional layers use filters to identify important features within the image. For mango detection, these features could include color, shape, and texture — key elements for distinguishing mangoes from other fruits or objects in the orchard (N. T. Anderson, K. B. Walsh, A. Koirala, et al., 2021; Koirala et al., 2020; Wang et al., 2019).

In the early layers of the CNN, basic features such as edges, colors, and simple textures are extracted. For mango detection, these might include the smooth, rounded edges of the mango or its distinct hues—green, yellow, or red—depending on its ripeness. As the data progresses through deeper layers, the network learns to identify more complex and abstract features, such as the skin patterns or curvature of the fruit, which help in differentiating mangoes from other objects, even in crowded environments or orchards (Annamalai & Suk Lee, 2003; Li et al., 2015; Payne et al., 2014; Payne et al., 2013).

Popular CNN architectures such as AlexNet (see Figure 1 and Figure 2), ZFNet, VGG-16, Darknet-19, and Darknet-53 have been successfully integrated into object detection frameworks like R-CNN, YOLO (Xia et al., 2023), and SSD (Maeda-Gutiérrez et al., 2020). These models can be swapped based on the specific needs of mango detection tasks. YOLO, for example, is widely used in agricultural settings due to its speed and efficiency in real-time fruit detection. YOLO's capability to quickly detect mangoes and their locations within an image makes it ideal for applications like real-time counting and ripeness assessment in mango orchards.

Figure 1. A typical architecture of AlexNet

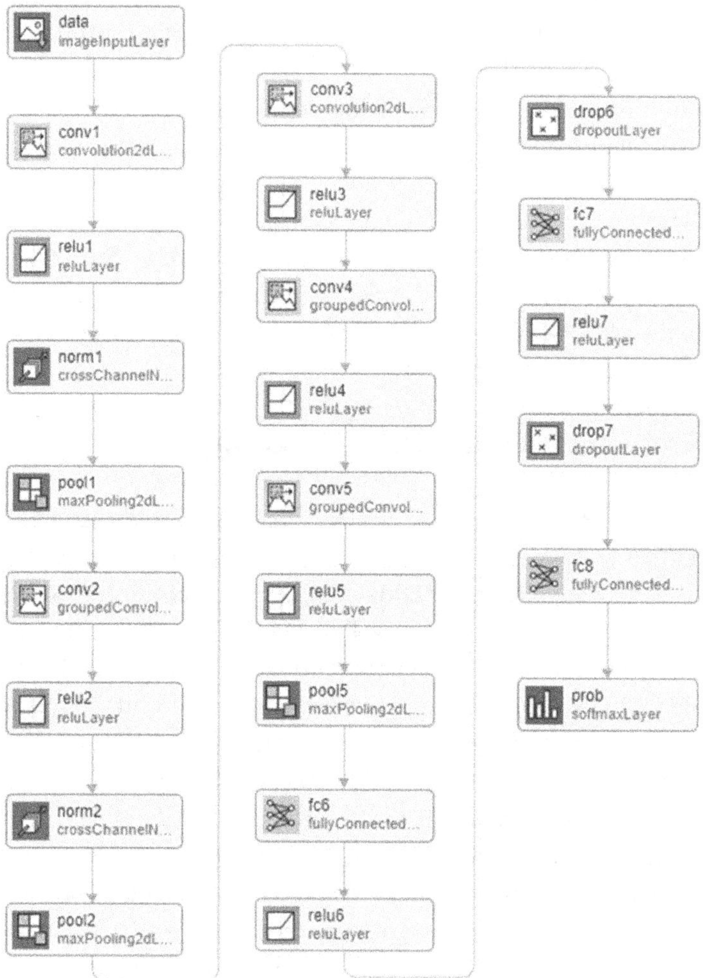

Earlier methods used a sliding window approach to classify objects within an image. This was computationally expensive, as it involved evaluating every possible region, which was slow and inefficient. For mango detection, this meant checking each small section of an orchard image for the presence of mangoes. The introduction of R-CNN (Regions with CNN Features) and its Selective Search algorithm

(Uijlings et al., 2013) helped by proposing regions likely to contain objects, thus reducing computational cost by focusing only on promising areas (Tu et al., 2020).

Although R-CNN with Selective Search made strides, it still faced limitations, especially with processing multiple region proposals for each image. Fast R-CNN addressed this by sharing convolutional feature maps across all region proposals, improving speed and efficiency in tasks like mango detection. However, Fast R-CNN still relied on external algorithms for region proposal, which led to the development of Faster R-CNN. This model integrated Region Proposal Networks (RPNs) (Girshick et al., 2016; Ren et al., 2015), automating the region proposal process, which enabled Faster R-CNN to autonomously identify areas in an image that likely contain mangoes. This improved both the speed and accuracy of mango detection within orchards.

Figure 2. The structure of AlexNet, consisting of 24 layers and 60.9 million learnable parameters.

	Name	Type	Activations	Learnable Sizes	State Sizes
1	data 227×227×3 images with 'zerocenter' nor...	Image Input	227(S) × 227(S) × 3(C) × 1(B)	-	-
2	conv1 96 11×11 convolutions with stride [4 4] a...	2-D Convolution	55(S) × 55(S) × 96(C) × 1(B)	Weig... 11 × 11 × 3... Bias 1 × 1 × 96	-
3	relu1 ReLU	ReLU	55(S) × 55(S) × 96(C) × 1(B)	-	-
4	norm1 cross channel normalization with 5 chann...	Cross Channel Norm...	55(S) × 55(S) × 96(C) × 1(B)	-	-
5	pool1 3×3 max pooling with stride [2 2] and pa...	2-D Max Pooling	27(S) × 27(S) × 96(C) × 1(B)	-	-
6	conv2 2 groups of 128 5×5 convolutions with st...	2-D Grouped Convo...	27(S) × 27(S) × 256(C) × 1(B)	Wei... 5 × 5 × 48 ... Bias 1 × 1 × 128...	-
7	relu2 ReLU	ReLU	27(S) × 27(S) × 256(C) × 1(B)	-	-
8	norm2 cross channel normalization with 5 chann...	Cross Channel Norm...	27(S) × 27(S) × 256(C) × 1(B)	-	-
9	pool2 3×3 max pooling with stride [2 2] and pa...	2-D Max Pooling	13(S) × 13(S) × 256(C) × 1(B)	-	-
10	conv3 384 3×3 convolutions with stride [1 1] an...	2-D Convolution	13(S) × 13(S) × 384(C) × 1(B)	Wei... 3 × 3 × 256... Bias 1 × 1 × 384	-
11	relu3 ReLU	ReLU	13(S) × 13(S) × 384(C) × 1(B)	-	-
12	conv4 2 groups of 192 3×3 convolutions with st...	2-D Grouped Convo...	13(S) × 13(S) × 384(C) × 1(B)	Wei... 3 × 3 × 192 ... Bias 1 × 1 × 192 ...	-
13	relu4 ReLU	ReLU	13(S) × 13(S) × 384(C) × 1(B)	-	-
14	conv5 2 groups of 128 3×3 convolutions with st...	2-D Grouped Convo...	13(S) × 13(S) × 256(C) × 1(B)	Wei... 3 × 3 × 192 ... Bias 1 × 1 × 128 ...	-
15	relu5 ReLU	ReLU	13(S) × 13(S) × 256(C) × 1(B)	-	-
16	pool5 3×3 max pooling with stride [2 2] and pa...	2-D Max Pooling	6(S) × 6(S) × 256(C) × 1(B)	-	-
17	fc6 4096 fully connected layer	Fully Connected	4096(C) × 1(B)	Weigh... 4096 × 92... Bias 4096 × 1	-
18	relu6 ReLU	ReLU	4096(C) × 1(B)	-	-
19	drop6 50% dropout	Dropout	4096(C) × 1(B)	-	-
20	fc7 4096 fully connected layer	Fully Connected	4096(C) × 1(B)	Weigh... 4096 × 40... Bias 4096 × 1	-
21	relu7 ReLU	ReLU	4096(C) × 1(B)	-	-
22	drop7 50% dropout	Dropout	4096(C) × 1(B)	-	-
23	fc8 1000 fully connected layer	Fully Connected	1000(C) × 1(B)	Weigh... 1000 × 40... Bias 1000 × 1	-
24	prob softmax	Softmax	1000(C) × 1(B)	-	-

One-stage detectors, such as YOLO and SSD, marked a significant advancement in object detection. Unlike R-CNN-based models that separate region proposal and classification, one-stage detectors predict object classes and bounding boxes in a single step, making them faster and more efficient for real-time applications (Jiang et al., 2018). YOLO is especially popular in commercial systems for monitoring mangoes in orchards due to its speed in processing images and accurately detecting mangoes at various stages of ripeness for harvest automation.

The progress of deep learning has revolutionized object detection by allowing models to automatically learn relevant features from raw image data, eliminating the need for manual feature engineering. From traditional segmentation methods to deep learning-based frameworks, models are now capable of recognizing an increasing variety of object classes in diverse environments. Innovations like Selective Search, Region Proposal Networks, and flexible feature extraction have made object detection faster, more efficient, and scalable. As deep learning techniques continue to evolve, one-stage detectors are setting new standards for performance and speed in object detection tasks.

2.2 MangoYOLO: A Successful Approach to Mango Detection and Counting

Object detection in agriculture, particularly for tree fruit yield estimation, is becoming increasingly important as machine vision systems are developed to automate tasks like fruit identification and counting. For mango detection, this involves several stages, including classification of potential fruit regions, localization of individual mangoes, and counting to estimate yield. Key steps in this process typically include:

1. Classification of Potential Fruit Regions: techniques such as color thresholding, key-point extraction, or convolutional filtering are used to distinguish mangoes from their surroundings.
2. Localization of Individual Objects: methods like blob segmentation, shape fitting, and bounding box regression are applied to precisely locate each mango.
3. Counting of Segmented Objects: once individual mangoes are localized, they are counted, providing vital input for yield estimation.

Since its introduction in 2015, the YOLO series has significantly advanced the field of object detection, evolving through various iterations to provide improved accuracy, speed, and computational efficiency. YOLOv8 (Neupane et al., 2024), released in 2023 by Ultralytics (Jocher et al., 2022), introduced innovative features like anchor-free detection, pseudo-ensembles, and mosaic augmentation, enhancing its ability to handle various tasks such as object detection, instance segmentation, and pose estimation. It also supports multiple model sizes ranging from nano to extra-large, with YOLOv8m achieving an impressive 50.2% mean Average Precision (mAP) on the COCO dataset, establishing its status as a state-of-the-art object detection model. Subsequent releases, YOLOv9 and YOLOv10, further boosted detection performance but increased computational complexity. YOLOv9 focused on resource-efficient design, while YOLOv10, developed by Tsinghua University (Wang et al., 2024), demonstrated superior object detection accuracy with higher resource demands.

Within this context, the MangoYOLO model (referred to Figure 3) was developed to address the specific challenges of mango fruit detection in orchards. Mangoes, with their unique shape and color variations, require a detection approach that can handle diverse lighting conditions and varying degrees of ripeness. The MangoYOLO framework builds upon YOLOv2(tiny) while optimizing for speed, accuracy, and computational efficiency. The innovative revisions include reducing the number of layers from 106 in YOLOv3 to just 33 in MangoYOLO, allowing for faster processing and less computational overhead while still achieving high accuracy.

Figure 3. MangoYOLO architecture (Koirala et al., 2019)

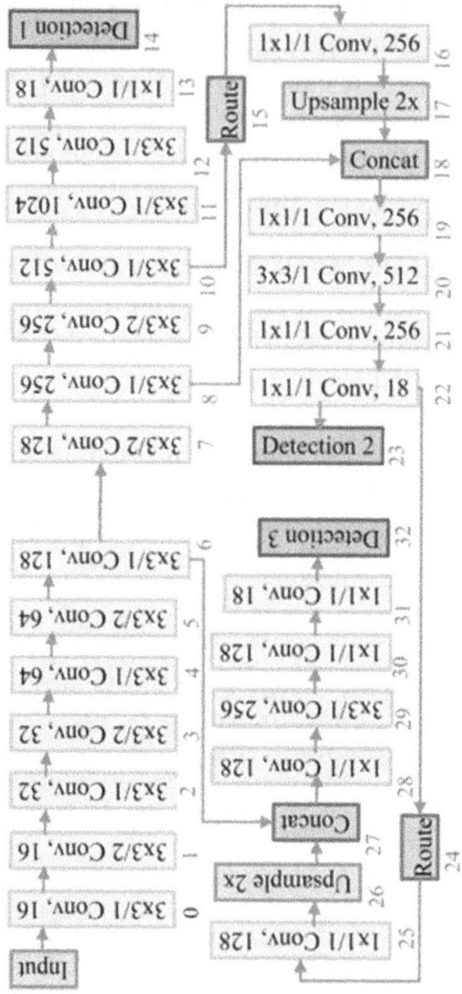

The MangoYOLO architecture focuses on detecting features at different scales to account for variations in mango size, especially considering that mangoes in real-world orchard conditions may be smaller or more obscured by foliage. By capturing features from early layers of the network, MangoYOLO is able to detect smaller or darker mangoes whose pixel information may otherwise be lost in deeper layers. The model also processes feature maps from intermediate layers (such as layers 6, 8, and 10) through up-sampling and merges them with feature maps of different resolutions, improving detection precision. Additionally, the first detection layer in MangoYOLO is positioned at layer 42, striking a balance between YOLOv2(tiny)'s layer 15 and YOLOv3's layer 82, offering an ideal trade-off for mango fruit detection.

The YOLO framework has also demonstrated its capacity to handle object detection efficiently by allowing adjustments to network resolution without requiring additional training. This flexibility is crucial for real-time applications in agriculture, where rapid and accurate detection is paramount. For mangoYOLO, the model was designed to handle images with a resolution of 512×512 pixels, which is appropriate for detecting mangoes in images with average fruit sizes around 43×48 pixels. At this resolution, MangoYOLO can detect mangoes at various scales, from smaller fruits to larger ones, without sacrificing performance.

The decision to replace max-pooling layers with convolutional layers was another key modification in MangoYOLO. In YOLOv3, max-pooling layers are typically used for dimensionality reduction, but convolutional layers are computationally more efficient and still perform a similar function. By integrating convolution layers into the architecture, MangoYOLO reduces the computational burden while maintaining the ability to extract relevant features. Furthermore, the removal of redundant layers, such as layer 13 in YOLOv2(tiny), contributes to further optimization, leading to faster processing times without sacrificing detection accuracy.

Unlike the deeper networks found in YOLOv3, which utilize residual blocks to combat the "vanishing gradient" problem in very deep architectures, MangoYOLO operates with a shallower network and does not require residual blocks. This design choice allows the network to focus on detecting mangoes at various scales with a reduced number of layers, making it faster and less computationally demanding, which is ideal for real-time applications like fruit detection in orchards.

The MangoYOLO architecture detects mangoes across three feature map scales (16×16, 32×32, and 64×64 pixels) for input images of 512×512 pixels, a resolution suitable for the dataset being used in this study. These scales are particularly well-suited to detecting mangoes in their various sizes and orientations within an orchard environment. By processing features at multiple scales, MangoYOLO is able to handle the challenges posed by fruit clusters, overlapping foliage, and varying fruit sizes.

In comparison with other object detection models such as Mask R-CNN and SSD, the trade-offs between speed, accuracy, and computational cost become apparent. While Mask R-CNN offers high accuracy, its detection speed is relatively slower, making it less suitable for real-time applications. YOLOv8, on the other hand, provides a more balanced trade-off between speed and accuracy, with the MangoYOLO model taking it a step further by optimizing the architecture for faster processing and higher detection efficiency.

In conclusion, the MangoYOLO model represents a significant advancement in object detection for agriculture, combining the best aspects of YOLO architecture with tailored optimizations for mango fruit detection. Through its reduced depth, multi-scale detection, and efficient feature map processing, MangoYOLO is capable of accurately identifying mangoes in complex orchard environments while maintaining high detection speeds. This makes it a powerful tool for automating processes in agriculture, particularly in mango yield estimation and other applications requiring real-time object detection.

2.3 Challenges and Progress in Mango Fruit Detection and Counting

Counting mangoes or other horticultural fruits on trees is very challenging because of the natural environment and the limits of current technology. Many factors affect detection and counting. One problem is occlusion, where fruits are hidden by leaves, branches, or even other fruits (Koirala et al., 2019; Wang et al., 2019). Trees with dense foliage make it even harder to see the fruits. Another issue is lighting. Sunlight and shadows change the way fruits look in images, making it harder for detection systems to work properly. For example, fruits may look overexposed under strong sunlight or unclear in low light.

The appearance of the fruit is also a big challenge. Mangoes come in different sizes, shapes, and colors, especially at different ripeness stages (N. T. Anderson, K. B. Walsh, A. Koirala, et al., 2021). These differences can confuse detection systems. Sometimes, the fruits look very similar to the leaves or background, which leads to misidentification. On top of that, the movement of tree branches and fruits caused by wind makes it difficult to keep detection accurate.

High-density clusters of mangoes also create issues. When many fruits grow close together, it is hard to tell one fruit from another, which can lead to counting errors. In orchards, scaling detection systems to large areas and maintaining speed and accuracy require a lot of resources. Environmental factors like rain, dust, or bad weather can also reduce image quality and make detection less reliable (Kestur et al., 2019).

Recent technology developments are helping solve these problems. Researchers have created advanced computer vision algorithms like YOLO and Faster R-CNN. These algorithms improve fruit detection accuracy and can work in real-time. Some studies suggest using multiple images from different viewpoints to get a better count. For example, Payne et al. (Payne et al., 2013) found that taking pictures from all four sides of a tree gave better results compared to only one or two sides. However, taking many images can cause problems like double counting, where the same fruit is detected multiple times. Researchers like Wang et al. (Wang et al., 2019) and Stein et al. (Stein et al., 2016) have used object tracking and depth information to fix this problem. They used tools like LiDAR to create 3D images of trees and track the fruits more accurately.

Another way to improve counting is by using correction factors for occluded fruits. This method calculates how many hidden fruits there are based on visible fruits. For example, Koirala et al. (Koirala et al., 2019) used this approach with their MangoYOLO system and achieved good results, estimating fruit counts with around 4.6% to 15.2% error.

In addition to better algorithms, autonomous systems like drones and robots are being used to capture orchard data. These machines are equipped with cameras and AI systems to process information more efficiently. Combining these tools with machine vision helps estimate fruit loads, predict yields, and even assist with harvesting (Mokrane et al., 2019; Okubo et al., 2024; Yıldırım & Ulu, 2023).

While progress is being made, there are still challenges. Scaling these technologies for large orchards and adapting them to different environments need further research. By continuing to optimize these methods, we can improve fruit detection and counting for better orchard management and productivity.

3. FRUIT SIZING

On-tree estimation of fruit size plays a critical role in various agricultural applications, such as predicting fruit maturity and harvest timing, estimating crop yield, and aiding in packing material and marketing decisions (e.g., tray inserts) (Jenni et al., 1998; Moreda et al., 2009). By measuring individual fruit sizes over time, researchers can estimate the rate of fruit expansion and understand how it responds to disease and environmental conditions (Jenni et al., 1996; Léchaudel & Joas, 2007; Mitchell, 1986; Ngouajio et al., 2003). Additionally, accurate fruit size estimation is essential for autonomous harvesting systems, as it helps control the harvest arm's approach and its grip on the fruit.

Size grading specifications vary depending on the commodity. For instance, weight is a commonly used grading specification, though some fruits are graded based on linear dimensions (Walsh & Subedi, 2014). The weight of a fruit can be estimated using an allometric relationship between its weight and linear dimensions, which can be measured non-destructively using methods such as calipers or machine vision techniques. Recent advancements in machine vision over the past decades, particularly in imaging hardware and the application of deep learning in image analysis, have significantly improved fruit sizing capabilities. Moreover, the expansion of broadband communication infrastructure (such as 5G) (Tang et al., 2021) and cloud computing further supports the development of sophisticated fruit sizing applications (Tang et al., 2021), ranging from handheld systems using mobile devices to specialized mobile imaging rigs for large-area assessments.

In orchards, assessing fruit size typically involves sampling, as it is not feasible to measure every fruit. A statistically valid sampling strategy is crucial for accurate size estimation, regardless of the measurement method used. Furthermore, farm management requires fruit size predictions weeks before harvest, but fruit can continue to grow until it is harvested. To address this, growth models are necessary to predict the size of fruit at harvest based on earlier measurements.

This section aims to cover topics such as (i) allometric relationships in fruit weight estimation, (ii) advancements in machine vision techniques, and the use of fruit growth models to forecast harvest weight. Case examples, including mango (Mangifera indica), are provided to illustrate the application of these methods. The ongoing development of these technologies, alongside continued advancements in data analysis and computational power, is expected to enhance fruit size estimation and its role in improving yield forecasting, harvest planning, and automation in orchard management.

3.1 Allometry: Relating Size to Weight in Fruit Estimation

Fruit weight can often be estimated based on the fruit's volume, accounting for its density. For fruits that approximate standard geometric shapes, common formulas linking lineal dimensions to volume are applied. For example, a spherical fruit follows the formula $V = \frac{4}{3}\pi r^3$, where r is the radius; an ellipsoid follows $V = \frac{4}{3}\pi ABC$, with A, B, and C being the semi-axes of the ellipsoid; a cylinder uses $V = \pi r^2 L$, where L is the length. These methods are particularly useful in determining the weight of uniform or geometrically simple fruits.

For mangoes, fruit size typically varies between 60 to 160 mm in length and 50 to 130 mm in width from stone hardening to harvest maturity, with daily growth rates of 0.3 mm in length and 0.2 mm in width during this period (Anderson et al., 2017). In Australia, mangoes are often packed in 7 kg trays, with the number of fruits per tray ranging from 10 to 22 depending on the average weight, which typically varies from 320 to 700 grams per fruit. As the tray size increases, the average fruit mass typically increases by 40 to 70 grams, and fruit length increases by 4 to 10 mm per tray size increment (Kernot et al., 1999).

A commonly used method to estimate mango fruit mass (M, in grams) involves the product of its length (L), width (W), and thickness (T), typically in centimetres, using the equation $M = kLWT$ (Spreer & Müller, 2011). Anderson et al. (Anderson et al., 2017) demonstrated this relationship's robustness across various growing conditions and growth stages, yielding a coefficient (k) ranging from 0.49 to 0.51 with an R-squared value of 0.97 and an RMSE of 28.7 grams. However, practical limitations arise when estimating fruit on a tree canopy, where orientation is uncontrollable, leading to variability in width and thickness as they are not consistently aligned in the image. Therefore, the allometric relationship can be simplified to with an R^2 of 0.93 (Anderson et al., 2017).

Various studies have empirically established relationships between fruit lineal dimensions or volume and fruit weight. For example, Marini et al. (Marini, 2001) used a quadratic equation to estimate apple fruit weight from diameter. While geometrically derived models follow physical principles, empirical models based on statistical relationships do not require the fruit to follow any specific geometric rule, although these models may be less universally applicable across different cultivars or environmental conditions.

Empirical relationships incorporate fruit density, which can vary with fruit shape, cultivar, and growing conditions. For instance, fruit density may decrease when voids form within the fruit or increase due to increased starch or woodiness content. Niklas (Niklas, 1994) and Misle et al. (Misle et al., 2013) have explored how fruit density can fluctuate during development, with studies indicating, for example, a 10% increase in pineapple fruit density in the 40 days before harvest. This variation poses challenges when applying a single model across different cultivars or growing conditions, as demonstrated by Yuan et al. (Yuan et al., 2020), who showed that the coefficient of the equation for kiwifruit weight needs to be adjusted according to the fruit's age due to changing density and shape.

Amaral and Walsh (Amaral et al., 2023) found that using a single model across different mango cultivars such as 'Honey Gold', 'Calypso', and 'Keitt' led to higher prediction errors (RMSE increased from 20.4 to 35.4 grams), highlighting the importance of cultivar-specific models. However, for certain applications, such as harvest management, these increased errors might be acceptable if the model provides a sufficiently accurate estimate (Schulze et al., 2015).

The precision of fruit weight estimation is also affected by the orientation of the fruit in the image. While machine vision typically relies on basic geometric formulas for volume estimation, image-based methods can offer additional parameters, such as fruit area, which may improve weight predictions. For instance, Utai et al. (Utai et al., 2019) explored the combination of fruit area with length and width measurements for improved mango fruit weight estimation.

In summary, allometric relationships between fruit weight and lineal dimensions offer a non-destructive way to estimate fruit weight while accounting for variations in fruit shape and density. These models, typically resulting in prediction errors of about 5% of average fruit mass, perform better when used within the conditions and cultivars for which they were developed. However, their accuracy decreases when applied to new fruit populations or under varying growing conditions. These models still represent a valuable tool in precision agriculture, especially when coupled with machine vision technologies.

3.2 Advancements in Mango Size Estimation Using Machine Vision

Mango size estimation, particularly in the context of pack-line fruit grading, has been extensively explored. Early methods focused on computer vision techniques using single (Yimyam et al., 2005) and multiple (Chalidabhongse et al., 2006) calibrated colour cameras for mango size estimation. These systems rely on maximizing contrast between the fruit and its background, with known camera-to-holder distances and estimation of fruit thickness from side views.

Traditional methods for measuring fruit dimensions, such as Vernier callipers or sizing rings, remain common, though they are labour-intensive and challenging, especially in the hot, humid field environment during mango fruiting seasons. As an alternative, machine vision systems mounted on farm vehicles can provide non-contact, efficient measurement of fruit size, facilitating the automation of this process. In recent years, there has been significant progress in the use of machine vision for fruit recognition and estimating fruit count on trees (Gongal et al., 2015; Payne & Walsh, 2014). Emerging machine learning technologies, particularly in object detection, have enabled high-performance fruit recognition and localization, contributing to the estimation of fruit number on trees (Gutiérrez et al., 2019; Qureshi et al., 2017; Sa et al., 2016). However, machine vision for fruit sizing requires unobstructed views of the fruit's full contour and precise measurements of camera-to-fruit distance. It is also important to note that while assessing fruit size for a sample is sufficient to estimate size distribution, the fruit visible in the outer canopy must be representative of the overall crop to be harvested.

For estimating mango fruit size on the tree, machine vision systems rely on the detection of fruit, object size in pixels, and accurate camera-to-fruit distance. Previous studies have focused on various techniques for fruit detection, including dense texture-based segmentation algorithms, colour and elliptical shape-based models, and colour and smoothness filters (Payne et al., 2014; Payne et al., 2013; Qureshi et al., 2017). More recently, deep learning methods like Faster Regional Convolutional Neural Networks (Faster R-CNN) have been applied to mango fruit detection (Sa et al., 2016; Stein et al., 2016). However, deep learning's high computational cost hampers its real-time application for tasks like in-field fruit sizing, keeping feature-based methods relevant. For instance, mango fruits often feature a distinct arc edge, making them suitable for rapid characterization using Histogram of Oriented Gradients (HOG) features (Dalal & Triggs, 2005). In our study in (Wang et al., 2017), we used colour thresholding and ellipse fitting, with an added cascade detection approach utilizing HOG features instead of Support Vector Machines (SVM) and Scale-Invariant Feature Transform (SIFT) features, to achieve a Root Mean Square

Error (RMSE) = 4.9 and 4.3 mm for estimated fruit length and width, respectively, relative to manual measurement.

4. FRUIT QUALITY MONITORING

Similar with yield estimation, non-destructively obtaining spatial and temporal variability of fruit quality on tree is also crucial to manage the orchard more precisely (Walsh & Subedi, 2014; Zude-Sasse et al., 2016). Hand-held near infrared spectroscopy (NIRS) instrument with the advantages of low-cost, non-destructive, rapid and on-site detection fits well for fruit quality assessment (K. Walsh et al., 2020). To precision orchard, forward from single instrument to internet of things (IoT) device for NIRS is a crucial direction. The hand-held NIRS instrument integrated with GPS sensor provides fruit quality properties and tree position simultaneously for fruit quality mapping (Walsh, 2023). Modern NIRS technology involves three sustaining pillars of instrument, chemometrics and model (Pasquini, 2018). Therefore, typical hand-held instrument, chemometrics, model and valuable field application were reviewed and discussed in this section.

4.1 Instrument

The development of the portable NIRS instrument for fruit quality outdoor use to maximize food potential of harvested fruit began in 1996. A portable NIRS instrument titled Fruit Tester 20 (Fantec, Kosai-city, Japan) was developed in 1998 using an interactance fibre optic probe and wavelength range of 600-1000 nm with size of 120×250×300 mm and weight of about 4 kg (Aoki et al., 2004). As for the second generation, a compact hand-held NIRS instrument named FQA-NIR GUN (Fantec, Kosai-city, Japan) was developed with weight of about 750 g and wavelength range of 600-1100 nm in 2003. In 1999, another portable NIRS instrument of K-BA100 was developed by Kubota Corporation (Osaka, Japan) with size of 300×118×240 mm and weight of 5 kg (Morimoto, 2002). An interactance feature probe with a ring illumination and a collector in the centre of the probe was designed for collecting the absorbed light scattered by the fruit tissue more efficiently. K-BA100 was upgraded and became more compact since 2019. Integrated Spectronics (Sydney, Australia) developed a Hand-held NIRS instrument named Nirvana integrated with GPS sensor in 2010. The upgrade generation of Nirvana named F750/F751 was developed by Flexi Instrument (Camas, WA, USA) in 2014. The predictive results of F750/F751 instruments including fruit quality index and tree location can be uploaded into the Fruit Map System (https://www.fruitmaps.com/) for mapping harvest decision map. It indicated a future direction for the hand-held NIRS instrument from a single instrument to an Internet of Things (IoT) instrument. In 2014, H100 series hand-held NIRS instruments were developed by Sunforest Company (Incheon, Korea) with wavelength range of 650-950 nm and weight of 400 g (Choi et al., 2017). The above instruments basically reflect a number of key nodes during development of hand-held NIRS fruit quality tester. The development features were summarized as follows:

1. Technologies for stable and precise measurement in the field. Field use requires consideration of conditional variations e.g. temperature, sunlight (Zhang & Yang, 2024). Kubota K-BA100R designed the automatic calibration system (ACS). ACS consisted of a wavelength calibration filter and a standard filter to calibrate the wavelength shift and output drift due to temperature variance or change of properties with the time (Morimoto, 2002). Felix F750/751 controlled an internal shutter on and off, light on and off to collects the white and dark references with every sample for calculating the absorbance spectrum (Sun, Subedi, & Walsh, 2020). To avoid the effects of outdoor temperature changes, Sunforest H100F will restart to collecting the white and dark refences for calibrating the instrument when the current temperature is 10°C higher than the start temperature (Sun, Du, et al., 2024).
2. IoT instrument. Fruit quality and the position of the affiliated tree requirement for precision orchard management. GPS sensor was integrated with Felix F750 and 751 devices for simultaneously providing fruit quality index and tree position. These data were subsequently uploaded into the support system for yielding the harvest decision map (K. B. Walsh et al., 2020). Bar camera was integrated with Sunforest H100F instrument for recording the tree position.
3. Miniaturization. The weight of the instruments developed from portable to hand-held and pocket-size. For example, the weights of K-BA100R, F750 and H100F hand-held NIRS instruments were 5 kg, 1.05kg and 400 g, respectively.

4.2 Chemometrics

NIRS is a secondary method requirement of chemometrics for model development between spectral signals and fruit quality attributes (K. B. Walsh et al., 2020). The state of art of chemometrics for NIRS and fruit quality assessment have been reviewed comprehensively in the past two years (Anderson & Walsh, 2022; Walsh et al., 2023; Zhang & Yang, 2024). So, the common knowledge was introduced from the commercial use perspective of hand-held NIRS instrument. The partial least squares (PLS) regression is widely accepted algorithm for model development with a range of standard pre-treatment methods of mean centering, standard normal variate (SNV) or multiplicative scattering correction (MSC) and derivative (K. Walsh et al., 2020; K. B. Walsh et al., 2020). The SNV, MSC and derivative methods generally fit well to a suitable wavelength range. For the specific instrument and fruit attributes, the suitable wavelength range was generally recommended. The wavelength ranges of 729-975 nm and 700-930 nm were recommended for F750 assessment of mango and olive dry matter content (DMC) (Sun, Subedi, & Walsh, 2020) and H100F estimation of pear soluble solid content (SSC) (Choi et al., 2017), respectively.

Heterogeneity of data often generates nonlinear relations between the spectral variables and fruit attributes along with the variation of seasons, regions and varieties. Locally weighted partial least squared regression (LWPLSR) and its extended method is an efficient prediction method (Lesnoff, 2024; Lesnoff et al., 2020). To fit well with the heterogeneity of spectral data, NIRS instrument companies also started provided artificial neural network (ANN) models to the users, for example, Felix instrument provided ANN models to F750 user in Australia. Convolution neural network (CNN) with the advantage of deep learning ability has been emerged as a promising evolution for assessment of fruit quality (Walsh et al., 2024). However, CNN requires a strong computer power and big data set, which are a challenge for hand-held NIRS instrument with the consideration of low-cost requirement. The big data set collection also needs a long period and sustainable work with cost input after the first instrument available.

Field use of NIRS is influenced by the variation of measurement conditions e.g. temperature, humidity and sunlight (Feng et al., 2011; Liu & Ying, 2015; Saranwong et al., 2003). External parameter orthogonalization (EPO), loading space standardization (LSS) and generalized least squares weighting (GLSW) were proposed for removing the influence of the external factor (Chen et al., 2005; Igne et al., 2019; Roger et al., 2003). Temperature and sunlight influence to the spectra was removed using EPO and GLSW in predicting SSC of table grape and navel orange using hand-held NIRS instrument (Sun, Guo, et al., 2024; Sun et al., 2023). Field use also requires these algorithms assembly with PLS in chemometrics software providing with the users.

4.3 Model

The big and qualified data set with a wider reference range is prior to model development, because the calibrated model will be used in assessment of new fruit populations originating from different seasons, orchards or cultivars (Subedi & Walsh, 2020; K. B. Walsh et al., 2020). Multi-source samples were adopted for the calibration and validation sets using the samples collected from different orchards and/or seasons e.g. apple (Peirs et al., 2005), kiwifruit (Afonso et al., 2022; Sarkar et al., 2020), avocado (Olarewaju et al., 2016; Subedi & Walsh, 2020; Wedding et al., 2013), mandarin (Guthrie et al., 2005), citrus (Cavaco et al., 2018; Pires et al., 2022; Torres et al., 2017), mango (Anderson et al., 2017), lemon (Serna-Escolano et al., 2024) and olive(Sun, Subedi, Walker, et al., 2020). Ncama et al. (Ncama et al., 2018) noted that the models could be used in various independent sets when the variability of calibration set was sufficiently wide and covered variation of samples in the external validation set. Similar results also were reported. For example, Olarewaju et al. (Olarewaju et al., 2016) noted that the predictive capacity of the model decreased using a single season model in predicting the next season samples. The root mean square error of prediction (RMSEP) of 3.8% was obtained for Hass avocado DMC compared with a merged population split into the calibration and validation sets with RMSEP of 2.5%. A single season model in predicting the fruit from the subsequent year (RMSEP = 2.6-5.0%) was lower than the three seasons model in predicting the samples come from a random split (RMSEP = 1.4%) (Wedding et al., 2013). The variability of orchard also causes the models to loss their predictive ability. The lower results were achieved using one orchard model in predicting the samples come from another orchard e.g. citrus (Cavaco et al., 2018; Guthrie et al., 2005; Pires et al., 2022) and lemon (Serna-Escolano et al., 2024). Anderson et al. noted that Mango DMC model impacted by ripening stage more than cultivar, season or region. Cultivar or physiological stage specific PLS model improved prediction relative to global model. Therefore, model robustness is generally achieved using representative samples covered the future samples (Anderson et al., 2020). The calibrated model should also be validated through different seasons, orchards and measurement conditions for meet the requirement of the practical use.

4.4 Application

The optimal harvest date (OHD) estimation as a typical practical use of NIRS technology is crucial to fruit quality and performance in postharvest storage (N. Anderson et al., 2021; K. B. Walsh et al., 2020). The indicators of fruit DMS or SSC to OHD have been successfully estimated for monitoring the variation during fruit maturity using hand-held NIRS instrument (K. Walsh et al., 2020). Zude et al. (Zude et al., 2006) reported the potential of NIRS for predicting accepted parameters enabling the determination of optimum harvest window for apple. Peirs et al. (Peirs et al., 2001) predicted the maturity of apple with a validation correlation of 0.90 and standard error of prediction of 7.4 days. The RMSEP depending on the cultivar varied between 5 to 9 days for individual fruit (Peirs et al., 2005). They noted that small errors will be achieved in the practical use via average values of multiple fruit in predicting OHD for a whole population. Bertone et al. (Bertone et al., 2012) reported that RMSEP computed on the averages of the predictions for apples from the same period was reduced to 1.0 day. Therefore, statistical results of averages and standard deviation (SD) reinforced the applicability of hand-held NIRS instrument application in orchard management (Afonso et al., 2022).

Miyamoto (Miyamoto, 2007) monitored the OHD of orange and diagnosed of fruit nutritional status for orchard management. The OHD of navel orange was 233 days after full bloom (DAFB) through hand-held NIRS instrument. The 55% of the sampling fruit qualified with the standard of SSC≥13% (Sun et al., 2024b). Harvest decision maps for describing the spatial distribution of citrus SSC along the orchard were attempted (Cavaco et al., 2016; Zude et al., 2008). Twenty trees with ten fruits for each tree were selected as the sampling trees for hand-held NIRS predicting mango DMC for mapping the harvest support map with 90% measurements above 14% DMC (Walsh et al., 2023; K. B. Walsh et al., 2020). To orchard management, averages reflect the quality level for the sampling block. The accuracy will be influenced by the uniformity of the block also. This requires a flexible definition of block with consideration of the variation of soil type, tree age, fruit cultivar, and flowering time (N. T. Anderson, K. B. Walsh, & D. Wulfsohn, 2021). Storage and fresh eating fruit requires different harvest standard levels. The OHD should be determined according to the detail purpose.

4.5 Future Perspective

In field fruit quality assessment through hand-held NIRS instrument has showed the potential, especially for fruit DMC and/or SSC attributes. The stable, low-cost instrument, chemometrics and robust model will promote the forward of NIRS field use. NIRS fusion of other sensors e.g. GPS, blue tooth and coupled with support decision system can accelerate the practical use to precision orchard. OHD estimation for orchard management is also promising. The cross fusion of NIRS and fruit industry e.g. tree status diagnose will be consider in the future research.

5. CONCLUSION

In conclusion, this chapter highlights the significant role of advanced technologies in improving the monitoring of fruit quality and quantity, with a particular focus on mango cultivation. From flowering to harvest, innovations such as machine vision, deep learning, and near-infrared (NIR) spectroscopy have provided farmers with precise tools to optimize production, enhance resource management, and promote sustainable farming practices.

Effective mango crop management relies on the transition from raw data to actionable knowledge. Accurate measurement of critical attributes, such as dry matter and fruit size, forms the foundation for informed decision-making. The development of digital measurement technologies and advancements in telecommunication networks offer new opportunities for real-time and scalable applications in mango production. This chapter emphasizes the integration of biological knowledge, advanced measurement techniques, and decision support systems to establish a structured framework for managing mango orchards.

Through mango-specific case studies, this chapter demonstrates the effectiveness of AI-driven technologies like MangoYOLO in overcoming field challenges such as occlusion and uneven lighting while achieving precise fruit detection and yield estimation. Similarly, the evolution of portable, IoT-enabled NIR spectroscopy devices showcases their ability to provide rapid, non-invasive quality assessments, enabling timely decisions regarding harvest and postharvest practices.

These advancements carry critical implications for the mango industry, addressing longstanding challenges such as climate variability, labour shortages, and logistical difficulties. By adopting these innovative technologies, mango producers can improve productivity, enhance fruit quality, and strengthen their competitive position in the market. Moving forward, integrating hybrid AI systems, scalable solutions, and real-time data-sharing platforms will further accelerate the modernization of mango cultivation, supporting the industry's sustainable growth and long-term development.

REFERENCES

Afonso, A. M., Antunes, M. D., Cruz, S., Cavaco, A. M., & Guerra, R. (2022). Non-destructive follow-up of 'Jintao'kiwifruit ripening through VIS-NIR spectroscopy–individual vs. average calibration model's predictions. *Postharvest Biology and Technology*, *188*, 111895. DOI: 10.1016/j.postharvbio.2022.111895

Amaral, M. H., McConchie, C., Dickinson, G., & Walsh, K. B. (2023). Growing degree day targets for fruit development of Australian mango cultivars. *Horticulturae*, *9*(4), 489. DOI: 10.3390/horticulturae9040489

Anderson, N., Walsh, K., Flynn, J., & Walsh, J. P. (2021). Achieving robustness across season, location and cultivar for a NIRS model for intact mango fruit dry matter content. II. Local PLS and nonlinear models. *Postharvest Biology and Technology*, *171*, 111358. DOI: 10.1016/j.postharvbio.2020.111358

Anderson, N., Walsh, K., Subedi, P., & Hayes, C. (2020). Achieving robustness across season, location and cultivar for a NIRS model for intact mango fruit dry matter content. *Postharvest Biology and Technology*, *168*, 111202. DOI: 10.1016/j.postharvbio.2020.111202

Anderson, N. T., Subedi, P. P., & Walsh, K. B. (2017). Manipulation of mango fruit dry matter content to improve eating quality. *Scientia Horticulturae*, *226*, 316–321. DOI: 10.1016/j.scienta.2017.09.001

Anderson, N. T., Underwood, J. P., Rahman, M. M., Robson, A., & Walsh, K. B. (2018). Estimation of fruit load in mango orchards: Tree sampling considerations and use of machine vision and satellite imagery [journal article]. *Precision Agriculture*. Advance online publication. DOI: 10.1007/s11119-018-9614-1

Anderson, N. T., & Walsh, K. B. (2022). The evolution of chemometrics coupled with near infrared spectroscopy for fruit quality evaluation. *Journal of Near Infrared Spectroscopy*, *30*(1), 3–17. DOI: 10.1177/09670335211057235

Anderson, N. T., Walsh, K. B., Koirala, A., Wang, Z., Amaral, M. H., Dickinson, G. R., Sinha, P., & Robson, A. J. (2021). Estimation of Fruit Load in Australian Mango Orchards Using Machine Vision. *Agronomy (Basel)*, *11*(9), 1711. https://www.mdpi.com/2073-4395/11/9/1711. DOI: 10.3390/agronomy11091711

Anderson, N. T., Walsh, K. B., & Wulfsohn, D. (2021). Technologies for Forecasting Tree Fruit Load and Harvest Timing—From Ground, Sky and Time. *Agronomy (Basel)*, *11*(7), 1409. https://www.mdpi.com/2073-4395/11/7/1409. DOI: 10.3390/agronomy11071409

Annamalai, P., & Suk Lee, W. (2003). *Citrus Yield Mapping System Using Machine Vision* St. Joseph, MI. https://elibrary.asabe.org/abstract.asp?aid=13701&t=5

Aoki, H., Tanabe, T., & Akinaga, T. (2004). Pre and postharvest data collecting system using the compact NIR analyzer. III International Symposium on Tropical and Subtropical Fruits 864, Assimakopoulos, F., Vassilakis, C., Margaris, D., Kotis, K., & Spiliotopoulos, D. (2024). Artificial Intelligence Tools for the Agriculture Value Chain: Status and Prospects. *Electronics (Basel), 13*(22), 4362. https://www.mdpi.com/2079-9292/13/22/4362

Bay, H., Ess, A., Tuytelaars, T., & Van Gool, L. (2008). Speeded-up robust features (SURF). *Computer Vision and Image Understanding, 110*(3), 346–359. DOI: 10.1016/j.cviu.2007.09.014

Bertone, E., Venturello, A., Leardi, R., & Geobaldo, F. (2012). Prediction of the optimum harvest time of 'Scarlet' apples using DR-UV–Vis and NIR spectroscopy. *Postharvest Biology and Technology, 69*, 15–23. DOI: 10.1016/j.postharvbio.2012.02.009

Cavaco, A., Antunes, M., Guerra, R., Rosendo, M., Pires, R., Brázio, A., Silva, L., Afonso, A., & Panagopoulos, T. (2016). Fruit quality survey techniques to optimize crop management in citrus orchards. VIII International Postharvest Symposium: Enhancing Supply Chain and Consumer Benefits-Ethical and Technological Issues 1194

Cavaco, A. M., Pires, R., Antunes, M. D., Panagopoulos, T., Brázio, A., Afonso, A. M., Silva, L., Lucas, M. R., Cadeiras, B., & Cruz, S. P. (2018). Validation of short wave near infrared calibration models for the quality and ripening of 'Newhall'orange on tree across years and orchards. *Postharvest Biology and Technology, 141*, 86–97. DOI: 10.1016/j.postharvbio.2018.03.013

Chalidabhongse, T., Yimyam, P., & Sirisomboon, P. (2006). 2D/3D Vision-Based Mango's Feature Extraction and Sorting. *9th International Conference on Control, Automation, Robotics and Vision*

Chen, Z.-P., Morris, J., & Martin, E. (2005). Correction of temperature-induced spectral variations by loading space standardization. *Analytical Chemistry, 77*(5), 1376–1384.

Choi, J.-H., Chen, P.-A., Lee, B., Yim, S.-H., Kim, M.-S., Bae, Y.-S., Lim, D.-C., & Seo, H.-J. (2017). Portable, non-destructive tester integrating VIS/NIR reflectance spectroscopy for the detection of sugar content in Asian pears. *Scientia Horticulturae, 220*, 147–153. DOI: 10.1016/j.scienta.2017.03.050

Dhonju, H. K., Walsh, K. B., & Bhattarai, T. (2024). Management information systems for tree fruit—1: A review. *Horticulturae*, *10*(1), 108. DOI: 10.3390/horticulturae10010108

Dorj, U.-O., Lee, M., & Yun, S. (2017). An yield estimation in citrus orchards via fruit detection and counting using image processing. *Computers and Electronics in Agriculture*, *140*, 103–112. https://doi.org/https://doi.org/10.1016/j.compag.2017.05.019. DOI: 10.1016/j.compag.2017.05.019

Feng, J., McGlone, A. V., Currie, M., Clark, C. J., & Jordan, B. R. (2011). Assessment of yellow-fleshed kiwifruit (actinidia chinensis 'hort16a') quality in pre-and post-harvest conditions using a portable near-infrared spectrometer. *HortScience*, *46*(1), 57–63. DOI: 10.21273/HORTSCI.46.1.57

Giménez-Gallego, J., Martinez-del-Rincon, J., González-Teruel, J. D., Navarro-Hellín, H., Navarro, P. J., & Torres-Sánchez, R. (2024). On-tree fruit image segmentation comparing Mask R-CNN and Vision Transformer models. Application in a novel algorithm for pixel-based fruit size estimation. *Computers and Electronics in Agriculture*, *222*, 109077. https://doi.org/https://doi.org/10.1016/j.compag.2024.109077. DOI: 10.1016/j.compag.2024.109077

Girshick, R., Donahue, J., Darrell, T., & Malik, J. (2016). Region-Based Convolutional Networks for Accurate Object Detection and Segmentation. *IEEE Transactions on Pattern Analysis and Machine Intelligence*, *38*(1), 142–158. DOI: 10.1109/TPAMI.2015.2437384 PMID: 26656583

Glorot, X., & Bengio, Y. (2010). Understanding the difficulty of training deep feedforward neural networks. Proceedings of the thirteenth international conference on artificial intelligence and statistics, Gongal, A., Amatya, S., Karkee, M., Zhang, Q., & Lewis, K. (2015). Sensors and systems for fruit detection and localization: A review. *Computers and Electronics in Agriculture*, *116*, 8–19. https://doi.org/http://dx.doi.org/10.1016/j.compag.2015.05.021

Guthrie, J., Walsh, K., Reid, D., & Liebenberg, C. (2005). Assessment of internal quality attributes of mandarin fruit. 1. NIR calibration model development. *Australian Journal of Agricultural Research*, *56*(4), 405–416. DOI: 10.1071/AR04257

Gutiérrez, S., Wendel, A., & Underwood, J. (2019). Ground based hyperspectral imaging for extensive mango yield estimation. *Computers and Electronics in Agriculture*, *157*, 126–135. https://doi.org/https://doi.org/10.1016/j.compag.2018.12.041. DOI: 10.1016/j.compag.2018.12.041

Hannan, M., Burks, T., & Bulanon, D. M. (2009). A machine vision algorithm combining adaptive segmentation and shape analysis for orange fruit detection. *Agricultural Engineering International: CIGR Journal.*

He, K., Zhang, X., Ren, S., & Sun, J. (2016). Deep residual learning for image recognition. *Proceedings of the IEEE conference on computer vision and pattern recognition*

He, L., Fang, W., Zhao, G., Wu, Z., Fu, L., Li, R., Majeed, Y., & Dhupia, J. (2022). Fruit yield prediction and estimation in orchards: A state-of-the-art comprehensive review for both direct and indirect methods. *Computers and Electronics in Agriculture, 195*, 106812. https://doi.org/https://doi.org/10.1016/j.compag.2022.106812 DOI: 10.1016/j.compag.2022.106812

Hussain, M. (2023). YOLO-v1 to YOLO-v8, the rise of YOLO and its complementary nature toward digital manufacturing and industrial defect detection. *Machines (Basel), 11*(7), 677. DOI: 10.3390/machines11070677

Igne, B., Allen, R., Gahkani, A., Kojic, D., Kovacs, Z., McCormick, R., & Mohan, S. (2019). Summary of the 2018 International Diffuse Reflectance Conference (IDRC) software shoot-out. In: SAGE Publications Sage UK: London, England.

Jenni, S., Stewart, K. A., Bourgeois, G., & Cloutier, D. C. (1996). Nondestructive volume estimation for growth analysis of eastern-type muskmelon ovaries. *HortScience, 31*(7), 1236–1236. DOI: 10.21273/HORTSCI.31.7.1236

Jenni, S., Stewart, K. A., Bourgeois, G., & Cloutier, D. C. (1998). Predicting yield and time to maturity of muskmelons from weather and crop observations. *Journal of the American Society for Horticultural Science, 123*(2), 195–201. DOI: 10.21273/JASHS.123.2.195

Jiang, Y., Li, C., Paterson, A. H., Sun, S., Xu, R., & Robertson, J. (2018). Quantitative Analysis of Cotton Canopy Size in Field Conditions Using a Consumer-Grade RGB-D Camera [Original Research]. *Frontiers in Plant Science, 8*(2233), 2233. Advance online publication. DOI: 10.3389/fpls.2017.02233 PMID: 29441074

Jocher, G., Chaurasia, A., Stoken, A., Borovec, J., Kwon, Y., Fang, J., Michael, K., Montes, D., Nadar, J., & Skalski, P. (2022). ultralytics/yolov5: v6. 1-tensorrt, tensorflow edge tpu and openvino export and inference. Zenodo.

Jrondi, Z., Moussaid, A., & Hadi, M. Y. (2024). Exploring End-to-End object detection with transformers versus YOLOv8 for enhanced citrus fruit detection within trees. *Systems and Soft Computing, 6*, 200103. https://doi.org/https://doi.org/10.1016/j.sasc.2024.200103

Kernot, I., Meurant, N., Holmes, R., MacLeod, N., Fullelove, G., & Bally, I. (1999). *Mango information kit*. Queensland Department of Primary Industries.

Kestur, R., Meduri, A., & Narasipura, O. (2019). MangoNet: A deep semantic segmentation architecture for a method to detect and count mangoes in an open orchard. *Engineering Applications of Artificial Intelligence*, *77*, 59–69. https://doi.org/https://doi.org/10.1016/j.engappai.2018.09.011. DOI: 10.1016/j.engappai.2018.09.011

Koirala, A., Walsh, K. B., Wang, Z., & Anderson, N. (2020). Deep Learning for Mango (Mangifera indica) Panicle Stage Classification. *Agronomy (Basel)*, *10*(1), 143. https://www.mdpi.com/2073-4395/10/1/143. DOI: 10.3390/agronomy10010143

Koirala, A., Walsh, K. B., Wang, Z., & McCarthy, C. (2019). Deep learning for real-time fruit detection and orchard fruit load estimation: Benchmarking of 'MangoYOLO' [journal article]. *Precision Agriculture*, *20*(6), 1–29. DOI: 10.1007/s11119-019-09642-0

Krizhevsky, A., Sutskever, I., & Hinton, G. E. (2012). Imagenet classification with deep convolutional neural networks. *Advances in Neural Information Processing Systems*.

Kuang, Y. C., Streeter, L., Cree, M. J., & Ooi, M. P. L. (May 2019). Evaluation of Deep Neural Network and Alternating Decision Tree for Kiwifruit Detection. 2019 IEEE International Instrumentation and Measurement Technology Conference (I2MTC)

Léchaudel, M., & Joas, J. (2007). An overview of preharvest factors influencing mango fruit growth, quality and postharvest behaviour. *Brazilian Journal of Plant Physiology*, *19*(4), 287–298.

LeCun, Y., Bengio, Y., & Hinton, G. (2015). Deep learning. *Nature*, *521*(7553), 436–444. DOI: 10.1038/nature14539 PMID: 26017442

LeCun, Y., Bottou, L., Bengio, Y., & Haffner, P. (1998). Gradient-based learning applied to document recognition. *Proceedings of the IEEE*, *86*(11), 2278–2324. DOI: 10.1109/5.726791

Lesnoff, M. (2024). Averaging a local PLSR pipeline to predict chemical compositions and nutritive values of forages and feed from spectral near infrared data. *Chemometrics and Intelligent Laboratory Systems*, *244*, 105031. DOI: 10.1016/j.chemolab.2023.105031

Lesnoff, M., Metz, M., & Roger, J. M. (2020). Comparison of locally weighted PLS strategies for regression and discrimination on agronomic NIR data. *Journal of Chemometrics*, *34*(5), e3209. DOI: 10.1002/cem.3209

Li, M., Chen, M., Zhang, Y., Fu, C., Xing, B., Li, W., Qian, J., Li, S., Wang, H., Fan, X., Yan, Y., Wang, Y., & Yang, X. (2015). Apple Fruit Diameter and Length Estimation by Using the Thermal and Sunshine Hours Approach and Its Application to the Digital Orchard Management Information System. *PLoS One*, *10*(4), e0120124. DOI: 10.1371/journal.pone.0120124 PMID: 25831065

Lin, T.-Y., Maire, M., Belongie, S., Hays, J., Perona, P., Ramanan, D., Dollár, P., & Zitnick, C. L. (2014). Microsoft coco: Common objects in context. Computer Vision–ECCV 2014: *13th European Conference*, Zurich, Switzerland, September 6-12, 2014, Proceedings, Part V 13

Liu, H., & Ying, Y. (2015). Evaluation of sugar content of Huanghua Pear on trees by visible/near infrared spectroscopy. *Guang pu xue yu Guang pu fen xi= Guang pu, 35*(11), 3078-3081.

Liu, W., Anguelov, D., Erhan, D., Szegedy, C., Reed, S., Fu, C.-Y., & Berg, A. C. (2016). SSD: Single Shot MultiBox Detector. In B. Leibe, J. Matas, N. Sebe, & M. Welling, *Computer Vision – ECCV 2016* Cham.

Lowe, D. G. (1999). Object recognition from local scale-invariant features. iccv, Luo, L., Tang, Y., Lu, Q., Chen, X., Zhang, P., & Zou, X. (2018). A vision methodology for harvesting robot to detect cutting points on peduncles of double overlapping grape clusters in a vineyard. *Computers in Industry*, *99*, 130–139. https://doi.org/ https://doi.org/10.1016/j.compind.2018.03.017

Maeda-Gutiérrez, V., Galvan-Tejada, C. E., Zanella-Calzada, L. A., Celaya-Padilla, J. M., Galván-Tejada, J. I., Gamboa-Rosales, H., Luna-Garcia, H., Magallanes-Quintanar, R., Guerrero Mendez, C. A., & Olvera-Olvera, C. A. (2020). Comparison of convolutional neural network architectures for classification of tomato plant diseases. *Applied Sciences (Basel, Switzerland)*, *10*(4), 1245. DOI: 10.3390/app10041245

Marini, R. P. (2001). Estimating mean fruit weight and mean fruit value for apple trees: Comparison of two sampling methods with the true mean. *Journal of the American Society for Horticultural Science*, *126*(4), 503–510. DOI: 10.21273/JASHS.126.4.503

Mimma, N.-E.-A., Ahmed, S., Rahman, T., & Khan, R. (2022). Fruits classification and detection application using deep learning. *Scientific Programming*, *2022*(1), 4194874. DOI: 10.1155/2022/4194874

Mirhaji, H., Soleymani, M., Asakereh, A., & Abdanan Mehdizadeh, S. (2021). Fruit detection and load estimation of an orange orchard using the YOLO models through simple approaches in different imaging and illumination conditions. *Computers and Electronics in Agriculture, 191,* 106533. https://doi.org/https://doi.org/10.1016/j.compag.2021.106533. DOI: 10.1016/j.compag.2021.106533

Misle, E., Kahlaoui, B., Hachicha, M., & Alvarado, P. (2013). Leaf area estimation in muskmelon by allometry. *Photosynthetica, 51*(4), 613–620. DOI: 10.1007/s11099-013-0062-x

Mitchell, P. (1986). Pear fruit growth and the use of diameter to estimate fruit volume and weight. *HortScience, 21*(4), 1003–1005. DOI: 10.21273/HORTSCI.21.4.1003

MIYAMOTO, K. (2007). Near infrared spectroscopy for bio-monitoring and bio-diagnosis of a fruit tree. *JOURNAL of the JAPANESE SOCIETY of AGRICULTURAL MACHINERY, 69*(3), 11-14.

Mokrane, A., Braham, A. C., & Cherki, B. (2019). UAV coverage path planning for supporting autonomous fruit counting systems. *2019 International Conference on Applied Automation and Industrial Diagnostics (ICAAID)*

Moreda, G. P., Ortiz-Cañavate, J., García-Ramos, F. J., & Ruiz-Altisent, M. (2009). Non-destructive technologies for fruit and vegetable size determination – A review. *Journal of Food Engineering, 92*(2), 119–136. https://doi.org/http://dx.doi.org/10.1016/j.jfoodeng.2008.11.004

Morimoto, S. (2002). Non-destructive near infrared spectrometers: Development of portable fruit quality meters. Near Infrared Spectroscopy, *Proceedings of the 10th International Conference*, NIR Publications, West Sussex, UK.

Ncama, K., Tesfay, S. Z., Fawole, O. A., Opara, U. L., & Magwaza, L. S. (2018). Non-destructive prediction of 'Marsh' grapefruit susceptibility to postharvest rind pitting disorder using reflectance Vis/NIR spectroscopy. *Scientia Horticulturae, 231*, 265–271.

Neupane, C., Walsh, K. B., Goulart, R., & Koirala, A. (2024). Developing Machine Vision in Tree-Fruit Applications—Fruit Count, Fruit Size and Branch Avoidance in Automated Harvesting. *Sensors (Basel), 24*(17), 5593. https://www.mdpi.com/1424-8220/24/17/5593. DOI: 10.3390/s24175593 PMID: 39275504

Ngouajio, M., Kirk, W., & Goldy, R. (2003). A simple model for rapid and nondestructive estimation of bell pepper fruit volume. *HortScience, 38*(4), 509–511. DOI: 10.21273/HORTSCI.38.4.509

Niklas, K. J. (1994). *Plant allometry: the scaling of form and process*. University of Chicago Press.

Ojala, T., Pietikainen, M., & Maenpaa, T. (2002). Multiresolution gray-scale and rotation invariant texture classification with local binary patterns. *IEEE Transactions on Pattern Analysis and Machine Intelligence*, *24*(7), 971–987. DOI: 10.1109/TPAMI.2002.1017623

Okubo, S., Tsunewaki, K., Kimura, T., Shimizu, H., Hiraguri, T., Shibasaki, A., Shimada, T., & Takemura, Y. (2024). Pear Blossom Counting System with Drones Using YOLO and Deep SORT. *2024 International Conference on Image Processing and Robotics (ICIPRoB)*

Olarewaju, O. O., Bertling, I., & Magwaza, L. S. (2016). Non-destructive evaluation of avocado fruit maturity using near infrared spectroscopy and PLS regression models. *Scientia Horticulturae*, *199*, 229–236.

Pasquini, C. (2018). Near infrared spectroscopy: A mature analytical technique with new perspectives–A review. *Analytica Chimica Acta*, *1026*, 8–36. DOI: 10.1016/j.aca.2018.04.004 PMID: 29852997

Payne, A., & Walsh, K. (2014). Machine vision in estimation of fruit crop yield. In S. Dutta Gupta & Y. Ibaraki (Eds.), *Plant image analysis: fundamentals and applications* (pp. 329-374). CRC Press. DOI: 10.1201/b17441-17

Payne, A., Walsh, K., Subedi, P., & Jarvis, D. (2014). Estimating mango crop yield using image analysis using fruit at 'stone hardening' stage and night time imaging. *Computers and Electronics in Agriculture*, *100*, 160–167. https://doi.org/http://dx.doi.org/10.1016/j.compag.2013.11.011. DOI: 10.1016/j.compag.2013.11.011

Payne, A. B., Walsh, K. B., Subedi, P. P., & Jarvis, D. (2013). Estimation of mango crop yield using image analysis – Segmentation method. *Computers and Electronics in Agriculture*, *91*, 57–64. https://doi.org/http://dx.doi.org/10.1016/j.compag.2012.11.009. DOI: 10.1016/j.compag.2012.11.009

Peirs, A., Lammertyn, J., Ooms, K., & Nicolaï, B. M. (2001). Prediction of the optimal picking date of different apple cultivars by means of VIS/NIR-spectroscopy. *Postharvest Biology and Technology*, *21*(2), 189–199. DOI: 10.1016/S0925-5214(00)00145-9

Peirs, A., Schenk, A., & Nicolaï, B. M. (2005). Effect of natural variability among apples on the accuracy of VIS-NIR calibration models for optimal harvest date predictions. *Postharvest Biology and Technology*, *35*(1), 1–13. DOI: 10.1016/j.postharvbio.2004.05.010

Pires, R., Guerra, R., Cruz, S. P., Antunes, M. D., Brázio, A., Afonso, A. M., Daniel, M., Panagopoulos, T., Gonçalves, I., & Cavaco, A. M. (2022). Ripening assessment of 'Ortanique'(Citrus reticulata Blanco x Citrus sinensis (L) Osbeck) on tree by SW-NIR reflectance spectroscopy-based calibration models. *Postharvest Biology and Technology*, *183*, 111750. DOI: 10.1016/j.postharvbio.2021.111750

Qureshi, W. S., Payne, A., Walsh, K. B., Linker, R., Cohen, O., & Dailey, M. N. (2017). Machine vision for counting fruit on mango tree canopies [journal article]. *Precision Agriculture*, *18*(2), 224–244. DOI: 10.1007/s11119-016-9458-5

Redmon, J., Divvala, S., Girshick, R., & Farhadi, A. (2016). You only look once: Unified, real-time object detection. *Proceedings of the IEEE conference on computer vision and pattern recognition*

Ren, S., He, K., Girshick, R., & Sun, J. (2015). Faster r-cnn: Towards real-time object detection with region proposal networks. *Advances in Neural Information Processing Systems*.

Reyes, J. F., Contreras, E., Correa, C., & Melin, P. (2021). Image analysis of real-time classification of cherry fruit from colour features. *Journal of Agricultural Engineering*, *52*(4). Advance online publication. https://doi.org/ DOI: 10.4081/jae.2021.1160

Roger, J.-M., Chauchard, F., & Bellon-Maurel, V. (2003). EPO–PLS external parameter orthogonalisation of PLS application to temperature-independent measurement of sugar content of intact fruits. *Chemometrics and Intelligent Laboratory Systems*, *66*(2), 191–204. DOI: 10.1016/S0169-7439(03)00051-0

Sa, I., Ge, Z., Dayoub, F., Upcroft, B., Perez, T., & McCool, C. (2016). DeepFruits: A Fruit Detection System Using Deep Neural Networks. *Sensors (Basel)*, *16*(8), 1222. DOI: 10.3390/s16081222 PMID: 27527168

Saranwong, I., Sornsrivichai, J., & Kawano, S. (2003). On-tree evaluation of harvesting quality of mango fruit using a hand-held NIR instrument. *Journal of Near Infrared Spectroscopy*, *11*(4), 283–293. DOI: 10.1255/jnirs.374

Sarkar, S., Basak, J. K., Moon, B. E., & Kim, H. T. (2020). A comparative study of PLSR and SVM-R with various preprocessing techniques for the quantitative determination of soluble solids content of hardy kiwi fruit by a portable Vis/NIR spectrometer. *Foods*, *9*(8), 1078. DOI: 10.3390/foods9081078 PMID: 32784804

Schulze, K., Nagle, M., Spreer, W., Mahayothee, B., & Müller, J. (2015). Development and assessment of different modeling approaches for size-mass estimation of mango fruits (Mangifera indica L., cv. 'Nam Dokmai'). *Computers and Electronics in Agriculture, 114*, 269–276. https://doi.org/https://doi.org/10.1016/j.compag.2015.04.013. DOI: 10.1016/j.compag.2015.04.013

Serna-Escolano, V., Giménez, M. J., Zapata, P. J., Cubero, S., Blasco, J., & Munera, S. (2024). Non-destructive assessment of'Fino'lemon quality through ripening using NIRS and chemometric analysis. *Postharvest Biology and Technology, 212*, 112870. DOI: 10.1016/j.postharvbio.2024.112870

Simonyan, K., & Zisserman, A. (2014). Very deep convolutional networks for large-scale image recognition. *arXiv preprint arXiv:1409.1556.*

Spreer, W., & Müller, J. (2011). Estimating the mass of mango fruit (Mangifera indica, cv. Chok Anan) from its geometric dimensions by optical measurement. *Computers and Electronics in Agriculture, 75*(1), 125–131. https://doi.org/http://dx.doi.org/10.1016/j.compag.2010.10.007. DOI: 10.1016/j.compag.2010.10.007

Srivastava, N., Hinton, G., Krizhevsky, A., Sutskever, I., & Salakhutdinov, R. (2014). Dropout: A simple way to prevent neural networks from overfitting. *Journal of Machine Learning Research, 15*(1), 1929–1958.

Stein, M., Bargoti, S., & Underwood, J. (2016). Image based mango fruit detection, localisation and yield estimation using multiple view geometry. *Sensors (Basel), 16*(11), 1915. DOI: 10.3390/s16111915 PMID: 27854271

Subedi, P. P., & Walsh, K. B. (2020). Assessment of avocado fruit dry matter content using portable near infrared spectroscopy: Method and instrumentation optimisation. *Postharvest Biology and Technology, 161*, 111078. DOI: 10.1016/j.postharvbio.2019.111078

Sun, X., Du, Y., Nawaz, M. A., Abobatta, W. F., Lyu, Q., Liu, J., Chen, Z., & Feng, S. (2024). Apple SSC estimation using hand-held NIRS instrument for outdoor measurement with ambient light correction. *Postharvest Biology and Technology, 217*, 113101. DOI: 10.1016/j.postharvbio.2024.113101

Sun, X., Guo, F., Liu, J., Chen, Z., Abobatta, W. F., Nawaz, M. A., & Feng, S. (2024). From lab to orchard use for models of hand-held NIRS instrument: A case for navel orange quality assessment considering ambient light correction. *Computers and Electronics in Agriculture, 219*, 108797. DOI: 10.1016/j.compag.2024.108797

Sun, X., Subedi, P., Walker, R., & Walsh, K. B. (2020). NIRS prediction of dry matter content of single olive fruit with consideration of variable sorting for normalisation pre-treatment. *Postharvest Biology and Technology*, *163*, 111140. DOI: 10.1016/j.postharvbio.2020.111140

Sun, X., Subedi, P., & Walsh, K. B. (2020). Achieving robustness to temperature change of a NIRS-PLSR model for intact mango fruit dry matter content. *Postharvest Biology and Technology*, *162*, 111117. DOI: 10.1016/j.postharvbio.2019.111117

Sun, X., Wang, Z., Aydin, H., Liu, J., Chen, Z., & Feng, S. (2023). First step for hand-held NIRS instrument field use: Table grape quality assessment consideration of temperature and sunlight chemometrics correction. *Postharvest Biology and Technology*, *201*, 112374. DOI: 10.1016/j.postharvbio.2023.112374

Szegedy, C., Liu, W., Jia, Y., Sermanet, P., Reed, S., Anguelov, D., Erhan, D., Vanhoucke, V., & Rabinovich, A. (2015). Going deeper with convolutions. *Proceedings of the IEEE conference on computer vision and pattern recognition*

Tan, K., Lee, W. S., Gan, H., & Wang, S. (2018). Recognising blueberry fruit of different maturity using histogram oriented gradients and colour features in outdoor scenes. *Biosystems Engineering*, *176*, 59–72. https://doi.org/https://doi.org/10.1016/j.biosystemseng.2018.08.011

Tang, Y., Dananjayan, S., Hou, C., Guo, Q., Luo, S., & He, Y. (2021). A survey on the 5G network and its impact on agriculture: Challenges and opportunities. *Computers and Electronics in Agriculture*, *180*, 105895. https://doi.org/https://doi.org/10.1016/j.compag.2020.105895. DOI: 10.1016/j.compag.2020.105895

Torres, I., Pérez-Marín, D., De la Haba, M.-J., & Sánchez, M.-T. (2017). Developing universal models for the prediction of physical quality in citrus fruits analysed on-tree using portable NIRS sensors. *Biosystems Engineering*, *153*, 140–148. DOI: 10.1016/j.biosystemseng.2016.11.007

Tu, S., Pang, J., Liu, H., Zhuang, N., Chen, Y., Zheng, C., Wan, H., & Xue, Y. (2020). Passion fruit detection and counting based on multiple scale faster R-CNN using RGB-D images. *Precision Agriculture*, *21*(5), 1072–1091. DOI: 10.1007/s11119-020-09709-3

Uijlings, J. R. R., van de Sande, K. E. A., Gevers, T., & Smeulders, A. W. M. (2013). Selective Search for Object Recognition [journal article]. *International Journal of Computer Vision*, *104*(2), 154–171. DOI: 10.1007/s11263-013-0620-5

Utai, K., Nagle, M., Hämmerle, S., Spreer, W., Mahayothee, B., & Müller, J. (2019). Mass estimation of mango fruits (Mangifera indica L., cv. 'Nam Dokmai') by linking image processing and artificial neural network. *Engineering in Agriculture, Environment and Food, 12*(1), 103–110. DOI: 10.1016/j.eaef.2018.10.003

Walsh, J., Neupane, A., Koirala, A., Li, M., & Anderson, N. (2023). The evolution of chemometrics coupled with near infrared spectroscopy for fruit quality evaluation. II. The rise of convolutional neural networks. *Journal of Near Infrared Spectroscopy, 31*(3), 109–125. DOI: 10.1177/09670335231173140

Walsh, J., Neupane, A., & Li, M. (2024). Evaluation of 1D convolutional neural network in estimation of mango dry matter content. *Spectrochimica Acta. Part A: Molecular and Biomolecular Spectroscopy, 311*, 124003. DOI: 10.1016/j.saa.2024.124003 PMID: 38354673

Walsh, K., McGlone, V., & Han, D. (2020). The uses of near infra-red spectroscopy in postharvest decision support: A review. *Postharvest Biology and Technology, 163*, 111139. DOI: 10.1016/j.postharvbio.2020.111139

Walsh, K., & Subedi, P. (2014). In-field monitoring of mango fruit dry matter for maturity estimation. *XXIX International Horticultural Congress on Horticulture: Sustaining Lives, Livelihoods and Landscapes (IHC2014)*: 1119

Walsh, K., & Wang, Z. (2018). Monitoring fruit quality and quantity in mangoes. In Galán Saúco, V., & Lu, P. (Eds.), *Achieving sustainable cultivation of mangoes* (pp. 313–338). Burleigh Dodds Science Publishing Limited., https://doi.org/, DOI: 10.19103/AS.2017.0026.14

Walsh, K. B. (2023). NIRS, fruit quality and Central Queensland University. *NIR News, 34*(7-8), 5–8. DOI: 10.1177/09603360231204541

Walsh, K. B., Blasco, J., Zude-Sasse, M., & Sun, X. (2020). Visible-NIR 'point'spectroscopy in postharvest fruit and vegetable assessment: The science behind three decades of commercial use. *Postharvest Biology and Technology, 168*, 111246. DOI: 10.1016/j.postharvbio.2020.111246

Wang, A., Chen, H., Liu, L., Chen, K., Lin, Z., Han, J., & Ding, G. (2024). Yolov10: Real-time end-to-end object detection. *arXiv preprint arXiv:2405.14458.*

Wang, Z., Walsh, K., & Koirala, A. (2019). Mango Fruit Load Estimation Using a Video Based MangoYOLO—Kalman Filter—Hungarian Algorithm Method. *Sensors (Basel), 19*(12), 2742. https://www.mdpi.com/1424-8220/19/12/2742. DOI: 10.3390/s19122742 PMID: 31216769

Wang, Z., Walsh, K. B., & Verma, B. (2017). On-tree mango fruit size estimation using RGB-D images. *Sensors (Basel)*, *17*(12), 2738. DOI: 10.3390/s17122738 PMID: 29182534

Wedding, B. B., Wright, C., Grauf, S., White, R., Tilse, B., & Gadek, P. (2013). Effects of seasonal variability on FT-NIR prediction of dry matter content for whole Hass avocado fruit. *Postharvest Biology and Technology*, *75*, 9–16. DOI: 10.1016/j.postharvbio.2012.04.016

Xia, X., Chai, X., Li, Z., Zhang, N., & Sun, T. (2023). MTYOLOX: Multi-transformers-enabled YOLO for tree-level apple inflorescences detection and density mapping. *Computers and Electronics in Agriculture*, *209*, 107803. https://doi.org/https://doi.org/10.1016/j.compag.2023.107803. DOI: 10.1016/j.compag.2023.107803

Yıldırım, Ş., & Ulu, B. (2023). Deep learning based apples counting for yield forecast using proposed flying robotic system. *Sensors (Basel)*, *23*(13), 6171. DOI: 10.3390/s23136171 PMID: 37448020

Yimyam, P., Chalidabhongse, T., Sirisomboon, P., & Boonmung, S. (2005). Physical properties analysis of mango using computer vision. *Proceeding of ICCAS*

Yuan, H., Martin, P., Fullerton, C., Gould, N., Hall, A., & Burdon, J. (2020). Refining the estimation of kiwifruit size from linear fruit dimensions. *Scientia Horticulturae*, *262*, 108878.

Zeiler, M. D., & Fergus, R. (2014). Visualizing and understanding convolutional networks. European conference on computer vision

Zhang, X., & Yang, J. (2024). Advanced chemometrics toward robust spectral analysis for fruit quality evaluation. *Trends in Food Science & Technology*, *104612*. Advance online publication. DOI: 10.1007/978-3-319-10590-1_53

Zhao, Y., Gong, L., Zhou, B., Huang, Y., & Liu, C. (2016). Detecting tomatoes in greenhouse scenes by combining AdaBoost classifier and colour analysis. *Biosystems Engineering*, *148*, 127–137. https://doi.org/https://doi.org/10.1016/j.biosystemseng.2016.05.001. DOI: 10.1016/j.biosystemseng.2016.05.001

Zhou, R., Damerow, L., Sun, Y., & Blanke, M. M. (2012). Using colour features of cv. 'Gala' apple fruits in an orchard in image processing to predict yield [journal article]. *Precision Agriculture*, *13*(5), 568–580. DOI: 10.1007/s11119-012-9269-2

Zinkevich, M., Weimer, M., Li, L., & Smola, A. (2010). Parallelized stochastic gradient descent. *Advances in Neural Information Processing Systems*, •••, 23.

Zude, M., Herold, B., Roger, J.-M., Bellon-Maurel, V., & Landahl, S. (2006). Non-destructive tests on the prediction of apple fruit flesh firmness and soluble solids content on tree and in shelf life. *Journal of Food Engineering*, *77*(2), 254–260. DOI: 10.1016/j.jfoodeng.2005.06.027

Zude, M., Pflanz, M., Kaprielian, C., & Aivazian, B. L. (2008). NIRS as a tool for precision horticulture in the citrus industry. *Biosystems Engineering*, *99*(3), 455–459. DOI: 10.1016/j.biosystemseng.2007.10.016

Zude-Sasse, M., Fountas, S., Gemtos, T. A., & Abu-Khalaf, N. (2016). Applications of precision agriculture in horticultural crops.

Chapter 7
AI-Driven Insights Into Revolutionizing Poultry Farming With Behavioral Monitoring and Ethical Management

Md. Sayem Kabir
https://orcid.org/0009-0001-7022-4187
American International University, Bangladesh

Tasnim Sultana Sintheia
https://orcid.org/0009-0007-3309-4348
American International University, Bangladesh

Kazi Tanvir
https://orcid.org/0009-0001-0612-6101
Vellore Institute of Technology, India

Dipta Gomes
https://orcid.org/0000-0003-3019-6051
American International University, Bangladesh

ABSTRACT

This chapter aims to create a system that uses AI and IoT technologies to monitor and understand chicken behaviors. It will use computer vision and machine learning to recognize common actions like pecking, scratching, resting, and feeding. The system will also track environmental factors such as temperature, humidity, and light to see how they affect behavior. Farmers will get real-time updates and insights

DOI: 10.4018/979-8-3373-0020-7.ch007

Copyright © 2025, IGI Global Scientific Publishing. Copying or distributing in print or electronic forms without written permission of IGI Global Scientific Publishing is prohibited.

through an easy-to-use dashboard or mobile app. Additionally, the system will keep improving over time by using advanced deep learning to make behavior detection more accurate and dependable. The system will use advanced deep learning models to detect chicken behaviors in real time. To ensure accurate results, it will rely on datasets with labeled examples of different chicken actions for precise training. It will also identify unusual behaviors, such as reduced feeding, and send timely alerts to help farmers take preventive steps. Based on the behavior analysis, the system will offer actionable suggestions to address potential issues.

INTRODUCTION

Chickens (Gallus gallus domesticus), domesticated birds descended from wild fowl(Lawal et al., 2020), exhibit a diverse range of physical traits across breeds, from medium to fairly large sizes, rounded bodies, and bare legs, with plumage spanning pure white, jet black, reddish tones, to earthy browns(Gebeyehu, 2021). These birds are opportunistic omnivores, using their sturdy legs and sharp beaks to scratch the ground and unearth a varied diet that includes seeds, insects, small reptiles, and even young mammals, showcasing their adaptability to different environments(Bobier, 2021). Despite their size and short wings, chickens possess the ability to fly short distances and defend themselves effectively, employing their beaks and claws in combat when threatened or challenged. These physical capabilities extend beyond mere survival, playing a critical role in their complex social dynamics, most notably in the formation of a fluid and ever-shifting hierarchy known as the "pecking order"(Strauss et al., 2022). This social structure emerges through constant interactions within the flock, where chickens establish and maintain dominance and group cohesion through a combination of aggressive and assertive behaviors(Tibbetts et al., 2022). Such behaviors include pecking at one another, chasing rivals, and striking imposing postures, all of which rely heavily on the effective use of their beaks and claws to signal rank and reinforce their standing within the group(Grandin & Deesing, 2022). This dynamic ranking system ensures that higher-status chickens secure priority access to vital resources while lower-ranking members adapt through submissive actions, creating a balance that sustains order and reduces conflict over time. Poultry behavior changes daily due to light exposure, feed availability, and environmental conditions. Circadian rhythms influence activity levels, while social interactions and stressors can alter feeding, resting, and aggression patterns(Neethirajan, 2024). Monitoring these changes helps optimize welfare and productivity in poultry farming. The pecking order within chicken flocks creates a structured hierarchy that determines which chickens get first access to essential resources such as food, water, and prime breeding areas, favoring those at the top of the social ladder.

Meanwhile, chickens lower in rank adopt submissive behaviors and patiently wait their turn, minimizing the risk of confrontations with their dominant counterparts. Far from being a fixed system, this social structure remains fluid and adaptable, evolving as the chickens continuously engage with one another, assert their positions, and challenge existing rankings through ongoing interactions over time.

Chickens play a very important role not only in their natural behaviors and social structures but also in the global economy, where they hold universal importance. Being omnivores with great foraging abilities, chickens are uniquely adapted and incredibly productive in the production of protein; thus, they become an integral part of food security and nutrition across the world(Papageorgiou et al., 2023). This has propelled the poultry industry to become one of the important economic sectors that generate multibillion-dollar revenues globally(Islam & Ali, 2024). The industry also underpins multi-millions of livelihoods, generating employment in farming, processing, distribution, and marketing. For instance, the chicken industry alone provides 1,375,952 jobs, $90.9 billion in wages, $449.5 billion in economic activity, and $36.7 billion in USA government revenues from its activities (Seeger et al., 2021).Indeed, chicken-related businesses bestow stable income upon numerous people and provide sustenance for innumerable families and communities across this great nation. So, chickens, by their very nature and economic value, play a part in sustaining ecosystems and human societies around the world. Their dual role as a key food source and a scientific resource highlights their multifaceted contributions to global development. Chickens help combat hunger and malnutrition, particularly in developing regions(Pius et al., 2021). With the poultry industry raking in billions of dollars and supporting millions of jobs around the world, it is very important that the chickens be kept healthy and productive to ensure these economic benefits. This is where research and monitoring of chicken behavior, food habits, disease factors become very important.

Understanding chicken behavior, particularly their social structure and *pecking order*, is a key area of focus. Studies explore how environmental factors like space, lighting, and temperature impact chicken behavior and how these factors influence flock dynamics and overall welfare. For instance, research looks into ways to reduce aggression and stress in crowded poultry environments by adjusting living conditions, which can lead to healthier chickens and higher productivity. Understanding the *pecking order* will help farmers improve flock management, reduce conflicts, and ensure that chickens have a balanced and healthy living environment (Bonnefous et al., 2022).

In addition to understanding chicken behavior, their food habits are another critical aspect of maintaining health and productivity. As omnivores, chickens have complex dietary needs that must be met to ensure optimal growth, egg production, and overall health(Givens, 2024). Research into how different feed types and nutritional supplements influence chicken well-being can help develop more efficient and sustainable feeding practices. Ethical and regulatory considerations play a crucial role in poultry nutrition, ensuring that feed ingredients are safe, humane, and environmentally responsible. Compliance with animal welfare regulations and sustainable farming guidelines helps prevent unethical practices while promoting transparency in the poultry industry(Bist et al., 2024). These practices not only improve the quality of chicken products but also reduce environmental impact, further enhancing the overall welfare of chickens and supporting the long-term sustainability of poultry farming.

Poultry behavior changes daily due to light exposure, feed availability, and environmental conditions. Circadian rhythms influence activity levels, while social interactions and stressors can alter feeding, resting, and aggression patterns. Such large-scale poultry farms demand a lot of manpower, precision, and resources in maintaining chicken behavior, food habits, and diseases. Real-time monitoring of challenges is becoming more imperative. The real-time detection system helps farmers to easily recognize and manage issues relating to flock dynamics, dietary needs, and disease prevention(George & George, 2023). It allows better management, as it reduces man-hours and costs involved with conventional methods of management. This will also always be able to highlight through this any disease or behavioral changes or nutritional deficits in time, ensuring flock healthiness and reflecting on productivity and farm economy(Åkerfeldt et al., 2021). Advances in poultry behavior research are deepening our understanding, with tools like AI-driven imaging and acoustic analysis revealing intricate details about flock stress and interactions. For instance, MSD Animal Health's Poultry Sense system delivers real-time insights into environmental and health metrics, empowering farmers with data-driven decisions. Likewise, Ancera's analytics platform monitors live production factors, minimizing losses and enhancing efficiency (Comes et al., 2024). Together, these efforts aim to simplify poultry farming, making it more sustainable and promising for farmers and the broader industry. However, real-time chicken behavior monitoring also raises ethical concerns. Continuous surveillance of animals, while beneficial for welfare and productivity, brings up questions about autonomy and potential stress caused by constant monitoring. Some argue that excessive reliance on technology may lead to treating chickens as mere data points rather than sentient beings with intrinsic value. Additionally, the use of AI-driven systems must ensure that interventions prioritize animal well-being rather than solely optimizing production outputs. Ethical poultry management should balance data-driven efficiency with humane treatment, ensuring that technology enhances, rather than compromises, the overall welfare of the flock.

In this study, different versions of YOLO models, a common model in object detection and real-time monitoring applications, are explored. Comparing multiple variations of the YOLO models will evaluate their effectiveness regarding chicken behavior monitoring. Related studies regarding this domain is discussed in Section 2. The basic architecture and working flow, structure of the YOLO models, are revealed in Section 3. In Section 4, the comparison of these models has been addressed by looking at various versions and variations of them. Finally, Section 5 will present conclusions and an outlook on future research objectives in this area.

Contributions of this study:

1. Real-Time Chicken Behavior Detection with YOLO Variants – Carried out a detailed analysis of using several YOLO versions on chicken identification in real-time.
2. Increased system accuracy in the recognition and categorization of poultry behavior – Created a new Behavioral Monitoring Framework to be AI based.
3. For Real-World Deployment – Proposed simple and time efficient models for farm applications.
4. Early Disease Detection Capabilities – AI views that deal with stress and disease precursors in the birds to take necessary action.
5. Achievement in Technology of automatic poultry health assessment – Founded the welfare monitoring system in relation to enhanced sustainable farming technology.

LITERATURE REVIEW

Traditional Way of Monitoring Chicken Behavior

In the early morning light, the old-fashioned way of keeping an eye on chicken behavior starts with a simple step: people walk into the chicken coop to watch and learn about the birds. This classic method involves regular visits where farmers or researchers write down what they see, like how the chickens eat, move, or rest in notebooks or on basic lists, using their experience and attention to detail. While this hands-on way builds a strong connection between the chickens and those caring for them, it has big downsides. It needs lots of people to do the watching, which costs money and doesn't work well for big farms. It also takes a lot of time, and since they can't watch all day, they might miss important clues about the chickens' health or happiness. Plus, every person sees things a bit differently, so the notes might not always match up, and the effort and cost make it clear that newer, smarter ways are needed.

Poultry Behavior Observation with YOLO Models

Recent advancements in poultry monitoring have leaned heavily on YOLO models, a family of object detection algorithms prized for their speed and precision. Several studies have harnessed different YOLO variants to track chicken behavior, count flocks, and assess welfare, revealing both their potential and their pitfalls.(Salihu et al., n.d.-a) pitted YOLOv5, YOLOv7, and YOLOv8 against each other, pairing them with DeepSORT for chicken tracking and counting. Their standout performer, YOLOv8 with dataset version V4, hit a mean Average Precision (mAP50) of 97.4% and an mAP50-95 of 75.4%, proving robust for general detection. Yet, it stumbled with small or crowded chickens, where overlapping bodies muddled results. Similarly,(Bist et al., 2023) trained multiple YOLOv6 versions to spot piling behavior in cage-free hens, with YOLOv6l_relu-PB leading the pack at an impressive mAP@0.50 of 98.9%. However, environmental factors like lighting and crowded scenes sparked detection errors and false positives, showing its limits in messy, real-world coops.

Other researchers have explored earlier YOLO iterations with mixed outcomes. (Sozzi et al., 2023) compared YOLOv4 and YOLOv4-tiny for detecting hens on the floor and identifying dust-bathing behavior. YOLOv4-tiny raced ahead with a blazing 208.5 frames per second (FPS), trouncing YOLOv4's sluggish 31.35 FPS, but both floundered at dust-bathing detection, scoring mAPs of just 28.2% and 31.6%, respectively. This weak performance contrasts sharply with Salihu and Bist's higher precision on simpler tasks, hinting that complex, dynamic behaviors trip up these models. (Triyanto et al., 2023)also tapped YOLOv4 with DeepSORT to track broiler movements for health and behavior insights, achieving solid detection but wrestling with slow processing times that demand hardware boosts for large-scale, real-time use. Meanwhile, (Kodaira et al., 2023) pushed YOLOv4 to near-perfection 99.9% mAP in controlled rooms to study hens' light color preferences, though its success leaned on a small sample and experimental setup, raising doubts about broader farm applications.

The YOLOv5 model pops up across studies with consistent real-time promise. (Wang et al., 2023) tested it alongside Efficientnet-YoloV3, YOLOv4-Tiny, and Faster-RCNN for behaviors like standing and grooming, with YOLOv5 clinching the top mAP and F1-scores at a zippy 55.55 FPS ideal for live monitoring. Efficientnet-YoloV3 and YOLOv4-Tiny, though, faltered in tricky nighttime scenes, while Faster-RCNN's competitive accuracy came at a snail-paced 3.54 FPS, ruling it out for practical use. Bhuiyan and (Bhuiyan & Wree, 2023) paired YOLOv5 with CNNs and IoT Tech, hitting 96.77% accuracy for stunned states and 99.17% for flock behavior, though they noted a lack of standardized models hampering wider adoption.

Comparing these YOLO variants, a pattern emerges: newer models like YOLOv8 (Salihu et al., n.d.-b) and YOLOv6 (Bist et al., 2024) excel in precision mAP50s of 97.4% and 98.9%, respectively edging out YOLOv4 and YOLOv5 in structured tasks. Yet, speed varies widely, with YOLOv4-tiny (208.5 FPS) outpacing YOLOv5 (55.55 FPS) and YOLOv4 (31.35 FPS), though it sacrifices accuracy on tougher behaviors like dust bathing. All models shine in controlled or simpler scenarios (e.g., Kodaira's 99.9% mAP), but crowded coops, environmental quirks, and dynamic actions expose cracks false positives, missed detections, or sluggish processing. For context, (Campbell et al., 2024) sidestepped YOLO entirely, using U-Net for broiler activity detection with a precision of 0.98 and F1-score of 0.94. While U-Net nailed segmentation, it struggled with clustering active behaviors, echoing YOLO's own woes with complex motion. Together, these findings spotlight YOLO's knack for real-time, high-accuracy detection in poultry care, but its stumbles with scalability, speed trade-offs, and tricky behaviors signal a need for tweaks or hybrid approaches to crack the toughest farm challenges.

This version groups all YOLO studies, compares their metrics (mAP, FPS, etc.), and contrasts their strengths and weaknesses in a conversational yet research-appropriate tone. It wraps up with a nod to a non-YOLO study (U-Net) to broaden the perspective.

Poultry Behavior Observation With Machine Learning Models

Machine learning has become a game changer in poultry farming, helping researchers and farmers keep an eye on chickens in smarter ways. Several studies have used different models to spot behaviors, check plumage conditions, and track flock health, each bringing its own strengths and struggles

(Campbell et al., 2024) rolled out ChickenNet, a setup built on Mask R-CNN with a ResNet-50 backbone, to check plumage conditions in laying hens. They used RGB images resized to 896 by 896 pixels and hit an amazing mean Average Precision (mAP) of 98.02% and an accuracy of 91.83%. That's a big win for figuring out if feathers are in good shape, which matters a lot for hen welfare. But here's the twist: adding extra depth info, like 3D data, didn't make it much better. One snag they ran into was that people labeling the images by hand didn't always agree, which threw some inconsistency into the mix. Meanwhile, (G. Li et al., 2020) also tapped Mask R-CNN but paired it with different ResNet backbones to catch preening behavior in Hy-Line Brown hens. Their best pick, ResNet101, scored a precision of 85.1%, a recall of 88.1%, and a mean intersection over union (MIOU) between 87.2% and 88.7%. While ChickenNet zoomed in on feather health, Li's work was all about behavior, showing how versatile Mask R-CNN can be. Both leaned on pre-trained weights from big datasets like COCO and ImageNet to get a head start, but they hit similar roadblocks: small datasets and the hassle of manual labeling held them back from being perfect.

Then there's (Teterja et al., 2024) who took a different tack with a video mosaicing and MobileNetV2 combo running on edge devices think small, portable tech you can use right in the barn. They aimed to spot behaviors like "Sleepy," "Eating," and "Drinking," landing at a decent 79.61% accuracy. This setup is cool because it supports sustainable farming by letting farmers check chickens on the spot without fancy setups. But it's not all smooth sailing. Getting it to work on different farms means tweaking it for each place's tech and environment like lighting or coop layout which takes extra work. Plus, the computing power it needs can slow things down, making real-time use tricky without beefier hardware.

Now, let's put these models side by side. ChickenNet's Mask R-CNN with ResNet-50 is a heavyweight champ for precision, nailing 98.02% mAP on plumage checks, outshining G. Li's ResNet101 setup, which topped out at 88.1% recall for preening. That gap makes sense feather conditions are static and easier to pin down compared to behaviors, which shift and wiggle in real time. Both Mask R-CNN versions lean on ResNet backbones, but ResNet-50's edge might come from its balance of depth and speed, while ResNet101 digs deeper into details at the cost of being a bit heavier to run. Teterja's MobileNetV2, though, plays a different game. Its 79.61% accuracy is lower, but it's built for lightweight, on-the-go use, not lab-perfect scores. Speed wise, MobileNetV2 is designed to be zippy on edge devices, though the studies don't give exact frames per second number to compare directly with Mask R-CNN. Still, Mask R-CNN's heft suggests it's slower, especially since it's typically used on bigger machines, not portable gadgets.

What ties these efforts together is their reliance on solid training data. ChickenNet and Li's work both used pre-trained weights to boost their start, but small datasets and human labeling quirks like different folks seeing the same feather or flap differently kept them from maxing out. Teterja's approach dodged some of that by focusing on video stitching and simpler behavior tags, but it trades precision for flexibility. Environmentally, all three face challenges: ChickenNet didn't gain much from depth data, hinting RGB might be enough for plumage, while Li's preening detection could falter under odd lighting or coop clutter, and Teterja's model needs tweaking for each farm's vibe. Another hurdle Mask R-CNN shines in controlled studies but might lag on big, chaotic farms, while MobileNetV2's portability could stretch further if the tech gets a boost.

Digging deeper, these models hint at bigger trends. ChickenNet's near perfect mAP shows machine learning can nail down specific traits like plumage with the right setup, but Li's slightly lower scores suggest behaviors are trickier chickens don't pose for the camera. MobileNetV2's lower accuracy isn't a flop; it's a trade-off for real-world grit, bringing AI out of the lab and into the coop. Limitations pile up across the board: manual labels mess with consistency, small datasets cap how much the models can learn, and real-time demands push hardware limits. Compared to each other, Mask R-CNN wins on accuracy for detailed tasks, while MobileNetV2 takes the crown for practical, on-site use. Together, they show machine learning's power in poultry care, whether it's spotting ruffled feathers or sleepy hens but also spotlight the need for bigger data, sharper labels, and faster gear to make them farm ready across the board.

Poultry Behavior Observation With Deep Learning Models

Deep learning has opened new doors in poultry farming, letting researchers and farmers track chicken behavior and health with cutting edge tools. Several studies have leaned on deep learning models to spot things like feather conditions and preening, while others have tried different tricks like sensor based systems. Let's dig into the deep learning efforts, compare how they stack up, and see where they fit in poultry care, with a quick nod to a non deep learning approach for contrast.

(Campbell et al., 2024) brought out ChickenNet, a system built on Mask R CNN with a ResNet 50 backbone, to check plumage conditions in laying hens. They used RGB images sized at 896 by 896 pixels and scored an awesome mean Average Precision of 98.02 % and an accuracy of 91.83 %. That's a big deal for keeping tabs on feather health, which ties straight to how well hens are doing. But here's the rub: tossing in extra depth info like 3D data didn't boost things much. A key hiccup they hit was that folks labeling the images by hand didn't always agree, leaving some wiggle room in the results. Jumping back to 2020, G. Li and his crew also used Mask R CNN but mixed in different ResNet backbones to catch preening behavior in Hy Line Brown hens. Their star player, ResNet 101, pulled off a precision of 85.1 %, a recall of 88.1 %, and a mean intersection over union between 87.2 and 88.7 %. ChickenNet zeroed in on feathers, while Li's work chased behavior, showing Mask R CNN can bend to different tasks. Both hitched their wagons to pre trained weights from big datasets like COCO and ImageNet to get rolling, but they bumped into the same walls: small datasets and the mess of manual labeling held them back from perfection.

Putting these deep learning models side by side, ChickenNet's Mask R CNN with ResNet 50 shines brighter for precision, clocking that 98.02 % mAP on plumage checks, outpacing G. Li's ResNet 101 setup, which topped out at 88.1 % recall for preening. That gap tracks feather conditions are steady and easier to pin down than behaviors that flutter and shift. Both lean on ResNet backbones, but ResNet 50 seems to strike a sweet spot between depth and speed, while ResNet 101 digs deeper into details, maybe at the cost of being a bit clunkier. Accuracy wise, ChickenNet's 91.83 % edges out Li's numbers, though we don't get exact speed stats like frames per second to see how fast they run head to head. Still, Mask R CNN's heft hints it's not the quickest, likely needing solid machines to crunch the numbers, which could slow it down in real time barn action.

What ties these two together is their hunger for good data. ChickenNet and Li's work both grabbed pre trained weights to jumpstart things, but small datasets meant they couldn't learn everything, and human labelers disagreed like one seeing a feather as fine while another calls it ruffled muddied the waters. Environment wise, they hit some snags too: ChickenNet didn't gain much from depth data, suggesting plain RGB images might do the trick for plumage, while Li's preening detection could stumble if the coop's lighting flickers, or stuff gets in the way. Scaling up's challenge Mask R CNN rocks in tight, controlled setups but might lag on sprawling, hectic farms with tons of birds. The reliance on hand done labels also means time and cost pile up, and if the dataset doesn't cover all the weird ways chickens act, the models miss out.

Zooming out, these deep learning runs tell a story. ChickenNet's near perfect mAP shows these tools can lock onto specific stuff like plumage when tuned right, but Li's lower scores flag that behaviors are tougher chickens don't freeze for the camera. Both prove Mask R CNN's muscle for poultry tasks, but they're not flawless. Small datasets cap how much they can learn, manual labels throw in hiccups and running them live might tax the gear too hard. Compared to each other, ChickenNet takes the crown for static checks like feathers, while Li's setup holds its own for trickier, moving targets like preening. Together, they spotlight deep learning's knack for detailed poultry monitoring, but the cracks data size, label quirks, and farm scale push for bigger datasets, sharper labeling, and maybe some hardware juice to make them barn ready across the board.

For a twist, (L. Li et al., 2024) went a different way with ISSA SVM, mixing the Sparrow Search Algorithm with a Support Vector Machine to spot violent chicken behaviors using inertial sensor data. This cut their feature list by 92.6 % and hit 94.27 % accuracy, beating out regular SVMs. It's a slick move for sensor-based tracking, sidestepping cameras altogether. But it's got its own headaches: sensors need to adapt to different chickens and setups, and battery life can't keep up forever. Unlike the vision-based Mask R CNN models, this one scale tricky for big farms sensors on every bird get pricey and fiddly. It's a cool contrast though: while ChickenNet and G. Li's work lean on eyes and images, ISSA SVM listens to motion, showing there's more than one way to crack the poultry puzzle. Still, the deep learning crew pulls ahead for precision on visual tasks, while ISSA SVM offers a leaner, sensor driven angle that might fit smaller, tech savvy setups better.

Wrapping it up, these studies paint a picture of deep learning's power in poultry care ChickenNet nailing feather checks, G. Li catching preening, and ISSA SVM as a sensor-based wildcard. Mask RCNN's high accuracy makes it a go to for detailed looks, but its data and speed limits keep it grounded. ISSA SVM skips the camera fuss but trades off with scalability woes. They all push the field forward, but nailing bigger data, cleaner labels, and real time grit could make them unbeatable for farms everywhere.

Table 1. Related studies summary table

Author(s)	Year	Model Used	Objective	Key Findings	Limitations
Christian Lamping et al.	2022	Mask R-CNN (ResNet-50)	Detect plumage conditions in laying hens	Achieved **mAP 98.02%** and **91.83% accuracy**	Inter-observer variability in manual annotations
Guoming Li et al.	2022	Mask R-CNN (ResNet-101)	Detect preening behavior among hens	Best precision **85.1%**, recall **88.1%**, and mIoU **87.2–88.7%**	Dataset size and manual labeling constraints
Suraj Salihu et al.	2023	YOLOv5, YOLOv7, YOLOv8 + DeepSORT	Tracking and counting chickens	YOLOv8 V4 had **mAP50 97.4%** and **mAP50-95 75.4%**	Poor performance on small, crowded objects
Ramesh Bahadur Bist et al.	2023	YOLOv6	Monitoring piling behavior in cage-free hens	YOLOv6l_relu-PB achieved **mAP@0 .50 of 98.9%**	Environmental factors caused detection errors
Sozzi et al.	2023	YOLOv4, YOLOv4-tiny	Detecting dust-bathing behavior	YOLOv4-tiny had **208.5 FPS** but low detection accuracy (**mAP 28.2%**)	Poor detection for complex dynamic behaviors
Campbell et al.	2024	U-Net + CFSGMM	Detect broiler chicken activities	High precision **0.98**, recall **0.90**, and F1-score **0.94**	Noisy clustering for highly active behaviors
Lihua Li et al.	2024	ISSA-SVM	Detect violent chicken behaviors using sensors	Accuracy **94.27%**, **feature set reduced by 92.6%**	Sensor adaptability and battery life issues
Fujie Wang et al.	2023	EfficientNet-YOLOv3, YOLOv4-Tiny, YOLOv5, Faster-RCNN	Detect various behaviors	YOLOv5 had best FPS **55.55**, high accuracy	Poor performance in nighttime scenarios
Kodaiera et al.	2023	YOLOv4	Study light color preferences of hens	**mAP 99.9%** in controlled conditions	Small sample size, limited generalization
Md Roman Bhuiyan et al.	2023	CNNs, YOLOv5	Analyze AI and IoT applications for tracking chicken health and behavior	**Accuracy 96.77% for stunned states, Accuracy 99.17% for flock behavior.**	Lack of standardized AI models for poultry welfare assessment.

continued on following page

Table 1. Continued

Author(s)	Year	Model Used	Objective	Key Findings	Limitations
Wiwit Agus Triyanto et al.	2023	YOLOv4	Study to track broiler movements for better health and behavior monitoring by AI.	**YOLOv4 with Deep SORT provided the best detection**	YOLOv4's high processing time limits its real-time tracking efficiency on large farms without hardware upgrades.
Dmitrij Teterja et al.	2024	MobileNetV2	Develop a video mosaicing and deep learning method for recognizing chicken behaviors to enhance animal welfare, health monitoring, farm management, and sustainable poultry farming.	**Accuracy 79.61%**	Adapting to different farms may need extra effort due to diverse technical and environmental factors.

METHODOLOGY

In this section, an overview is provided of the data collection process, the pre-processing techniques applied to the dataset, the models trained, and the evaluation metrics utilized for the analysis.

Dataset Description

The dataset of images of chicken behaviors for analysis uploaded to Roboflow was developed to further the comparative studies of YOLO models for recognition and analysis of chicken behaviors. It consists of 1,026 images and has an average of 1.3 annotations per image with a total of 1,311 labels created. The images have an average of 0.23 MP, and the average size of the images is 360x640 pixels which allows use of the models in mobile applications. The dataset comprises of 21 behavior classes from the basic instinctive behavior (a portion of the images includes: feeding: 141, self-preening: 127, walking: 104) to physiological and health indicators including breathing: 90, bird flu: 61, and other images such as abnormal chickens: 45, and dead chickens: 59. It is also enriched by a few appearances of allo-preening (24), wing and leg stretching (13) and wing stretching (6) which are considered rare behaviors. This dataset has a collection of a variety of behavioral expressions including that of social interactions (fighting: 41, menacing: 30) as well as activities indicative of environmental conditions, thus can be used for general behavioral studies as well as in-depth assessment of health conditions. There is sufficient class balancing to allow proper cross-training of deep learning models.

Figure 1. Methodology diagram

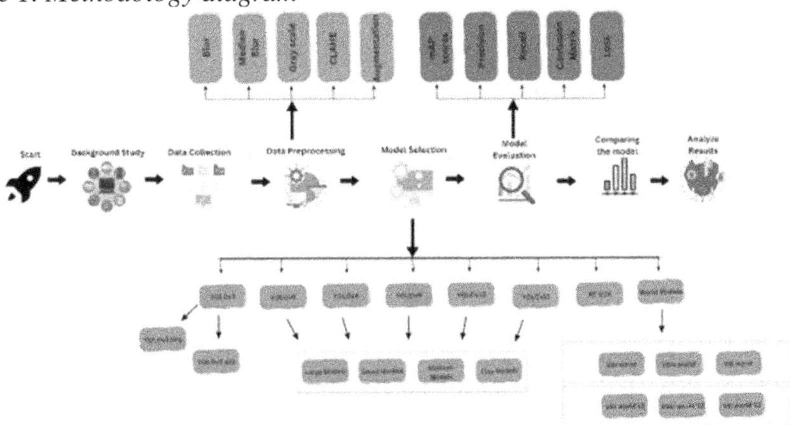

Preprocessing

In this study, we employed a number of preprocessing techniques to upscale the image quality and enhance the performance of our models with the help of Albumentations Library. Each of these methods is explained in detail as follows:

Blur: The Blur function is a technique that applies a Gaussian blur to images in the dataset. This means it slightly softens or smooths the image by spreading out the pixels, which helps in reducing noise (random unwanted details) and very small patterns that might not be important. By doing this, the model learns to focus on bigger, more general shapes and structures rather than tiny details, making it more reliable. In this dataset, the blur effect is applied only 1% of the time, and the blur size can be anywhere between three to seven pixels. This method is useful in pre-processing images, meaning it prepares the data in a way that makes the AI model better at recognizing objects even if they appear blurred or slightly hidden (occluded). Ultimately, this can improve the model's accuracy and performance by teaching it to focus on important features rather than unnecessary details.

Median Blur: Median Blur is a technique used to remove noise from images, making them clearer and improving their quality. It works by taking the median value (middle value) of the surrounding pixels in a small area of the image, which helps in filtering out unwanted noise while keeping the important details. This method is especially useful for handling "salt and pepper" noise, which appears as random white and black spots on an image—something common in real-world photos. One of the biggest advantages of Median Blur is that it preserves the edges of objects in an image, unlike other blurring techniques that might make them look too smooth or unclear. In this dataset, the filter is applied to only 1% of the images, meaning that it is rarely used but still helps improve overall image quality. The blur effect can range between three to seven pixels, which ensures that the images look natural and realistic while still benefiting from noise reduction. By using this method, the AI model can work with cleaner and sharper images, leading to better recognition and improved performance.

To gray: The To Gray function is used to convert an image from color to grayscale, meaning it removes all colors and turns the image into different shades of gray. It does this by averaging the values of the pixels, which helps simplify the image. In many cases, color information is not very important, and the AI model needs to focus more on shapes, edges, and structures rather than colors. For example, if an object's shape is the key detail for recognition, color might just be an extra distraction. By using a weighted average method, the function ensures that different colors contribute to brightness in the same way as in the original image, keeping the balance of light and dark areas. Even though the image looks grayscale, it is still stored in three channels (Red, Green, and Blue) so that the format remains the same, allowing the model to process it without issues. This conversion makes the AI more adaptable and efficient, especially in situations where color differences are minimal or not important, ensuring that it learns from the most important features of the image rather than relying on color details.

CLAHE: CLAHE (Contrast Limited Adaptive Histogram Equalization) is a technique used to improve the contrast of an image, making details clearer, especially in images that look faint or low in contrast. It works by dividing the image into smaller sections, called tiles, and then adjusting the brightness and contrast of each tile separately. This method ensures that even areas with poor lighting or dull colors become more visible. Unlike normal histogram equalization, CLAHE limits the contrast enhancement in each tile to prevent over-brightening or excessive contrast, which could make parts of the image look unnatural. The clip limit (ranging from 1.0 to 4.0) controls how much contrast is increased, while the tile grid size (8x8) determines how small or large each section is when adjusting the brightness. This technique is only applied 1% of the time to ensure that the images still look natural while benefiting from enhanced details. By using CLAHE, the AI model can better recognize fine details in images, even when they are dark, dull, or low in contrast, leading to improved accuracy in object detection and analysis.

Dataset Augmentation

Data augmentation is a method applied in machine learning in order to increase the volume and diversity of the original data set by performing certain transformations on the available information (Dey et al., 2024). This is very important for the improvement of models' performance especially for a case when the data set is small or unbalanced so that overfitting is avoided, and model generalization is enhanced. In this case, data augmentation methods were used to boost the class representations and balance all 21 behavior classes, which expanded the database from 1,026 to 2,100 images, consisting of 100 images per class. The augmentation methods used consisted of flipping lengthwise, rotation from -15° to +15°, as well as shearing ±10° horizontally and vertically. Saturation range -25% to +25%, brightness -15% to +15%, and exposure -10% to +10% were also modifications that were made. Also, noise was added to less than 0.1% of pixels, which created diversity without distorting data. These augmentations help in enhancing the model robustness and generalization.

Loss Functions

Binary cross entropy (BCE): BCE is a kind of loss function for classification tasks that object detection uses, by measuring a difference between the predicted probability of the object being present and true class label, 0 in case of no object, 1 in the case of an object(Hurtik et al., 2022). BCE works with a comparison of the predicted probability with a true label and calculates log-loss for every class. This loss helps train the models to optimize predictions in multi-class detection tasks, as is the case with YOLOv3 and YOLOv5, where the detection task involves both localization and classification.

$$L = -(y\log(p) + (1-y)\log(1-p))$$

Where:

- y is the actual label (0 or 1),
- p is the predicted probability of the object being present.

Intersection over Union: IoU stands for Intersection over Union. It is a metric to assess the accuracy of the object detection model, comparing the predicted bounding box to the ground truth bounding box(Zheng et al., 2020). It does so by taking the area of overlap between the predicted box and the actual box, then dividing that by the area of the union of them. IoU is a measure of how well predicted bounding boxes overlap with the ground truth objects. It is used in object detection, part of the loss function to be optimized for getting the best bounding box prediction at training time. The applications can be found in YOLOv3, YOLOv5, YOLOv8, YOLOv9, YOLOv10, and YOLOv11 models.

$IoU = Area\ of\ Union/Area\ of\ Intersection$

Complete Intersection over Union (CIoU): Complete Intersection over Union is a more advanced metric that has been used to assess the quality of bounding box predictions in object detection tasks(Du et al., 2021). Different from traditional IoU, which considers only overlap, CIoU takes into account more geometric factors: distance between the center points of predicted and ground truth boxes, aspect ratio consistency, and box alignment. This leads to much better bounding box localization, especially when the predicted box is far from the ground truth box. The CIoU has started to find a place in the more modern YOLO incarnations, such as YOLOv8, YOLOv9, YOLOv10, and YOLOv11, which are dedicated to improving detection accuracy, especially on hard scenes with small or overlapped objects.

Where:

- is the squared Euclidean distance between the center points of the predicted and ground truth boxes.
- is the diagonal length of the smallest enclosing box.
- is the aspect ratio consistency factor.
- α is a weight factor for balancing the aspect ratio term.

Generalized Intersection over Union (GIoU): Generalized Intersection over Union generalizes the classic IoU metric, which is limited for non-overlapping bounding boxes(Jiang et al., 2023). GIoU includes a penalty term from the smallest bounding box containing both the prediction and the ground truth bounding box in the case of nonoverlapping boxes for better localization. Mathematically, this can be described as:

This metric further encourages a better alignment and positioning of bounding boxes, especially useful in cases with sparse objects or occlusions. GIoU has found applications on modern object detection frameworks such as YOLOv4 and YOLOv5, since it helps increase precision and robustness during the detection process.

Distribution Focal Loss: Distribution Focal Loss is an advanced loss function, enhancing the model's hard-example-oriented ability, especially for dense and occluded object detection tasks(X. Li et al., 2020). It works by applying a focal mechanism that shifts focus toward the most challenging samples, such as high-class imbalance. The core behind DFL is to adjust the weight of each sample's contribution to the loss concerning its difficulty in classification. DFL extends the traditional Focal Loss by taking into account the distribution of predicted class probabilities for improved bounding box regression.

Where:

- is the predicted probability for class t
- α is the balancing factor to adjust the importance of rare classes
- γ is the focusing parameter to adjust the rate at which easy examples are downweighed

AdamW Optimizer: The AdamW optimizer is based on the Adam algorithm but introduces weight decay (L2 regularization) in a more decoupled way, which allows for better regularization without interfering with the adaptive learning rates(Tanvir et al., 2024). In object detection tasks, AdamW improves convergence speed and generalization by penalizing weights that are too large. The key difference with respect to Adam is that weight decay here is imposed directly on the parameter update and not as a part of the gradient computation(Zhang et al., 2021). This is more effective for regularization, preventing overfitting in deep networks.

Where:

- is the parameter at time t
- is the first moment estimate (mean of gradients)
- is the second moment estimate (variance of gradients)
- λ is the weight decay coefficient
- η is the learning rate
- ϵ is a small constant to prevent division by zero

Model Descriptions

YOLOv3 Architecture Description

Figure 2. YOLOv3 base architecture

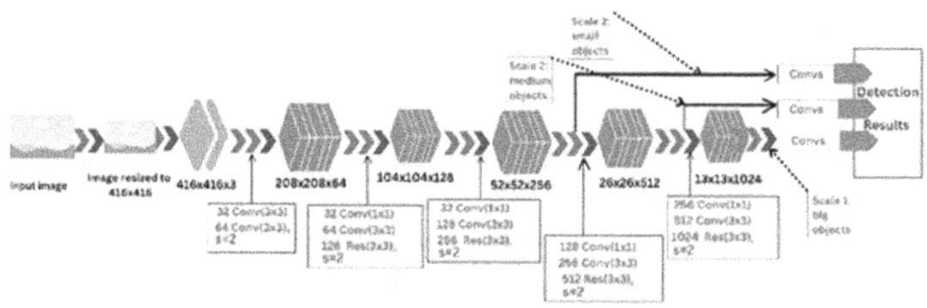

YOLOv3 is a real-time object detection model, released in 2018, with a balance of speed and accuracy. Compared with the previous versions like YOLOv1 and YOLOv2, the new model improves the detection of small objects using multi-scale predictions(F. Li et al., 2024). YOLOv3 detects objects at three feature map levels, capturing fine details of small objects at higher resolutions and larger objects at lower resolutions. It uses anchor boxes for precise object localization and Darknet-53 as a backbone for efficient feature extraction. The YOLOv3-SPP variant adds a Spatial Pyramid Pooling layer to enhance multi-scale feature aggregation. On the COCO dataset, YOLOv3 achieves 57.9% mAP at IoU 0.5 and runs at 30 FPS on a Titan X GPU, outperforming Faster R-CNN and Mask R-CNN in speed. However, it struggles with fine-grained object detection in crowded scenes and is less precise for small or complex objects.

Figure 3. YOLOv5 base architecture

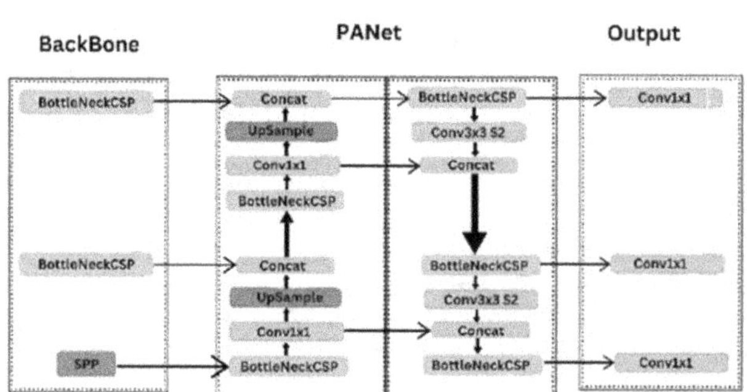

YOLOv5 Architecture Description

YOLOv5 is a real-time object detection model released in 2020, focused on fast training and inference with lightweight architecture to achieve high accuracy(Mahaur et al., 2023). YOLOv5 uses an AutoAnchor algorithm to adaptively generate and optimize anchor boxes, improving detection precision. Its backbone, modified CSPDarknet53 with SPPF, enhances computation efficiency and feature extraction. YOLOv5 achieves 66.8% mAP at IoU=0.5 on the COCO dataset, with up to 140 FPS on a high-performance GPU, outperforming YOLOv3 in speed and accuracy. It has four variants: YOLOv5n (lightweight, 1.9M parameters), YOLOv5s (balanced, 7.2M parameters), YOLOv5m (robust, 21.2M parameters), and YOLOv5l (high-performance, 46.5M parameters). All variants use IoU and Binary Cross-Entropy losses for localization and classification. YOLOv5 is open-source and supports mobile deployment.

Figure 4. YOLOv8 base architecture

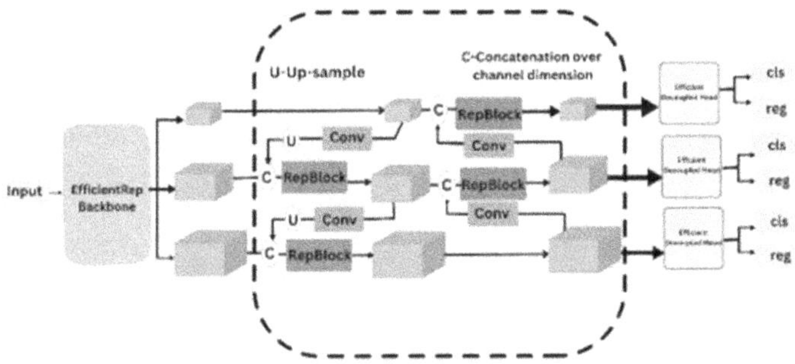

YOLO v8 Architecture

YOLOv8, introduced by Ultralytics in January 2023, is a high-performance model designed for object detection, segmentation, classification, tracking, and pose estimation. It features a custom CSPDarknet53 backbone with the C2f module, improving small object detection and feature fusion. YOLOv8 uses an anchor-free design and decouples the head into separate modules for better performance in complex scenes. With Complete IoU and Distribution Focal Loss, YOLOv8 achieves 53.9% mAP at IoU=0.5 and 280 FPS on an NVIDIA A100. It has four variants: YOLOv8n (1.8M parameters), YOLOv8s (11.2M), YOLOv8m (25.9M), and YOLOv8l (62.3M), balancing speed, accuracy, and computational cost for various applications. This is mainly because the changes have been different for the scaled forms with increased layer depth, scaling in channels, and feature aggregations in order to create YOLO variants(Diwan et al., 2023). Free to use, open-sourced, and being actively developed, high deploy ability ensures, for a wide application area, top performances that include real-time tasks.

Figure 5. YOLOv9 base architecture

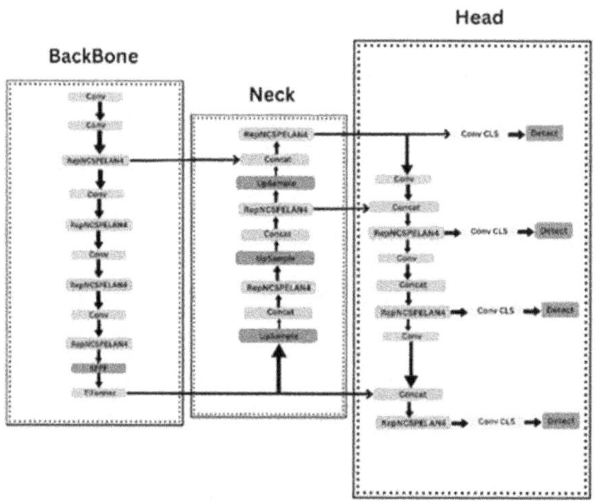

YOLO v9 Architecture

YOLOv9 introduces two groundbreaking features: Programmable Gradient Information (PGI) and General Efficient Layer Aggregation Network (GELAN), which enhance learning flow and efficiency, particularly for large and diverse datasets like MS COCO. YOLOv9 is faster, more accurate, and delivers superior generalization and detection precision, excelling in complex and challenging scenarios. It comes in four variants: YOLOv9n (1.7M parameters, 120 layers), optimized for edge deployment; YOLOv9s (10.8M, 152 layers), balancing speed and accuracy; YOLOv9m (26.5M, 248 layers), which offers high robustness and precision for demanding tasks; and YOLOv9c, the largest variant with 65M parameters and 372 layers, designed for maximum performance in fine-grained detection tasks. YOLOv9 significantly improves small object detection and performs better in difficult environments with improved feature aggregation. PGI optimizes gradient usage during model training, enhancing learning efficiency, while GELAN boosts feature aggregation, resulting in improved detection capabilities under various conditions. These advancements make YOLOv9 well-suited for real-time applications and challenging tasks requiring exceptional accuracy. All variants use Complete Intersection over Union (CIoU) loss for accurate localization and apply Distribution Focal Loss to enhance both bounding box prediction and classification performance. While YOLOv9 offers remarkable improvements, the increased network complexity, particularly in the Compact variant, demands more computation, making it less suitable for constrained devices. However, YOLOv9 excels in generalizing to diverse environments, even in cases involving occluded or densely packed objects. Overall, YOLOv9 represents a significant leap forward in object detection, providing outstanding performance across a wide range of applications and offering flexibility for multiple deployment scenarios. The largest variant, YOLOv9c, has 65 m parameters integrated with 372 layers for maximum performance on tasks requiring the most fine-grained detection and accuracy (Farid et al., 2023). All variants make use of Complete Intersection over Union loss for localization, whereas Distribution Focal Loss has been applied in order to improve both the quality of the bounding box prediction and classification. While YOLOv9 advances the field, network complexity rises, especially within its Compact variant, requiring more computation for constrained devices. This solution will generalize well in almost any diverse environment that may have occluded objects or very congested detection. YOLOv9 is a huge leap in object detection, doing an amazing job in a wide range of applications, flexible for deployment multiple times.

Figure 6. YOLOv10 base architecture

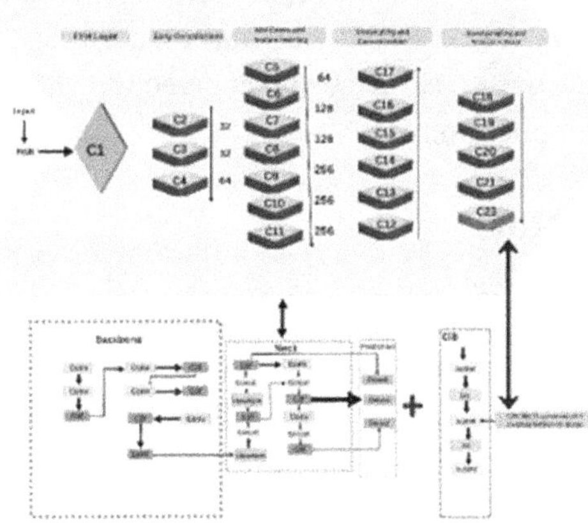

YOLOv10 Architecture

YOLOv10, announced by Tsinghua University in May 2024, achieves an optimal balance between accuracy and computational efficiency. It introduces Consistent Dual Assignment for richer supervision and removes Non-Maximum Suppression during inference to reduce computation. Architectural innovations like Parallel Split-Attention and Compact Inverted Bottleneck Blocks enhance attention mechanisms and efficiency, while Scaled Residual Connections improve information flow. YOLOv10 outperforms YOLOv9 and YOLOv8 in accuracy with reduced latency. It offers four variants: Nano (1.6M parameters, 110 layers), Small, Medium, and Large, catering to various operational needs, with YOLOv10n ideal for edge applications. YOLOv10s has 12 million parameters with 140 layers, a sweet balance between speed and accuracy(Paul & Machavaram, 2024). It is, therefore, well-suited for general-purpose applications in real-time. With 28 million parameters on 230 layers, the YOLOv10m variant is powerful in taking on complex tasks when high accuracy is needed. Its largest variant, called YOLOv10l, comprises 65 million parameters with 370 layers that provide state-of-the-art performance for high precision applications, albeit at computational demand. For object detection, YOLOv10 uses Complete Intersection over Union loss. For precision and efficiency of bounding box prediction and classification, Distribution Focal Loss. Variants, such as YOLOv10l, will be prohibitive in computational time, hence impossible to deploy on resource-constrained platforms, including mobile and edge devices. Even a finely tuned model will likely be far outperformed by YOLOv10 in environments with very high clutter or occlusion. But with the best accuracy, speed, and versatility, YOLOv10 raised a new bar in real-time object detection and was particularly suited for a wide variety of applications.

YOLOv11 Architecture

Ultralytics introduced YOLOv11 on September 30, 2024, a real-time object detector with considerable improvements compared to the evolution of YOLO versions(Sapkota et al., 2024). The network design introduces transformer-based architectures and self-attention mechanisms that work much better than ever for object detection tasks in multi-object and complicated scenarios. Innovations like Consistent Dual Assignments during training eliminate Non-Maximum Suppression at inference, streamlining detection in YOLOv11 with improved computational efficiency. These enable YOLOv11 to perform better in small object detection, occlusion handling, and low contrast than previous versions, alongside improvements in accuracy and processing speed. The main variants of YOLOv11 include Nano: YOLOv11n, Small: YOLOv11s, Medium: YOLOv11m, and Large: YOLOv11l, each tuned for different computational and application needs. YOLOv11n, with 2.0 million parameters and 115 layers, is the lightest and fastest, optimized for edge and mobile deployments. YOLOv11s, with 14 million parameters and 150 layers, strikes a balance between speed and accuracy for general-purpose detection tasks. YOLOv11m, with 30 million parameters and 240 layers, is more robust for higher accuracy needs. YOLOv11l, with 70 million parameters and 390 layers, provides the best performance in precise applications. YOLOv11 uses CIoU loss for localization, DFL for bounding box regression, and multitask learning for classification, pose estimation, and orientation detection. Its transformer-based backbone improves feature extraction and generalization, enabling better performance in challenging scenes. However, its architectural complexity, especially in the larger variants, limits deployment on resource-constrained platforms. Even smaller versions may struggle with complex scenes involving heavy occlusions or high object density. Despite these challenges, YOLOv11 sets the standard for object detection, instance segmentation, pose estimation, and orientation detection in a wide range of applications, proving invaluable for real-time use. This version also improves robustness in difficult conditions like low-light environments and dynamically changing scenes. The balance between performance and flexibility allows YOLOv11 to adapt to diverse use cases across industries such as healthcare, autonomous driving, and surveillance. It's engineered to handle various tasks with increased accuracy, allowing faster decision-making in high-stakes scenarios. With its highly optimized architecture, YOLOv11 provides a practical solution for environments requiring both high efficiency and exceptional accuracy across multiple platforms and devices, maintaining seamless integration and scalability, while enabling innovation in complex environments.

Figure 7. YOLOv11 base architecture

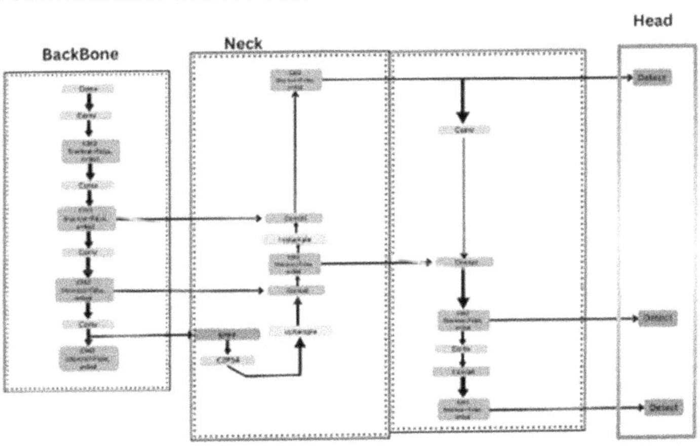

YOLO World Models

The YOLO World models, optimized for large-scale and diverse environments, come in six main variants: YOLOv8 Large World, YOLOv8 Large World v2, YOLOv8 Medium World, YOLOv8 Medium World v2, YOLOv8 Small World, and YOLOv8 Small World v2. Architectural refinement and loss functions are the main differences among them. The YOLOv8 Large world model uses advanced CSPDarknet-based backbones with SPPF layers, utilizing CIoU and BCE Loss for balanced localization and classification accuracy. The YOLOv8 Large World v2 version further improves this with Split-Attention modules, Scaled Residual Connections, and Scaled Weight Shortcuts, together with Distribution Focal Loss for enhanced bounding box precision and learning of class distribution. Compared with this, YOLOv8 Medium World and YOLOv8 Medium World v2 reduce the complexity of parameters while maintaining multiscale feature processing with Compact Inverted Bottlenecks for better computational efficiency in mid-tier tasks; the variant, v2, adds enhanced Parallel Split-Attention for cluttered scenarios. YOLOv8 Small World is for edge efficiency, incorporating lightweight residual blocks and Depthwise-Separable Convolutions optimized with GIoU Loss for minimal computational demands(Ma et al., 2024). YOLOv8 Small World v2, on the other hand, introduces a shallower attention mechanism with fewer layers but integrates a new Attention Augmented Feature Pyramid, making it perfect for ultra-low latency tasks on resource-constrained platforms. Across all models v2 iterations offer superior accuracy and generalization and are thus better suited for high-occlusion and dense environments, while the earlier versions balance speed and performance for real-time applications.

Figure 8. YOLO world base architecture

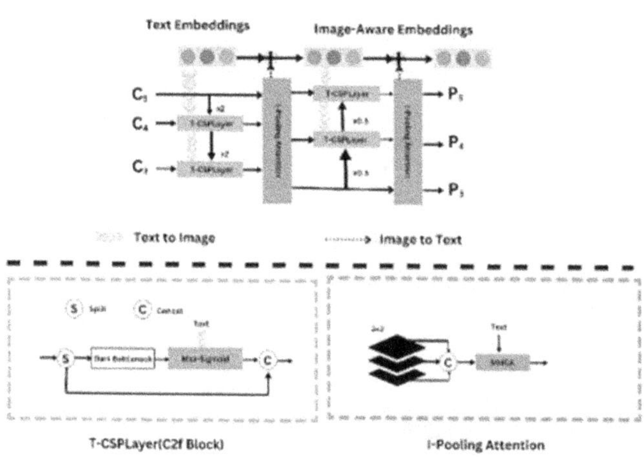

Model Evaluation Parameters

mAP50(B)

mAP50(B) stands for the mean Average Precision at IoU = 0.5 with bounding box confidence and is a common metric for object detection(Su et al., 2024). This gives the model's capability in terms of correct object localization and classification, given the IoU threshold of 0.5. In such a case, predictions are taken to be correct when there is an IoU $\geq$ 0.5 between the predicted and ground truth bounding boxes. This is an average of APs calculated over all classes, with a class's AP being the area under its PR curve. It is mathematically defined as:

mAP50-95(B)

mAP50-95(B) provides an overall measure of model performance based on precision and recall over a set of more conservative localization criteria(Natij et al., 2024). It is the average AP calculated over different IoU thresholds at a step size of 0.05 (0.5,0.55…,0.95). This metric considers the quality of object localization for the model and is considered the gold standard in modern benchmarks for object detection. Mathematically, it can be written as:

Validation Distribution Focal Loss

The object detection validation loss helps as an important metric, showing the performance of the model on unseen data and hence in finding out overfitting or underfitting problems. It usually consists of a sum of bounding box losses, classification losses, and sometimes object confidence losses. The total validation loss may be represented as a weighted sum of these individual losses: On the other hand, Distribution Focal Loss (dfl_loss) is a special loss that is intended to do a better bounding box regression prediction by predicting a full probability distribution over possible coordinates rather than fixed coordinates. Hence, this can help be more robust against the localization uncertainty and quantization error. The dfl_loss is calculated as a negative log-likelihood between the prediction and ground truth distributions(Kataria, 2023). All these combined into one metric together represent a more complete assessment of model performance in both the accuracy of classification and localization toward more accurate object detection.

Confusion Matrix

The confusion matrix is important in the evaluation of model performance for object detection, as it summarizes the relation of predicted and actual object detections. It categorizes the predictions based on four main components: TP, FP, FN, and TN, whereby not only classification accuracy but spatial precision of bounding boxes will be analyzable. It is also taken into consideration by the IoU metric whether or not, in object detection, a predicted bounding box is taken by the matrix to sufficiently overlap with the ground truth. Many other different key metrics can then be computed from this confusion matrix, including precision, recall, and F1-score(Naidu et al., 2023). These encompass the model's capability for correctly detecting objects while reducing the number of false alarms and missing detections. In turn, this converts the confusion matrix into one of the most informative pieces in object detection model performance assessment and refining.

Precision, Recall & F1-Score

Precision and recall have been both basic metrics, which define the performance of an object detection model-especially about the exact classification and proper detection(Zeng et al., 2022). The precision of a model actually refers to the correct rate of the positive predicted values. It can be measured by the ratio between the true positive detection outputs out of the total positive predications: TP + FP. It refers to how many of the objects that the model identified are actually correct. Recall, on the other hand, is about how well the model is at detecting all relevant objects inside an image. This will be calculated as a ratio of the true positive detected, TP, against all the total actual objects which is TP + FN. It gives the share of actual objects which have been correctly detected. Where precision is about minimizing false positives, recall concerns the reduction of false negatives. The two are probably the most important metrics in object detection, as they make for a good trade-off between trying to find as many objects as possible-appropriately, high recall-and having the right ones detected-appropriately, high precision. Most applications tend to use the F1-score, or the harmonic mean of precision and recall, since it gives a single measure that balances both. Precision and recall together give the overall effectiveness of a model in detecting and classifying objects while limiting errors(Tanim et al., 2024).

RESULTS AND ANALYSIS

We trained a lot of deep learning object detection models on this dataset for analysis. Here is a table with all the description of the models below:

Table 2. Results

Model		Time	Training Loss			Validation Loss			mAP-score		Precision	Recall
			Box	Class	DFL	Box	Class	DFL	50 (B)	50-95(B)		
V3	**T**	**1125**	**0.58**	**0.439**	**1.03**	**1.14**	**1.10**	**1.47**	**0.755**	**0.538**	**0.848**	**0.693**
	spp	4865	0.44	0.354	1.04	1.12	1.18	1.95	0.723	0.547	0.794	0.634
V5	N	1222	0.66	0.755	1.06	1.11	1.05	1.36	0.805	0.581	0.869	0.691
	S	1289	0.51	0.321	0.95	1.07	1.06	1.34	0.785	0.602	0.813	0.744
	M	**1864**	**0.47**	**0.281**	**0.94**	**1.05**	**1.01**	**1.37**	**0.803**	**0.607**	**0.734**	**0.781**
	L	2883	0.48	0.305	0.96	1.07	1.05	1.45	0.797	0.615	0.813	0.699
V8	N	1207	0.57	0.629	1.01	1.07	1.03	1.38	0.796	0.595	0.808	0.713
	S	**1403**	**0.46**	**0.278**	**0.93**	**1.07**	**1.03**	**1.40**	**0.795**	**0.601**	**0.877**	**0.731**
	M	2099	0.42	0.250	0.92	1.04	0.97	1.44	0.791	0.601	0.838	0.719
	L	3095	0.46	0.288	0.97	1.06	1.01	1.49	0.806	0.609	0.797	0.736
V9	C	3229	0.44	0.938	1.01	1.08	0.99	1.43	0.802	0.601	0.841	0.706
	S	1794	0.49	0.282	1.02	1.05	1.03	1.55	0.772	0.580	0.806	0.750
	T	1593	1.00	1.597	1.35	1.28	1.71	1.64	0.517	0.352	0.523	0.497
	M	**2631**	**0.45**	**0.263**	**0.96**	**1.04**	**0.95**	**1.49**	**0.796**	**0.608**	**0.828**	**0.735**
V10	N	1545	1.09	1.524	1.96	2.14	2.22	2.67	0.739	0.565	0.735	0.684
	S	**1701**	**0.95**	**0.509**	**1.86**	**2.25**	**2.12**	**2.84**	**0.797**	**0.599**	**0.838**	**0.690**
	M	2364	2.09	3.504	2.91	2.68	3.95	3.29	0.345	0.238	0.445	0.295
	L	3469	2.08	3.401	2.97	2.78	3.90	3.56	0.378	0.257	0.382	0.388
V11	N	1298	0.55	0.594	0.99	1.06	1.10	1.36	0.764	0.578	0.773	0.736
	S	1494	1.11	1.740	1.45	1.37	1.87	1.62	0.451	0.293	0.442	0.428
	M	**2310**	**0.48**	**0.291**	**0.96**	**1.05**	**1.02**	**1.40**	**0.777**	**0.598**	**0.781**	**0.723**
	L	2856	1.20	1.857	1.54	1.44	2.04	1.72	0.413	0.272	0.494	0.416
World Model	8SW	2465	1.11	1.837	1.42	1.37	1.93	1.61	0.429	0.290	0.394	0.484
	8SWV2	2246	0.49	0.299	0.94	1.01	0.90	1.35	0.796	0.607	0.795	0.737
	8MW	3268	0.56	0.341	1.01	1.04	0.96	1.37	0.789	0.604	0.816	0.733
	8MWV2	**3389**	**0.43**	**0.253**	**0.92**	**1.06**	**0.83**	**1.46**	**0.811**	**0.617**	**0.845**	**0.727**
	8LW	4464	1.05	1.805	1.38	1.29	1.84	1.56	0.471	0.331	0.364	0.497
	8LWV2	4345	1.07	1.850	1.42	1.36	2.00	1.65	0.441	0.303	0.429	0.468
Rt-dtr	**L**	**4671**	**1.22**	**2.216**	**1.40**	**0.41**	**0.73**	**0.46**	**0.728**	**0.559**	**0.838**	**0.719**

Variants of YOLOv3 Model Analysis

In this work, we compare the performance of YOLOv3 and YOLOv3-SPP over a variety of metrics. YOLOv3 showed higher precision (0.84868), recall (0.69333), and mAP@50 (0.75551); hence, more accurate object detection with fewer false positives, compared to YOLOv3-SPP, which showed slightly lower precision (0.79491), recall (0.6349), and mAP@50 (0.72313). While YOLOv3-SPP showed a marginal improvement in mAP@50-95 (0.54766 vs. 0.53805) and better box localization during validation, as indicated by the lower box loss of 1.12673 versus 1.14542, longer training time, 4865.18 seconds versus 1125.61 seconds, and weaker recall, make it less efficient. Therefore, YOLOv3 is the better performing model, with the best performance in terms of detection accuracy, speed, and efficiency, and is more suitable for real-time applications.

Variants of YOLOv5 Model Analysis

Among all the YOLOv5 models, YOLOv5 Nano has the best metrics/ mAP50 of 0.80538 and metrics/mAP50-95 of 0.58173, indicating its good generalization capability. However, with high training losses, it has a train/box_loss of 0.66093, a train/cls_loss of 0.75593, and a train/dfl_loss of 1.06165, which means this model is not efficient in learning, though it takes the least time in training: 1222.57 seconds. Among these, the best precision is obtained by the YOLOv5 Large model: 0.81385, competitive metrics/mAP50: 0.79781, and metrics/mAP50-95: 0.6154, but it also had the longest training time: 2883.14 seconds and higher validation losses: val/box_loss = 1.07844, val/cls_loss = 1.05434. YOLOv5 Medium offers balanced performance, with lower training losses such as train/box_loss = 0.47183, train/cls_loss = 0.28099, and train/dfl_loss = 0.94066, while offering competitive metrics/mAP50 = 0.8039, metrics/mAP50-95 = 0.60723, and faster training in 1864.86 seconds in total. YOLOv5 Small performs moderately in the middle across metrics with a value of metrics/mAP50 = 0.78593 and metrics/mAP50-95 = 0.60261 and didn't excel in any categories. If considering all the parameters, YOLOv5 Medium will be efficient by balancing accuracy, training time, and lower losses as a whole, making it the best for robust and practical deployment.

Variants of YOLOv8 Model Analysis

Among all the parameters, in general, the YOLOv8 Nano model has the shortest training time: 1207.87 seconds, but with bigger training losses: train/box_loss = 0.57603, train/cls_loss = 0.62937, train/dfl_loss = 1.01251, which reflects inefficiency of convergence. Its metrics/mAP50 (0.79627) and metrics/mAP50-95

(0.59589) are competitive but not the highest. YOLOv8 Small has a great balance, with very strong performance in precision (0.87779), very competitive metrics/mAP50 (0.79529), and metrics/mAP50-95 (0.60141), with moderate training time of 1403.15 seconds. YOLOv8 Medium has the lowest training losses, at train/box_loss = 0.42451, train/cls_loss = 0.25006, and train/dfl_loss = 0.92772, decent precision of 0.8381, and strong metrics/mAP50-95 of 0.60158, finishing training in 2099.74 seconds. The YOLOv8 Large model, though having the highest metrics/mAP50 (0.80646) and metrics/mAP50-95 (0.60901), has the longest training time at 3095.05 seconds and has a slightly higher training loss than the medium and small models. Based on all metrics, YOLOv8 Small emerges as the best choice, combining high precision, competitive mAP scores, and moderate training time, making it suitable for practical deployment.

Variants of YOLOv9 Model Analysis

YOLOv9 Tiny has the least training time among the models of YOLOv9, which is 1593.19 seconds. However, it suffers from the lowest performance regarding metrics/ mAP50 0.51715 and metrics/mAP50-95 0.35255 coupled with remarkably high training losses: train/box_loss = 1.00153, train/cls_loss = 1.59712, train/dfl_loss = 1.3588. YOLOv9 Small trains faster, taking 1794.15 seconds but with slightly higher training losses, train/box_loss = 0.49592 compared to medium and large, while realizing a moderate precision of 0.80617 and recall of 0.75051. YOLOv9 Medium gives a great balance of performance and losses with the highest metrics/mAP50-95 of 0.60845 while being very competitive in precision 0.82845 and recall 0.73551 at a moderate training time of 2631.32 seconds. The balanced losses of YOLOv9C have the best precision of 0.84183 and metrics/mAP50 of 0.80211, though this has the longest training time at 3229.15 seconds. Thus, the YOLOv9 Medium is the best because it provides a very good trade-off between accuracy and training efficiency, with its moderate losses making it useful in practical applications.

Variants of YOLOv10 Model Analysis

A comparison of the performance of different YOLO v10 models, such as Large, Medium, Nano, and Small, indicates that YOLO v10 Small outperforms its peers in all major metrics: highest precision, 0.83886; recall, 0.69056; mAP50, 0.7979; and mAP50-95, 0.5992, while having the lowest training losses, box: 0.95965, class: 0.50903, DFL: 1.86929, and validation losses, which essentially translates to good generalization. Competitive YOLO v10 Nano-precision of 0.7352, recall of 0.68405, mAP50 of 0.73961, and mAP50-95 of 0.56517-offers a sweet trade-off between accuracy and efficiency but lags a bit behind v10 Small by the loss metric.

Contrary to this, YOLO v10 Large and Medium had to struggle with considerably higher losses, such as box loss: 2.08015 for Large and 2.09336 for Medium, while for detection metrics, they showed considerably lower results: where Large reached only 0.38219 precision, and Medium reached 0.44541, really underlining their inefficiency despite longer training up. Here, the smaller learning rate for YOLO v10 Small and Nano is 1.72e-05, which allows for stable convergence, whereas the larger one for Large and Medium is 0.00043. These results confirm YOLO v10 Small as the most effective and reliable model for object detection tasks, having combined high accuracy, efficient training, and strong generalization capabilities.

Variants of YOLOv11 Model Analysis

The medium variant, YOLO v11 Medium, performed better across most metrics in the YOLO v11 variants' performance evaluation-Large, Medium, Nano, and Small. The maximum precision was 0.78021, and it had the highest recall: 0.72312, mAP50: 0.7773, and mAP50-95: 0.59875 while being lower for training losses: box 0.48864, class: 0.29085, DFL: 0.96266 and validation losses: box: 1.05377, class: 1.02557, DFL: 1.40952. YOLO v11 Nano had comparable performance, slightly lower metrics: precision 0.77392, recall 0.73622, mAP50 0.76401 with slightly higher losses and, therefore, is also an efficient alternative. While large and small models of YOLO v11 were characterized by lower accuracy and higher losses, the smallest model reached the minimum value of precision-0.4425 and maximum value of train losses among these models. The lower learning rates of Medium and Nano, 1.72e-05, contributed to their superior stability and convergence. Thus, the most balanced and efficient model for accurate and quick object detection in this comparison is YOLO v11 Medium.

Variants of YOLO World Model Analysis

In the given comparison, the YOLO v8 Medium World v2 model was better, with the best metrics for both precision 0.84576 and mAP: mAP50: 0.81158 and mAP50-95: 0.61756, indicating that the model detects objects better than others. This model gave the lowest training losses box: 0.43245, class: 0.25392, DFL: 0.92908 and the lowest validation losses: box: 1.06479, class: 0.8396, DFL: 1.46001, reflecting in its robust optimization and generalization. While the YOLO v8 Small World v2 model showed close results in precision 0.79594 and recall 0.73762, its mAP metrics and losses were a bit worse. Models like YOLO v8 Large World and its v2 variant showed higher losses and poorer precision 0.36431–0.42983 and mAP metrics, which means they are not as good. The medium world v2 model has a learning rate consistency of 1.72e-05, hence stable and reinforcing the best performance for the most accurate object detection in an efficient manner in this dataset.

Rt dtr Model Analysis

The precision-recall curve of the RT DTR model showed high performance for the classes, with an average precision (mAP@0.5) of 0.741 across all classes. High precision values were recorded for specific behaviors such as "crouch chicken," "drinking," and "walking." On the other hand, some conditions were challenging for the model, such as "feeding" and "leg stretching," since they exhibited low precision. The F1-Confidence curve showed that at the confidence threshold of 0.677, the overall F1 score is 0.76, which means that the model is fairly balanced between precision and recall in the dataset. The training and validation losses showed gradual improvement; the metrics have stabilized, reflecting a well-optimized learning process. Overall, this model provides a very reliable base for automatic behavior analysis, with further possible improvements in those categories where it performed poorly.

Among all the models have been compared YOLOv3, YOLOv5 Medium, YOLOv8 Small, YOLOv9 Medium, YOLO v10 Small, YOLO v11 Medium, YOLO v8 Medium World v2 performed the best among their families. The mAP values alongside precision, recall and loss curves have been added below for all these models:

DISCUSSION

The study successfully analyzed chicken behavior using various YOLO object detection models, evaluating their effectiveness in real-time monitoring for poultry farming. Results indicate that while newer YOLO versions, such as YOLOv10 and YOLOv11, improve accuracy and speed, challenges remain in detecting occluded and small objects and adapting to varying environmental conditions. The research underscores the economic and agricultural significance of automated poultry monitoring, offering benefits such as early disease detection, stress management, and enhanced resource allocation. The findings suggest that integrating multi-modal data, improving model generalization, and refining detection architectures can further advance poultry behavior analysis. Future developments should focus on enhancing model robustness, incorporating self-attention mechanisms, and leveraging multi-camera fusion to address occlusion challenges, ultimately promoting efficiency and sustainability in large-scale poultry farming.

Figure 9. Analyzing multiple evaluation parameters

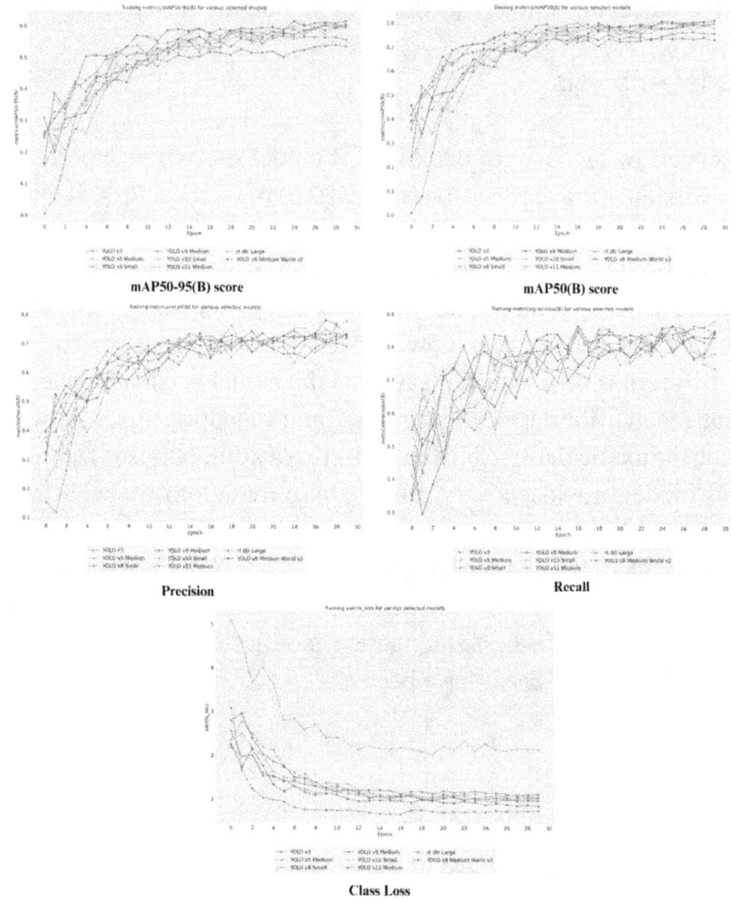

mAP50-95(B) score

mAP50(B) score

Precision

Recall

Class Loss

Figure 10. Confusion matrix analysis

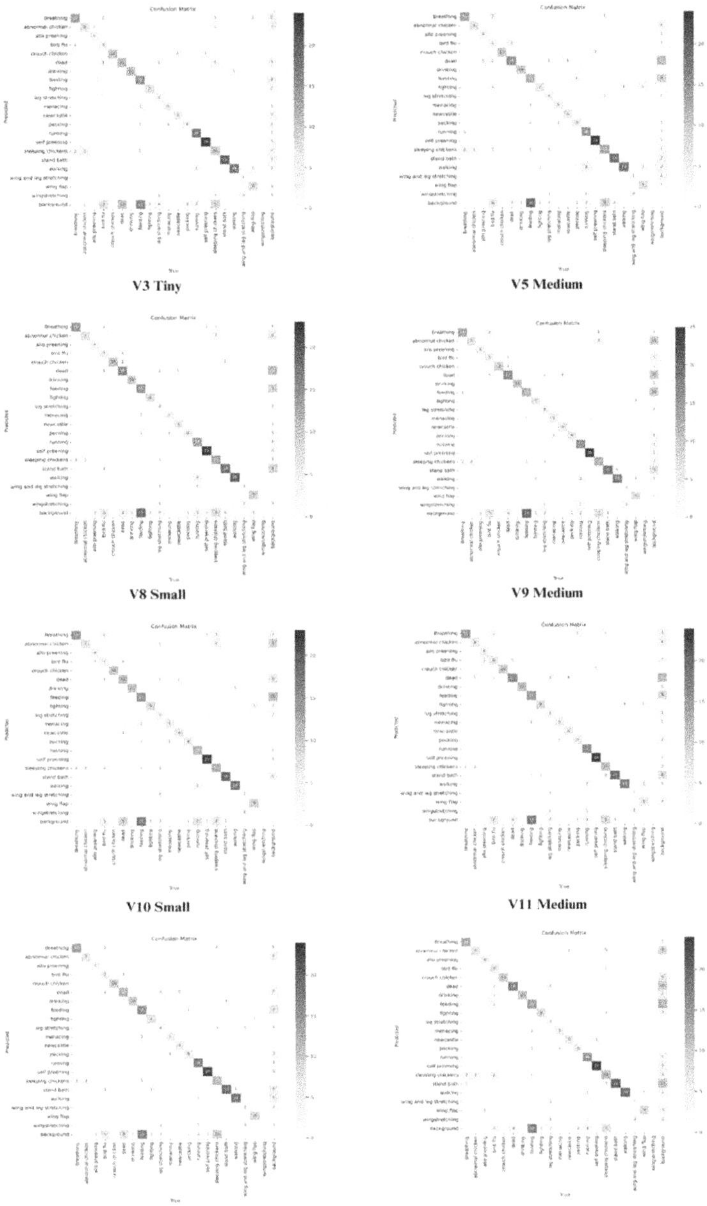

V3 Tiny **V5 Medium**

V8 Small **V9 Medium**

V10 Small **V11 Medium**

V8 Medium World V2 **rt dtr Large**

Figure 11. Some examples of the model's output overall

Future Directions

1. **Improved Model Generalization**
    - o To improve the current training data, more conditions have to be added such as different lighting, different breeds of the animal and more complexity in the environment.
    - o Explore the relationship between transformers and self-attention to enhance object detection in context of high movement rates.
2. **Occlusion and Crowded Scene Handling**
    - o Create architectures which are halfway between CNNs and attention-based models in order to enhance the detection performance at crowded scenes.
    - o Use of multi camera fusion can help to deal with the occlusion challenges in large poultry farms.
3. **Multi-Modal Data Fusion**

- o Make use of data from sensors such as thermal and environmental sensors as well as motion sensors, in conjunction with vision sensors.
- o Combine audio, thermal imagery, and inertial sensor data to provide more forms of learning for improving behavioral profiling.

4. **Real-Time Adaptability and Edge Computing**
   - o While the models can be used on cloud, the best models should be optimized for deployment on low power edge devices for on-site real-time monitoring.
   - o Introduce mechanisms for updating models by the received data about the environment and using these data for modifying the models.
5. **Automated Disease Detection and Early Warning Systems**
   - o Include more disease patterns to the datasets, for instance poor appetite, lameness and lack of energy.
   - o Develop algorithms that can identify behaviors that are peculiar and characteristic of diseases.

Therefore, considering these concerns, the next research can help to improve the poultry farming efficiencies that are also responsible, feasible, and sustainable. Thus the development of the next generation of fully automatic systems using artificial intelligence will remain critical in enhancing productivity, welfare, and food availability throughout the world.

CONCLUSION

In this paper, chicken behavior is analyzed comprehensively by employing popular scientific real-time object detection models, known as YOLO models. Therefore, the work presented in the research can be considered to have met its objectives through using deep learning approaches toward smart and automated CHicken behaviour monitoring – SCHoMo. The comparative cardinality of these quantitative YOLO models—ranging from YOLOv3 to YOLOv11 and further to YOLO World variants—has thus offered insights into the relative merits and drawbacks of the architectural configurations for detecting, categorising and analysing the behaviours of chickens. The studies show that although YOLO series including the YOLOv10 and YOLOv11, offer some striking enhancements in terms of accuracy and speed, the problem of occlusion, detecting small-sized objects, and changes in the environment is still unresolved.

Furthermore, the use of real-time monitoring technologies to poultry farming offers a chance for increased efficiency, decreased cost of operation and better outcomes on animal welfare. Automated behaviour recognition systems will go a long way to enhance early disease detection, stress & resource management in Large Scale Integrated Poultry Farms. More development of these models is needed to improve their predictive capacity with better datasets, optimization methods and the utilization of biological knowledge from a range of disciplines to provide better solutions for the management of poultry.

REFERENCES

Åkerfeldt, M. P., Gunnarsson, S., Bernes, G., & Blanco-Penedo, I. (2021). Health and welfare in organic livestock production systems—A systematic mapping of current knowledge. *Organic Agriculture, 11*(1), 105–132. DOI: 10.1007/s13165-020-00334-y

Bhuiyan, M. R., & Wree, P. (2023). Animal Behavior for Chicken Identification and Monitoring the Health Condition Using Computer Vision: A Systematic Review. *IEEE Access, 11*, 126601–126610. IEEE Access. DOI: 10.1109/ACCESS.2023.3331092

Bist, R. B., Bist, K., Poudel, S., Subedi, D., Yang, X., Paneru, B., Mani, S., Wang, D., & Chai, L. (2024). Sustainable poultry farming practices: A critical review of current strategies and future prospects. *Poultry Science, 103*(12), 104295. DOI: 10.1016/j.psj.2024.104295 PMID: 39312848

Bist, R. B., Subedi, S., Yang, X., & Chai, L. (2023). A Novel YOLOv6 Object Detector for Monitoring Piling Behavior of Cage-Free Laying Hens. *AgriEngineering, 5*(2), 2. Advance online publication. DOI: 10.3390/agriengineering5020056

Bobier, C. A. (2021). What Would the Virtuous Person Eat? The Case for Virtuous Omnivorism. *Journal of Agricultural & Environmental Ethics, 34*(3), 3. Advance online publication. DOI: 10.1007/s10806-021-09860-5

Bonnefous, C., Collin, A., Guilloteau, L. A., Guesdon, V., Filliat, C., Réhault-Godbert, S., Rodenburg, T. B., Tuyttens, F. A. M., Warin, L., Steenfeldt, S., Baldinger, L., Re, M., Ponzio, R., Zuliani, A., Venezia, P., Väre, M., Parrott, P., Walley, K., Niemi, J. K., & Leterrier, C. (2022). Welfare issues and potential solutions for laying hens in free range and organic production systems: A review based on literature and interviews. *Frontiers in Veterinary Science, 9*, 952922. Advance online publication. DOI: 10.3389/fvets.2022.952922 PMID: 35990274

Campbell, M., Miller, P., Díaz-Chito, K., Hong, X., McLaughlin, N., Parvinzamir, F., Martínez Del Rincón, J., & O'Connell, N. (2024). A computer vision approach to monitor activity in commercial broiler chickens using trajectory-based clustering analysis. *Computers and Electronics in Agriculture, 217*, 108591. DOI: 10.1016/j.compag.2023.108591

Comes, J. D. G., Poniman, M., van Oosten, L., Doets, K., de Cloe, S., Geertsema, C., & Pijlman, G. P. (2024). Infectious clone of a contemporary Tembusu virus and replicons expressing reporter genes or heterologous antigens from poultry viruses. *Biotechnology Journal, 19*(1), 2300254. DOI: 10.1002/biot.202300254 PMID: 37750498

Dey, S. K., Thakur, U., Upadhyay, A., & Kabir, Md. S., Sintheia, T. S., & Raju, V. (2024). Improving Guava Disease Detection using Inception Mobilenet. *2024 International Conference on Advances in Computing, Communication and Applied Informatics (ACCAI)*, 1–7. DOI: 10.1109/ACCAI61061.2024.10601750

Diwan, T., Anirudh, G., & Tembhurne, J. V. (2023). Object detection using YOLO: Challenges, architectural successors, datasets and applications. *Multimedia Tools and Applications*, *82*(6), 9243–9275. DOI: 10.1007/s11042-022-13644-y PMID: 35968414

Du, S., Zhang, B., Zhang, P., & Xiang, P. (2021). An Improved Bounding Box Regression Loss Function Based on CIOU Loss for Multi-scale Object Detection. *2021 IEEE 2nd International Conference on Pattern Recognition and Machine Learning (PRML)*, 92–98. DOI: 10.1109/PRML52754.2021.9520717

Farid, A., Hussain, F., Khan, K., Shahzad, M., Khan, U., & Mahmood, Z. (2023). A Fast and Accurate Real-Time Vehicle Detection Method Using Deep Learning for Unconstrained Environments. *Applied Sciences (Basel, Switzerland)*, *13*(5), 5. Advance online publication. DOI: 10.3390/app13053059

Gebeyehu, S. T. (2021, May 17). *Genetic mapping on local Swedish chicken breeds* [Second cycle, A2E]. SLU, Dept. of Animal Breeding and Genetics (until 231231). https://stud.epsilon.slu.se/16684/

George, D. A. S., & George, A. S. H. (2023). Optimizing Poultry Production Through Advanced Monitoring and Control Systems. *Partners Universal International Innovation Journal*, *1*(5), 5. Advance online publication. DOI: 10.5281/zenodo.10050352

Givens, D. I. (2024). Animal board invited review: Dietary transition from animal to plant-derived foods: Are there risks to health? *Animal*, *18*(9), 101263. DOI: 10.1016/j.animal.2024.101263 PMID: 39121724

Grandin, T., & Deesing, M. J. (2022). Genetics and animal welfare. In Grandin, T. (Ed.), *Genetics and the Behavior of Domestic Animals* (3rd ed., pp. 507–548). Academic Press., DOI: 10.1016/B978-0-323-85752-9.00013-5

Hurtik, P., Tomasiello, S., Hula, J., & Hynar, D. (2022). Binary cross-entropy with dynamical clipping. *Neural Computing & Applications*, *34*(14), 12029–12041. DOI: 10.1007/s00521-022-07091-x

Islam, M. T., & Ali, A. (2024). Sustainable green energy transition in Saudi Arabia: Characterizing policy framework, interrelations and future research directions. *Next Energy*, *5*, 100161. DOI: 10.1016/j.nxener.2024.100161

Jiang, K., Itoh, H., Oda, M., Okumura, T., Mori, Y., Misawa, M., Hayashi, T., Kudo, S.-E., & Mori, K. (2023). Gaussian affinity and GIoU-based loss for perforation detection and localization from colonoscopy videos. *International Journal of Computer Assisted Radiology and Surgery*, *18*(5), 795–805. DOI: 10.1007/s11548-022-02821-x PMID: 36913126

Kataria, S. (2023). *Robust Speaker Recognition using Perceptual and Adversarial Speech Enhancement* [Johns Hopkins University]. http://jhir.library.jhu.edu/handle/1774.2/68958

Kodaira, V., Siriani, A. L. R., Medeiros, H. P., De Moura, D. J., & Pereira, D. F. (2023). Assessment of Preference Behavior of Layer Hens under Different Light Colors and Temperature Environments in Long-Time Footage Using a Computer Vision System. *Animals (Basel)*, *13*(15), 15. Advance online publication. DOI: 10.3390/ani13152426 PMID: 37570237

Lawal, R. A., Martin, S. H., Vanmechelen, K., Vereijken, A., Silva, P., Al-Atiyat, R. M., Aljumaah, R. S., Mwacharo, J. M., Wu, D.-D., Zhang, Y.-P., Hocking, P. M., Smith, J., Wragg, D., & Hanotte, O. (2020). The wild species genome ancestry of domestic chickens. *BMC Biology*, *18*(1), 1. Advance online publication. DOI: 10.1186/s12915-020-0738-1 PMID: 32050971

Li, F., Yan, H., & Shi, L. (2024). Multi-scale coupled attention for visual object detection. *Scientific Reports*, *14*(1), 11191. DOI: 10.1038/s41598-024-60897-8 PMID: 38755252

Li, G., Hui, X., Lin, F., & Zhao, Y. (2020). Developing and Evaluating Poultry Preening Behavior Detectors via Mask Region-Based Convolutional Neural Network. *Animals (Basel)*, *10*(10), 10. Advance online publication. DOI: 10.3390/ani10101762 PMID: 32998372

Li, L., Wang, Z., Hou, W., Zhou, Z., & Xue, H. (2024). A Recognition Method for Aggressive Chicken Behavior Based on Machine Learning. *IEEE Access, 12*, 24762–24775. IEEE Access. DOI: 10.1109/ACCESS.2024.3365552

Li, X., Wang, W., Wu, L., Chen, S., Hu, X., Li, J., Tang, J., & Yang, J. (2020). Generalized Focal Loss: Learning Qualified and Distributed Bounding Boxes for Dense Object Detection. *Advances in Neural Information Processing Systems, 33*, 21002–21012. https://proceedings.neurips.cc/paper/2020/hash/f0bda020d2470f2e74990a07a607ebd9-Abstract.html

Ma, N., Wu, Y., Bo, Y., & Yan, H. (2024). Chili Pepper Object Detection Method Based on Improved YOLOv8n. *Plants*, *13*(17), 17. Advance online publication. DOI: 10.3390/plants13172402 PMID: 39273886

Mahaur, B., Mishra, K. K., & Kumar, A. (2023). An improved lightweight small object detection framework applied to real-time autonomous driving. *Expert Systems with Applications*, *234*, 121036. DOI: 10.1016/j.eswa.2023.121036

Naidu, G., Zuva, T., & Sibanda, E. M. (2023). A Review of Evaluation Metrics in Machine Learning Algorithms. In Silhavy, R., & Silhavy, P. (Eds.), *Artificial Intelligence Application in Networks and Systems* (pp. 15–25). Springer International Publishing., DOI: 10.1007/978-3-031-35314-7_2

Natij, Y., El Karch, H., Maafiri, A., & Mezouari, A. (2024). Evaluating the Performance of YOLO Object Detectors for Plant Disease Detection. *2024 11th International Conference on Wireless Networks and Mobile Communications (WINCOM)*, 1–6. DOI: 10.1109/WINCOM62286.2024.10656918

Neethirajan, S. (2024). *Rethinking Poultry Welfare—Multifactorial and Digital Approaches to Enhance Animal Well-Being* (SSRN Scholarly Paper 5034840). Social Science Research Network. DOI: 10.2139/ssrn.5034840

Papageorgiou, M., Goliomytis, M., Tzamaloukas, O., Miltiadou, D., & Simitzis, P. (2023). Positive Welfare Indicators and Their Association with Sustainable Management Systems in Poultry. *Sustainability (Basel)*, *15*(14), 14. Advance online publication. DOI: 10.3390/su151410890

Paul, A., & Machavaram, R. (2024). *Advancements in Capsicum Detection in Thermal Greenhouse Imaging: Transitioning from Yolov10 to Sam-Dino* (SSRN Scholarly Paper 4917658). Social Science Research Network. https://papers.ssrn.com/abstract=4917658 DOI: 10.2139/ssrn.4917658

Pius, L. O., Strausz, P., & Kusza, S. (2021). Overview of Poultry Management as a Key Factor for Solving Food and Nutritional Security with a Special Focus on Chicken Breeding in East African Countries. *Biology (Basel)*, *10*(8), 8. Advance online publication. DOI: 10.3390/biology10080810 PMID: 34440042

Salihu, S., Jauro, S. S., & Salihu, M. N. (n.d.-a). *Enhancing Poultry Welfare and Production through Advanced Chicken Detection, Tracking and Counting: A Deep Learning Approach.*

Salihu, S., Jauro, S. S., & Salihu, M. N. (n.d.-b). *Enhancing Poultry Welfare and Production through Advanced Chicken Detection, Tracking and Counting: A Deep Learning Approach.*

Sapkota, R., Meng, Z., Churuvija, M., Du, X., Ma, Z., & Karkee, M. (2024). *Comprehensive Performance Evaluation of YOLO11, YOLOv10, YOLOv9 and YOLOv8 on Detecting and Counting Fruitlet in Complex Orchard Environments* (arXiv:2407.12040). arXiv. https://doi.org//arXiv.2407.12040DOI: 10.48550

Seeger, R. M., Hagerman, A. D., Johnson, K. K., Pendell, D. L., & Marsh, T. L. (2021). When poultry take a sick leave: Response costs for the 2014–2015 highly pathogenic avian influenza epidemic in the USA. *Food Policy*, *102*, 102068. DOI: 10.1016/j.foodpol.2021.102068

Sozzi, M., Pillan, G., Ciarelli, C., Marinello, F., Pirrone, F., Bordignon, F., Bordignon, A., Xiccato, G., & Trocino, A. (2023). Measuring Comfort Behaviours in Laying Hens Using Deep-Learning Tools. *Animals (Basel)*, *13*(1), 1. Advance online publication. DOI: 10.3390/ani13010033 PMID: 36611643

Strauss, E. D., Curley, J. P., Shizuka, D., & Hobson, E. A. (2022). The centennial of the pecking order: Current state and future prospects for the study of dominance hierarchies. *Philosophical Transactions of the Royal Society B: Biological Sciences, 377*(1845). *Philosophical Transactions of the Royal Society of London. Series B, Biological Sciences*, *1845*(1845), 20200432. Advance online publication. DOI: 10.1098/rstb.2020.0432 PMID: 35000437

Su, S., Tang, Z., & Zhu, Y. (2024). High-density foreground object detection in optical remote sensing images via semantic fusion and box alignment. *The Visual Computer*, *40*(6), 4355–4371. DOI: 10.1007/s00371-023-03086-4

Tanim, S. A., Shrestha, T. E., Tanvir, K., & Kabir, Md. S., Mridha, M. F., & Haq, M. K. (2024). Single-Level Fusion for Enhancing Meat Quality Classification with Explainable AI. *2024 IEEE International Conference on Computing, Applications and Systems (COMPAS)*, 1–6. DOI: 10.1109/COMPAS60761.2024.10796775

Tanvir, K., & Kabir, Md. S., Anik, Z. H., Hasan, Md. R., Tushar, MD. T., & Rahman, R. B. (2024). Forest Fire Detection Using Ensemble Deep Learning Model With XAi. *2024 IEEE International Conference on Computing, Applications and Systems (COMPAS)*, 1–6. DOI: 10.1109/COMPAS60761.2024.10796582

Teterja, D., Garcia-Rodriguez, J., Azorin-Lopez, J., Sebastian-Gonzalez, E., Nedić, D., Leković, D., Knežević, P., Drajić, D., & Vukobratović, D. (2024). A Video Mosaicing-Based Sensing Method for Chicken Behavior Recognition on Edge Computing Devices. *Sensors (Basel)*, *24*(11), 11. Advance online publication. DOI: 10.3390/s24113409 PMID: 38894200

Tibbetts, E. A., Pardo-Sanchez, J., & Weise, C. (2022). The establishment and maintenance of dominance hierarchies. *Philosophical Transactions of the Royal Society B: Biological Sciences, 377*(1845). *Philosophical Transactions of the Royal Society of London. Series B, Biological Sciences, 1845*(1845), 20200450. Advance online publication. DOI: 10.1098/rstb.2020.0450 PMID: 35000449

Triyanto, W. A., Adi, K., & Suseno, J. E. (2023). Detection and Tracking of Broiler Flock Movements in The Chicken Coop using YOLO. *E3S Web of Conferences, 448*, 02064. DOI: 10.1051/e3sconf/202344802064

Wang, F., Cui, J., Xiong, Y., & Lu, H. (2023). Application of deep learning methods in behavior recognition of laying hens. *Frontiers in Physics (Lausanne), 11*, 1139976. Advance online publication. DOI: 10.3389/fphy.2023.1139976

Zeng, N., Wu, P., Wang, Z., Li, H., Liu, W., & Liu, X. (2022). A Small-Sized Object Detection Oriented Multi-Scale Feature Fusion Approach With Application to Defect Detection. *IEEE Transactions on Instrumentation and Measurement, 71*, 1–14. *IEEE Transactions on Instrumentation and Measurement*. Advance online publication. DOI: 10.1109/TIM.2022.3153997

Zhang, G., Kenta, N., & Kleijn, W. B. (2021). *Extending AdamW by Leveraging Its Second Moment and Magnitude* (arXiv:2112.06125). arXiv. https://doi.org//arXiv.2112.06125DOI: 10.48550

Zheng, Z., Wang, P., Liu, W., Li, J., Ye, R., & Ren, D. (2020). Distance-IoU Loss: Faster and Better Learning for Bounding Box Regression. *Proceedings of the AAAI Conference on Artificial Intelligence, 34*(07), 07. Advance online publication. DOI: 10.1609/aaai.v34i07.6999

Chapter 8
Will Quantum Communication and Processing Accelerate the Progress of Automated and Sustainable Farming?

Pirunthavi Wijikumar
https://orcid.org/0009-0001-9818-6880
University of Vavuniya, Sri Lanka

Jahan Hassan
https://orcid.org/0000-0002-0939-2106
Central Queensland University, Australia

Anwaar Ul-Haq
https://orcid.org/0000-0002-5145-7276
Central Queensland University, Australia

Biplob Rakshit Ray
https://orcid.org/0000-0002-3016-1695
Central Queensland University, Australia

Rukshani Puvanendran
https://orcid.org/0000-0003-3481-5294
University of Vavuniya, Sri Lanka

ABSTRACT

Quantum computing is revolutionizing computation and communication, with its

DOI: 10.4018/979-8-3373-0020-7.ch008

Copyright © 2025, IGI Global Scientific Publishing. Copying or distributing in print or electronic forms without written permission of IGI Global Scientific Publishing is prohibited.

integration into drones transforming agrotech and other industries. The Internet of Quantum Drones (IoQDs) enables ultra-secure communication, rapid data processing, and precise navigation through quantum computing, quantum communication, and quantum sensors. This integration tackles challenges in intelligent farming, such as real-time decision-making, ecological complexity, and efficient data management. IoQDs support applications in precision agriculture, disaster management, environmental monitoring, and defense using quantum-based image processing and machine learning. This chapter explores advancements, case studies, and IoQD integration with 6G and satellites, ensuring low-latency global connectivity. Challenges—scalability, cost, hardware downsizing, and regulations are analyzed alongside potential solutions. A case study on IoQD-based weed detection demonstrates its agricultural potential. Future directions highlight quantum technologies' transformative role in drone networks across industries.

INTRODUCTION

In recent years, many academic studies and industry innovations have focused on drones, commonly called unmanned aerial vehicles (UAVs) (Ghamari, Rangel, Mehrubeoglu, Tewolde, & Sherratt, 2022). The UAV system is very useful to industry application as it can cover a large geographical (like broadacre) area to achieve various objectives, like collecting data and spraying herbicides, in a short amount of time. Further, drones are capable of live streaming, real-time video capture, and object carrying in addition to flying ("Security analysis of drones systems: Attacks, limitations, and recommendations", 2020). Some examples of drone applications are military and rescue operations (Ghamari et al., 2022), (Al-Adhami, Hasoun, Gbashi, & Hassan, 2022), (Ko et al., 2021), (Zolanvari, Jain, & Salman, 2020), precise agricultural monitoring (Gopal Dutta, 2020), (Yashika Gupta, 2022), (Wadod & Mohammed, 2023), biodiversity (Barnas et al., 2020), (Millner et al., 2023), (Behjati et al., 2021), environmental and natural disaster monitoring (Dambal, Mohadikar, Kumbhar, & Guvenc, 2019), (Jabbar et al., 2024), and road and safety maintenance (Mustafa & Alizadeh, 2024), (Alsamhi, Ma, Ansari, & Gupta, 2019), (Hemmati, Zarei, & Souri, 2023), (Tiurlikova, Stepanov, & Mikhaylov, 2019).

One of key aspect to use UAV's capability is to ensure fast and reliable drone communication to transfer data that takes place between drone-to-drone (D2D), drone-to-base station (D2BS), drone-to-network (D2N), and drone-to-satellite (D2S) over the communication channels (P. P. Kumar, Kumar, Muthu, Rajendirane, & Samuel, 2022). The other key aspects are the processing of drone data within feasible timeline. Current drone networks face significant limitations in processing speed and efficiency due to the constraints of computing technologies. These limitations often result in delays when processing complex algorithms and large datasets, creating computational bottlenecks in real-time decision-making and data analysis (Haider et al., 2022).

To cover all aspect of application scenarious, the drone communication needs to: internet and intranet based. Drones that connect via the internet utilize the Internet of Things (IoT) to create the Internet of Drones (IoD), which enables data transfer and control without the need for human interaction. While internet-based communication offers rapid data speeds and global access, it is not practical or available in many working scenarios, including remote areas, crises, wildlife monitoring and military missions (Khan, Jhanjhi, Brohi, & Nayyar, 2020). An efficient deployment mechanism is required for this type of application in order to establish communication between the UAVs and the base station (Shafiee et al., 2021). The intranet drone communication addresses the gap by transferring data using a local network without the need for an internet connection.

Multiple wireless channels and network protocols/technologies used by commercial drone communications systems like Radiofrequency (RF), Bluetooth, cellular, Wi-Fi, ZigBee, LoRa, and satellite (A. Sharma et al., 2020), (Khan et al., 2020), (Arya, Yang, & Wang, 2023) to enable UAVs to connect to base stations (Ghazali, Teoh, & Rahiman, 2021). The use of a particular type of communication and processing technologies depends on criteria such as cost, payload weight, compatibility, speed, power requirements, and bandwidth (A. Sharma et al., 2020).

The effectiveness and dependability of drone operations are greatly impacted by the numerous drawbacks of employing communication modalities for drone communication (Maher et al., 2020). Current drone networks have limitations such as processing speed and efficiency due to the constraints of classical computing. These limitations often result in delays when processing complex algorithms and large datasets, creating computational bottlenecks in real-time decision-making and data analysis (Haider et al., 2022). In terms of encryption and security, current drone networks rely on classical encryption methods that are increasingly vulnerable to sophisticated cyber-attacks, data breaches, and hacking. Data transmission and communication are often limited by classical data transmission rates and bandwidth constraints, which can lead to data loss or delays due to network congestion or interference (Fouda, Ibrahim, Güvenç, & Ghosh, 2019).

Furthermore, managing and coordinating large fleets of drones is rather challenging due to the limitations of classical computing infrastructure and the complexity of simultaneously coordinating several autonomous systems (Chiti, Picchi, & Pierucci, 2022). Such challenges include scalability, communication, data processing, real-time decision-making, security, and energy efficiency. While contemporary 5G or 6G technology offers large device connectivity, low latency, and high data rates (Maher et al., 2020), it necessitates the broad installation of 5G base stations and internet connectivity, which can be difficult in isolated or rural locations (Khan et al., 2020).

Leveraging quantum computing and communication methods, such as quantum entanglement, drone systems and networks can achieve instantaneous data transmission over long distances, thereby minimizing data loss and transmission delays and ensuring more reliable communication between drones (Liu et al., 2020). Additionally, the ability of quantum computing to process exponentially larger datasets at much faster speeds significantly improves real-time data analysis and decision-making capabilities, leading to more efficient overall drone operations. Furthermore, quantum algorithms excel in solving complex optimization problems, enhancing adaptive path planning and resource allocation within these networks (A. Kumar, Augusto de Jesus Pacheco, Kaushik, & Rodrigues, 2022). Quantum computing also introduces quantum cryptography, which provides near-unbreakable encryption and strengthens security protocols, making unauthorized access and cyberattacks more challenging (Ralegankar et al., 2022). Furthermore, it stated that, in terms of navigation and sensing, quantum computing offers improved scalability by effectively managing large datasets and complex computations, which facilitates better coordination across extensive drone networks. Drone operations will be redefined and new possibilities will become available as this technology advances.

In this study, we investigate the revolutionary potential of quantum computing and quantum communication in developing drone networks, with a focus on resolving restrictions in classical drone systems related to communication and processing efficiency (Chen, 2021), (Al-Hraishawi, Rehman, Razavi, & Chatzinotas, 2024). By examining several quantum-based methodologies, we hope to provide a thorough understanding of how the Internet of Quantum Drones (IoQDs) might improve drone operations in a variety of applications by providing fast, secure, and efficient data processing and communication. Our research not only focuses on the gaps and limitations in standard UAV systems, but also emphasizes the advantages of quantum technologies in overcoming processing challenges and enabling instantaneous communication over vast distances.

Our contribution include:

- A comprehensive review of the most recent advances in quantum-enabled drone networks, including communication protocols, processing limits, and potential applications.
- A detailed comparison of classical and quantum-based drone systems, highlights quantum technologies' advantages in terms of communication and processing efficiency.
- A discussion on future research directions, challenges, and practical implications of deploying IoQDs.

The structure of the paper is as follows: Section 2 presents the study's background, including a case study on current drone technologies in smart farming, and a brief introduction to quantum technology. Section 3 details the methodology adopted for this survey. Section 4 explores key findings on drone communication and processing from existing literature. Section 5 delves into the applications of IoQDs. Section 6 focuses on the advancements of IoQDs. Section 7 proposes a comprehensive case study of IoQD based weed detection in smart farming. Section 8 discusses the challenges and limitations associated with IoQDs while outlining potential future directions. Finally, the study concludes in Section 9 by summarizing the findings.

BACKGROUND

Current Drone Technologies in Smart Farming: Case Study

The drone-based agricultural weed detection system uses contemporary technology for communication to ensure seamless operations. Recent improvements in the drone industry, such as high-resolution imaging sensors, autonomous navigation systems, and sophisticated communication protocols, have drastically changed methods for farming (Zhang., 2024). The Figure 1 demonstrates a drone system which has been currently used for precise weed detection and management in smart agriculture. The system includes several technologies in the subsequent stages:

- Autonomous Flight Path Planning and Navigation: The latest drones have been equipped with GPS-based navigation systems and AI-driven flight planning techniques that improve flight paths, providing complete agricultural coverage while avoiding overlap. LiDAR sensors and infrared imaging are used to operate obstacle identification and avoidance systems, which improve autonomous navigation.
- Data Acquisition with High Resolution Imaging: During its flight, drone employs multispectral, hyperspectral, and RGB imaging devices to acquire pictures with excellent resolution, allowing for accurate identification of weed locations.
- Data Transmission: Data is then transmitted to a base station via one of several communication technologies, including 4G/5G cellular networks for fast and reliable data transfer in connected areas, RF for dependable communication in non-cellular regions, and LoRa and satellite communication for operations in remote or isolated areas.

- Data Processing with Machine Learning Techniques: Once the data arrives at the base station, it is processed using machine learning techniques to produce thorough weed maps.
- Targeted Spraying: Weed maps, which contain geolocation data pinpointing weed-infested areas, are saved locally and then sent to the spraying drone. The weed coordinates are then transmitted to the spraying drone using the same data communication methods. Finally, the spraying drone uses geolocation data from the base station for determining the best flight path, ensuring precise navigation to each recognized weed patch. Advanced sensors allow the drone to identify and avoid obstructions while hovering precisely over weed areas. Herbicides are sprayed in a targeted and efficient manner, which reduces chemical waste and environmental damage. Throughout the spraying process, communication technologies including 4G/5G, RF, and satellite connectivity enable real-time updates, ensuring precision and agility during operations.
- Real-time Monitoring: The entire system is fully connected with IoT systems, allowing for real-time monitoring of spraying activities.

Figure 1. Weed detection framework for smart farming using current drone technologies

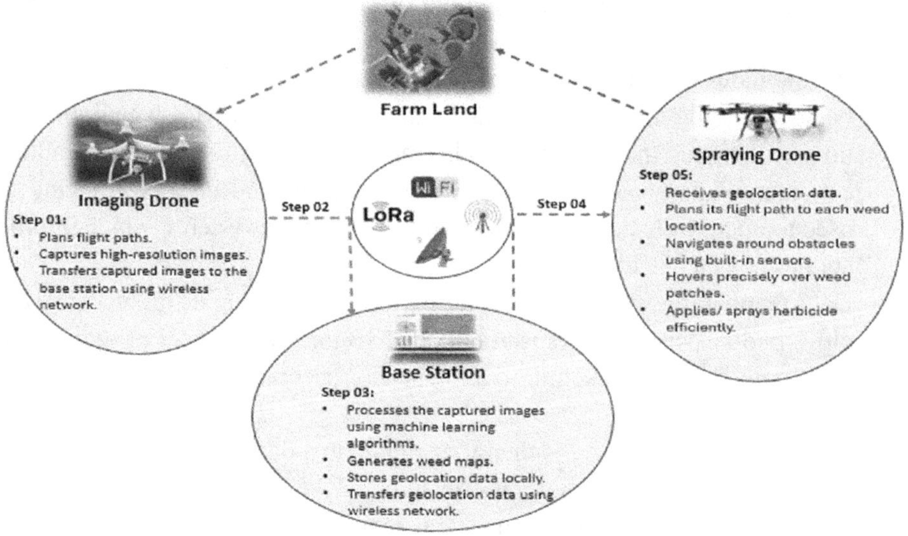

Introduction to Quantum Technology

Quantum communication is fundamentally different from traditional communication in the way information is represented, transmitted, and processed. While classical communication depends on bits (binary 0s and 1s), quantum communication uses qubits, which can exist in superpositions of states (Hasan, Chowdhury, Saiam, & Jang, 2023). To illustrate it further, we can say that in classical communication, information is represented using bits that take values of either 0 or 1, mathematically belonging to the set $\{0,1\}$. In contrast, quantum communication uses Qubits, which are represented as vectors in a two-dimensional complex vector space. A Qubit $|\psi\rangle$ can be expressed as

$|\psi\rangle = \alpha|0\rangle + \beta|1\rangle$

where $|0\rangle = (1)$

0 and $|1\rangle = (0)$

1 are the basis vectors, and α and β are complex numbers that satisfy the normalization condition $|\alpha|2 + |\beta|2 = 1$. This normalization ensures that the Qubit remains in a valid quantum state. Unlike classical bits, which are strictly either 0 or 1, Qubits can exist in a superposition, meaning they can be simultaneously be $|0\rangle$ and $|1\rangle$. For example, a common superposition state is created using the Hadamard gate, yielding $|\psi\rangle = 1 \sqrt{2} (|0\rangle + |1\rangle)$, demonstrating the unique capability of qubits to represent multiple possibilities at the same time.

In its essence, quantum technology is able to process information in ways through the use of microscopic physical systems including photons, atoms, ions, and superconductors that are not conceivable in the traditional macroscopic world (Cheng, Qi, Sun, Liu, & Xie, 2024). Quantum computers have the potential to solve complicated problems exponentially more rapidly than classical computers by utilizing quantum bits, or qubits (A. Kumar et al., 2022). The field of quantum technology, which is still in its early stages of development, is based on the application of specific quantum systems' properties, such as quantum entanglement, quantum superposition, and quantum tunnelling—for practical purposes (Sidhu et al., 2021).

In many different industries, quantum computing has the potential to solve large-scale computing challenges more quickly and affordably (Kantsepolsky, Aviv, Weitzfeld, & Bordo, 2023). Drone networks can benefit from the cutting-edge capabilities of quantum computing combined with the adaptability and dynamism of UAVs through the integration of quantum communication. Quantum communication is another form that represents a paradigm shift in safe information transfer, offering previously unattainable degrees of security.

The Internet of Quantum Drones (IoQDs) and drone-to-satellite based applications depends on reliably connectivity and heavy data processing in real time. Therefore, quantum computation can excel and address shortcomings of drone-based application using traditional communication and computation technologies (Chiti et al., 2022), (Ali et al., 2023). Integrating quantum computing into drone communication offers numerous advantages, including enhanced data processing, secure communication, efficient resource management, improved machine learning models, instantaneous data transfer, advanced simulation capabilities, and better scalability (A. Kumar et al., 2022). For instance, drones equipped with quantum sensors can be useful for a wide range of tasks, such as spectrum measurements and surface albedo assessment, human-made structures (such as tunnels or underground residences), and delivery-based systems collision detection (A. Kumar et al., 2021). Quantum technology represents a significant step forward in the evolution of drone technology, promising to overcome many of the limitations of current systems and unlocking new potential for innovation and performance.

The use of quantum technology into smart farming systems has enormous potential for tackling numerous critical difficulties related to modern farming. IoQDs provide for more effective, reliable, and extensible smart farming systems. Major applications for quantum technology in smart farming consist of:

- Secure Communication: Quantum key distribution (QKD) provides impenetrable security for transferring critical data related to agriculture between drones and base stations, assuring data confidentiality and safety (Ralegankar et al., 2022).
- Advanced Data Processing: The use of quantum computing speeds up big data processing for crop health monitoring, soil evaluation, and yield prediction, which are critical for implementing precision agriculture (A. Kumar et al., 2022), (C. Maraveas et al., 2024).
- Optimization of Resources: Quantum techniques enable more precise utilization of resources models for distribution of water, spraying process of pesticide, and application of fertilizer, minimizing wastage while increasing efficiency, sustainability and production (C. Maraveas et al., 2024), (Mukhamedieva et al., 2024).
- Real-Time Decision Making: Quantum-enhanced systems can provide immediate data into the health of crops, outbreaks of diseases, and watering needs, allowing farmers to make real-time decisions (C. Maraveas et al., 2024).

The combination of IoT technology and quantum technology can significantly improve the efficiency of autonomous smart farming systems, rendering them more reliable, expandable, and adaptable to changing environments. Quantum technologies are a significant step ahead in the growth of drone-based smart farming, opening up new opportunities for advancement and sustainability in precise farming.

METHODOLOGY USED FOR THE SURVEY

This section details the methodologies that underpin the survey presented in this paper. Systematic literature review methodology was adapted for gathering the articles and thematic grouping for their analysis in this survey. Figure 2 represents how the review was undertaken from generating themes to presenting the final findings. For this survey, articles of recent six years (2018 to 2024) were considered. In step one, thematic method was used to identify main survey topics and their associated keywords as presented in a mind map in Figure 3. Three main themes were identified as listed below and illustrated in the figure.

- Standard drone technologies and applications
- Current quantum technologies
- Advancements of IoQDs and applications

Figure 2. Flow chart showing the process of systematic literature review on IoQD

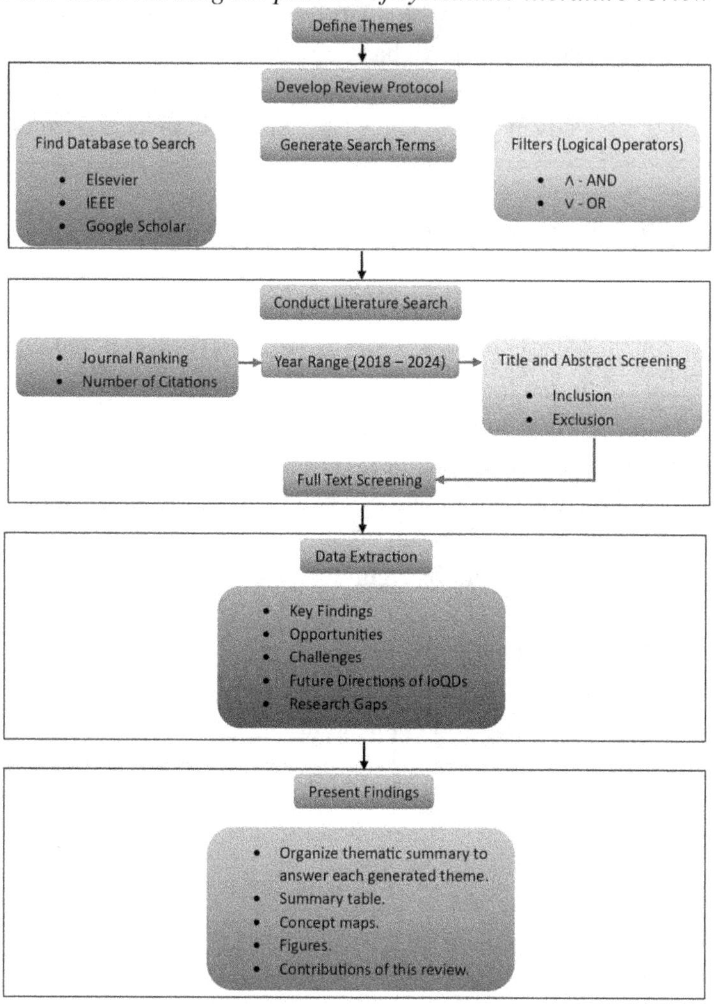

Based on the above themes, keywords were generated, related to the survey's aims and objectives, which are opportunities and challenges of IoQD systems and application, as illustrated in Figure 3.

Figure 3. The mind map underpinning the literature search of the survey

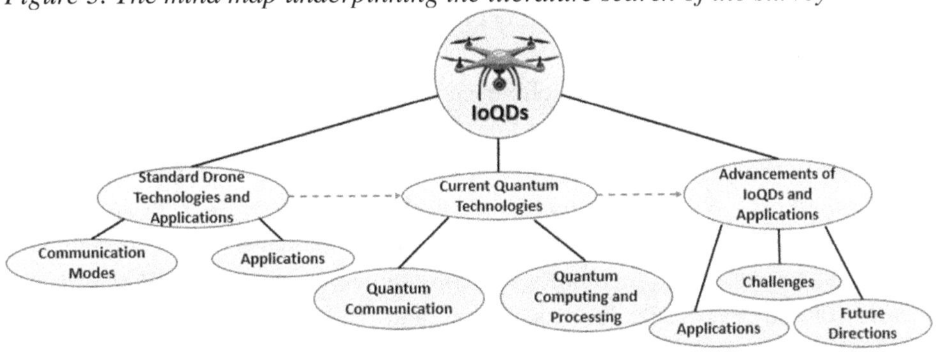

These keywords or terms of Figure 3 were searched using logical operators ("AND" and "OR") in three databases (Elsevier, IEEE and Google Scholar). Relevant articles were then exported to Microsoft Excel for additional processing and filtering. The acquired findings underwent further review for relevancy utilizing journal ranking, titles, abstracts, and citation counts. The publications that had not pertained to the focus of the current survey have been excluded. Specifically, papers with a wide range of topics and relevance to general quantum computing, or optimization of the drone and AI had been excluded. An analysis had been conducted on a final data set of publications.

One paper may appear in more than one key research area due to the similarity in some of the keywords/phrases used in each key search area. As a result, a duplicate check was carried out to exclude similar papers across the major research fields. The appropriate literature that was gathered for each important study area was thoroughly examined to extract relevant information such as the main contribution of each paper to knowledge, the research methodology, the problem that was solved, and its limitations and gaps. The extracted information then critically analyzed and thematically organized before marking in-depth and critical analysis in each theme to report finding of this survey.

LITERATURE REVIEW

Current State of Drone Communication and Processing

The field of drone networks has seen significant advancements in recent years, driven by improvements in hardware, software, and wireless communication technologies. Modern drones are equipped with advanced sensors, high-resolution cameras, GPS, and various other payloads, enabling a wide range of applications (Ralegankar et al., 2022). The versatility of drone networks has led to their adoption in various sectors.

The integration of artificial intelligence (AI) and machine learning (ML) into drone networks has led to increased levels of autonomy (A. Kumar et al., 2021). AI-powered drones can perform tasks such as obstacle avoidance, target recognition, and autonomous navigation without constant human intervention. Despite these advancements, current drone networks face several challenges. Security remains a critical concern, with vulnerabilities in communication channels posing risks of interception and unauthorized access. Latency and bandwidth limitations can impact real-time decision-making and data transmission, particularly in densely populated or remote areas. Additionally, the computational power of onboard systems is often constrained, affecting the drones' ability to process large volumes of data and perform complex tasks autonomously.

These technologies promise to enhance security, increase data processing capabilities, and improve overall network efficiency, but their practical implementation and scalability remain areas of ongoing research and development. Scaling drone networks to manage large fleets remains a significant challenge (A. Kumar et al., 2022). Current systems can handle small to medium-sized fleets effectively, but large-scale operations are often hampered by computational and communication limitations (Chiti et al., 2022). Despite these advancements, fully autonomous operations are still limited by the computational capabilities of onboard processors and the complexity of real-world environments (B. Wang et al., 2018). The limited energy capacity of batteries restricts f light time and operational range, necessitating frequent recharges or battery swaps. Advances in battery technology and energy-efficient propulsion systems are being explored to address this issue, but significant improvements are still needed to support extended missions (Ko et al., 2021). Overall, while current drone networks have made significant strides in functionality and application, ongoing advancements are necessary to overcome existing limitations and fully realize their potential in future technological landscapes.

Quantum Technology

Quantum Communication

Rapid developments in quantum communication technologies, which offer unprecedented levels of information transmission efficiency and security, define the quantum era. These technologies are anticipated to completely transform computers, secure communication, and a host of other fields as they develop and become more integrated with current systems. Space overcomes the range constraint of existing ground-based networks, enabling quantum communication applications at global scales (Sidhu et al., 2021).

Quantum communication would significantly enhance the security of drone networks through QKD to create secure communication channels by exchanging encryption keys through QKD's specialized laser and photon detector technology (NASA, 2020). Any attempt to eavesdrop on the communication would be detected, as measuring a quantum system disturbs it (Conrad et al., 2023). The integration of QKD with classical communication networks has been a significant leap forward (Chiti et al., 2022). These hybrid systems offer a practical approach to deploying quantum-secure communication within existing infrastructure. These networks may carry out difficult tasks more securely and reliably by creating new opportunities for sensitive missions such as military operations (Ishveena, 2022), surveillance, and delivery of confidential information applications (A. Kumar et al., 2022), (Hill et al., 2017).

Moreover, drone technology could be revolutionized by quantum entanglement, which offers immediate computation and communication, enhanced coordination, enhanced navigation, secure communication, extremely sensitive detection, and advanced processing capabilities (Cheng et al., 2024), (Chiti et al., 2022), (Al-Hraishawi et al., 2024). With their unprecedented performance and reliability, entangled drones have an array of purposes in the environmental, commercial, and military domains.

Leveraging quantum entanglement could potentially allow for instantaneous communication over long distances (Ishveena, 2022). Each user receives one photon of entangled photons for each pair of particles; a loss of entanglement between the photons would notify the users of any eavesdropping. Using optical fibres is one of the most popular ways to transmit quantum-encrypted messages. However, a significant portion of photons in f ibres scatter before arriving at their intended destination. If quantum information travels through the atmosphere, more photons will survive (Michael, 2021). Although practical applications for instantaneous communication are still theoretical, advancements in this f ield could revolutionize how drones coordinate and share data. Future entangled drone networks are becoming closer, even though there are still obstacles to overcome in terms of technological development, scalability, and integration. These obstacles are being met by continuous breakthroughs in quantum research and technology (A. Kumar et al., 2022).

Quantum Computing and Data Processing

Quantum technologies are transforming drone systems by providing unprecedented computing power. Quantum computing accelerates difficult optimization issues including f light path planning, collision avoidance, and resource allocation (Khoshnoud, Esat, de Silva, & Quadrelli, 2019), (Chen, 2021). Quantum computers can process massive amounts of data at unprecedented speeds by using qubits, which can be in several states simultaneously (A. Kumar et al., 2022). Google revealed "Willow," a quantum computing chip that is said to solve problems in five minutes that would take standard supercomputers ten septillion years (Chris, 2024). This capacity transforms high-performance computing (HPC) operations, including real-time analysis of massive datasets, making it critical for applications like disaster management, precision agriculture, and wildlife monitoring (A. Kumar et al., 2022), (Chiti et al., 2022).

Drones may utilize quantum optimization techniques to improve critical operational factors including altitude, trajectory, and energy efficiency, leading to longer flight durations and increased reliability (Mozaffari, 2018), (Rahmadya, 2021). Furthermore, quantum computing makes it easier to develop resource allocation algorithms that can effectively manage drone networks' processing and communication resources. This is especially useful for quantum-enhanced swarming activities, where drones must communicate securely and make coordinated choices in real-time (Chiti et al., 2022), (Ali et al., 2023). Quantum image processing allows drones to acquire, process, and analyze high-resolution imagery more accurately, leading to advanced applications in environmental monitoring and smart agriculture (A. Kumar et al., 2021).

Quantum machine learning methods improve autonomous decision-making in drones, allowing for real-time analysis and secure data transfer between drones and ground control stations (GCS) (Chiti et al., 2022), (Ali et al., 2023). Quantum-enhanced machine learning enables drones to adapt to dynamic settings with better precision (Caruso, Chessa, Escolar, Barba, & López, 2021). Integrating quantum computing with drone swarms can optimize flight trajectories, reduce energy consumption, and enhance mission efficiency (Chiti et al., 2022). The combination of quantum computing and drone technology improves communication and security. Quantum-safe cryptographic algorithms enable secure data transfer, while satellite-based quantum communication systems enhance the reach and robustness of drone networks (Chen, 2021), (Chiti et al., 2022). These improvements pave the way for the Internet of Quantum Drones (IoQDs), a network of drones that process quantum data in real time for vital missions.

Despite its transformative potential, quantum computing confronts problems such as scalability, qubit stability, and high error rates (A. Kumar et al., 2022). Quantum error correction and qubit technologies are making viable and scalable quantum systems a reality (Sidhu et al., 2021). Future drone systems are anticipated to rely on hybrid frameworks that integrate classical and quantum computing. Classical systems are reliable and scalable, whereas quantum computing provides unrivaled benefits in optimization, data processing, and security. This symbiotic connection will drive the next generation of autonomous drones, enabling resilient, adaptive, and efficient operations across varied applications (Chen, 2021), (Ali et al., 2023).

Quantum Navigation and Sensing

Quantum technologies are redefining navigation and sensing capabilities by overcoming traditional system limitations such as signal interference, environmental influences, and calibration difficulties. Classical systems, like GPS and radar, frequently suffer with reliability under adverse conditions, such as signal obstruction, bad weather, or deliberate signal spoofing (Ralegankar et al., 2022). Quantum technologies use superposition and entanglement to overcome limitations, resulting in revolutionary improvements (Al-Hraishawi et al., 2024). Quantum sensors, such as gravimeters and magnetometers, provide high sensitivity and precision, enabling applications like single-photon imaging and gravitational wave detection (Al-Hraishawi et al., 2024). Quantum accelerometers and gyroscopes offer precise inertial navigation without external signals, making them appropriate for signal-deprived environments (Al-Hraishawi et al., 2024).

The integration of quantum sensing with classical systems provides a viable opportunity for hybrid solutions that combine the best of both approaches. Quantum-enhanced inertial navigation systems can provide dependable navigation in locations with weak or unavailable GPS signals (Ralegankar et al., 2022). Quantum sensors can improve precision, stability, and environmental resistance, making them useful for crucial applications including drone navigation, environmental sensing, and aerospace operations (Ralegankar et al., 2022), (Al-Hraishawi et al., 2024).

Quantum sensors achieve remarkable measurement sensitivity by using quantum states of matter, such as entangled particles or superpositions. Atomic interferometry and quantum-enhanced spectroscopy provide precise measurements beyond classical constraints (Eckert et al., 2006). Quantum sensors, unlike classical sensors, require minimum calibration since they compare relative environmental changes using unchanging atomic properties (Ralegankar et al., 2022). Quantum sensors offer advantages such as scalability, precision, accuracy, and resilience to environmental perturbations, making them transformational instruments for several sectors (Sidhu et al., 2021), (Bova, Goldfarb, & Melko, 2021).

As quantum sensing advances, its applications are increasing into different domains such as environmental monitoring, aerospace, defense, and medical imaging (Hill et al., 2017). Quantum sensing technology is rapidly paving the way for quantum-enhanced drone networks, which prioritize precision and efficiency. To fully realize the potential of these technologies, substantial hurdles must be overcome, such as improving hardware design, establishing error correction methods, and guaranteeing seamless interface with current systems (Sidhu et al., 2021).

Future breakthroughs in quantum sensing and navigation technologies are projected to change a wide range of industries by providing unparalleled accuracy, dependability, and efficiency. These breakthroughs, which enable robust solutions such as quantum-enhanced navigation systems and hybrid classical-quantum networks, have the potential to remove existing restrictions and uncover groundbreaking capabilities across essential applications.

APPLICATIONS OF IOQDS

While the potential applications of quantum technology in drone systems are promising, there are several challenges to address. These include the development of practical and scalable quantum hardware, integration with existing drone systems, and the need for further research to fully realize the benefits of quantum technologies. Ongoing advancements in quantum technology and its integration with drone systems are expected to drive innovation and expand the capabilities of drones in various fields. Figure 4 depicts the possible applications of the IoQD across a variety of industries, emphasizing the transformative impact of quantum communication and computing on important domains. Here's an overview of how quantum communication may affect each area depicted in the diagram:

- Disaster, Wildlife & Environmental Monitoring: Quantum communication facilitates instantaneous data transmission between monitoring equipment, allowing for swift responses to natural disasters and improving the precision of wildlife tracking and environmental evaluations (A. Kumar et al., 2021). IoT sensors with quantum enhancements may make it possible to identify and react to natural disasters like earthquakes, floods, and fires more precisely and quickly (Budker et al., 2022), (Crawford et al., 2021), (Zaiser et al., 2016).
- Geographical Mapping: Quantum communication promotes more accurate data gathering and transmission between devices in geographical mapping, hence improving the quality of geographic information systems (GIS) (Ralegankar et al., 2022), (A. Kumar et al., 2021). Applications in resource management, environmental preservation, and urban planning may particularly benefit from this level of precision (Al-Hraishawi et al., 2024), (Hill et al., 2017).
- Shipping & Delivery: Quantum communication could improve logistics by allowing extremely secure data sharing between delivery vehicles and tracking systems (A. Kumar et al., 2021). Throughout the logistics network, sensitive data, including shipment location and customer information, is protect-

ed using quantum-based cryptography (A. Kumar et al., 2022), (Hill et al., 2017).

- Smart Health: IoQD has the potential to improve remote health monitoring and diagnosis by enabling the safe and reliable transfer of patient data across medical equipment (A. Kumar et al., 2022). By guaranteeing data integrity and security, quantum technologies can also improve predictive analysis and medical imaging (Al-Hraishawi et al., 2024), (Hill et al., 2017).
- Internet Services: Quantum communication has the potential to transform the structure of internet services by improving encryption and data integrity (A. Kumar et al., 2022). This can be critical for private browsing, data sharing, and other online services that prioritize the safety of information. IoQD has the potential to contribute to the development of more robust and secure internet infrastructure, as well as improved cyber threat protection (Chiti et al., 2022), (Ali et al., 2023).
- Secure Confidential Message Communication: IoQD's utilization in secure communication channels relies on quantum cryptography, which enables potentially unbreakable encryption (Ralegankar et al., 2022). This can be used to safeguard sensitive communications in the government, finance, and other industries where data breaches have serious effects (A. Kumar et al., 2022), (Ralegankar et al., 2022), (Hill et al., 2017).
- Defense, Military & Surveillance: In defense, quantum communication improves data security, allowing for more dependable surveillance, monitoring, and decision-making procedures (A. Kumar et al., 2022), (Hill et al., 2017). Quantum technologies offer safe command and control communications, which is crucial in high-risk defense missions (Ishveena, 2022).
- Searching & Rescue Operations: Quantum-based IoT devices could dramatically improve search and rescue operations by delivering extremely precise location data (Ralegankar et al., 2022). IoQD-enabled sensors may interact in real time, even in remote or hazardous environments, allowing rescuers to locate and aid individuals more efficiently (Conrad et al., 2023).
- Gaming & Entertainment: IoQD has the potential to improve user experience in the gaming and entertainment industries by enabling faster, more secure interactions across several platforms (Ralegankar et al., 2022). Quantum communication ensures minimal latency, which enhances the quality of online gaming and virtual reality (VR) experiences that need real-time data transfer.
- Smart Agriculture: In agriculture, quantum-enhanced IoT devices can improve resource efficiency by delivering precise, real-time data on soil health, crop development, and meteorological conditions (A. Kumar et al., 2021),

(Al-Hraishawi et al., 2024). This can lead to more informed, data-driven decisions that increase yields and reduce resource use (Eckert et al., 2006).

Figure 4. IoQDs and its applications

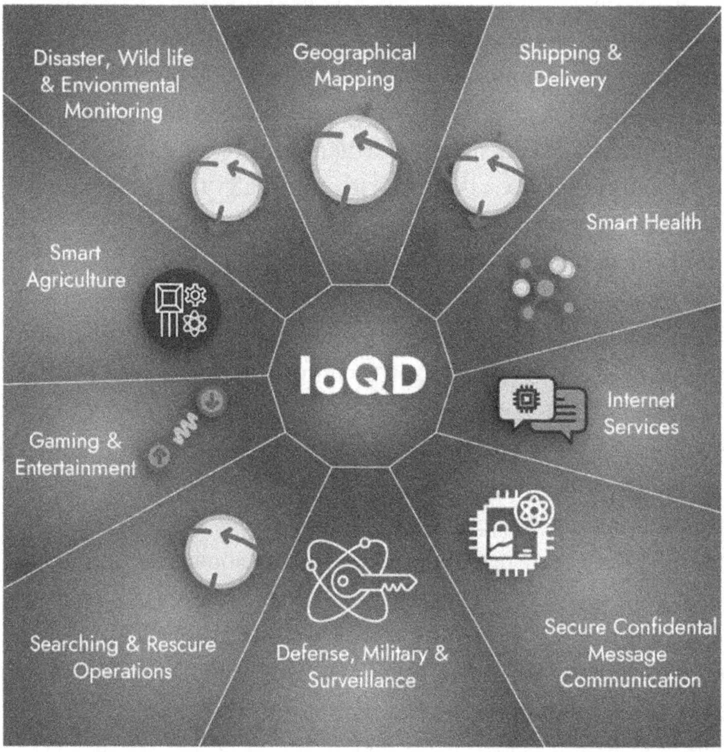

Table 1 provides a brief summary of the benefits of IoQDS in different applications.

Table 1. Benefits of IoQDs in different sectors

Sector	Applications & Benefits
Disaster, Wildlife, & Environmental Monitoring	• Real-time data transmission (A. Kumar et al., 2021). • Enhancing disaster response (A. Kumar et al., 2021). • Improving the precision of wildlife tracking and environmental evaluations (Budker et al., 2022), (Crawford et al., 2021), (Zaiser et al., 2016).
Geographical Mapping	• Accurate data gathering and transmission (Ralegankar et al., 2022), (A. Kumar et al., 2021). • Improving the quality of geographic information systems (GIS) (Ralegankar et al., 2022), (A. Kumar et al., 2021). • Precise environmental preservation, urban planning, and resource management (Al-Hraishawi et al., 2024), (Hill et al., 2017).
Shipping & Delivery	• Improving logistics by allowing extremely secure data sharing (A. Kumar et al., 2022), (Hill et al., 2017).
Smart Health	• Facilitating secure and reliable remote health monitoring and diagnosis (A. Kumar et al., 2022). • Enabling the safe and reliable transfer (A. Kumar et al., 2022). • Improving predictive analysis and medical imaging (Al-Hraishawi et al., 2024), (Hill et al., 2017). • Ensuring data integrity (Al-Hraishawi et al., 2024), (Hill et al., 2017).
Internet Services	• Strengthens encryption and data integrity (A. Kumar et al., 2022). • Improving cybersecurity (Chiti et al., 2022), (Ali et al., 2023). • Enabling secure browsing, private data sharing, and cyber threat protection (Chiti et al., 2022), (Ali et al., 2023).
Secure Confidential Communication	• Create unbreakable encryption (Ralegankar et al., 2022). • Safeguarding sensitive government, finance, and corporate communications effects (A. Kumar et al., 2022), (Ralegankar et al., 2022), (Hill et al., 2017).
Defense, Military & Surveillance	• Enhancing secure military communication, reliable surveillance, and decision-making (A. Kumar et al., 2022), (Hill et al., 2017).
Searching & Rescue Operations	• Providing highly accurate real-time location data (Ralegankar et al., 2022). • Allowing efficient search and rescue efforts (Conrad et al., 2023).
Gaming & Entertainment	• Improving gaming and VR experiences. • High-speed and secured data transfer across platforms (Ralegankar et al., 2022).
Smart Agriculture	• Improving resource efficiency by delivering precise, real-time data (A. Kumar et al., 2021), (Al-Hraishawi et al., 2024).

Military and Surveillance

Drones' capacity to deliver real-time intelligence, improve situational awareness, and reduce threats to human forces has made them essential in military and surveillance missions. QKD offers unprecedented security for military communications by enabling secure key exchange (Ghamari et al., 2022). This ensures that sensitive military communications remain confidential, even in the face of potential quantum computing threats. By utilizing quantum entanglement and superposition, quantum communication systems can establish secure channels in military networks and command systems that are resistant to interception and tampering (Ralegankar et al., 2022). This is crucial for maintaining the integrity and confidentiality of military operations and strategic information.

Sensors like quantum gravimeters and magnetometers can be used for enhanced detection of submarines, aircraft, and other assets by measuring gravitational and magnetic anomalies with unprecedented accuracy (Al-Hraishawi et al., 2024). Quantum-enhanced imaging technologies can improve surveillance and reconnaissance by providing higher resolution and contrast in imaging systems. Quantum imaging techniques, such as quantum-enhanced radar and Lidar, can detect objects and gather detailed information in various environmental conditions, including low visibility and adverse weather (Jabbar et al., 2024), (Pereira et al., 2020). In addition, quantum sensors and accelerometers can improve the navigation and positioning capabilities of military drones (Al-Hraishawi et al., 2024). Quantum-enhanced inertial navigation systems provide precise measurements of acceleration and rotation, enabling accurate navigation in GPS-denied environments such as underground or underwater operations (Ralegankar et al., 2022).

This capability is crucial for autonomous drones performing reconnaissance or tactical missions in challenging conditions. The use of quantum encryption can safeguard communication channels in real-time, enhancing the security of drone operations and preventing unauthorized access (A. Kumar et al., 2022). Integrating quantum-enhanced sensors into drone systems can significantly improve their sensing capabilities. Quantum imaging and sensing technologies can enhance the ability of drones to detect and identify targets, measure environmental conditions, and gather intelligence with greater precision (Al-Hraishawi et al., 2024).

In particular, authors of (A. Kumar et al., 2022) suggested QW which is a new field of application made possible by IoQDs. QW is a subfield of quantum technologies that uses the internet and quantum technologies for military applications that impact defense, intelligence, and military capabilities in all types of warfare; the development of new military doctrines, strategies, and attack capabilities; and issues related to peace and ethics. Based on their views, by offering more precise navigation, extremely secure communication, and computational power, IoQDs will enhance the current military's capabilities in intelligence, surveillance, target acquisition, and reconnaissance. IoQDs also create opportunities to form a private network for territory surveillance. They emphasized that IoQDs will enable faster and more secure data transmission and processing, which will aid in the completion of any significant operation.

In military and surveillance applications, drones have become indispensable tools for enhancing operational effectiveness and situational awareness (Al-Hraishawi et al., 2024). Equipped with advanced sensors and high-resolution cameras, drones provide real-time intelligence, reconnaissance, and surveillance capabilities, enabling detailed monitoring of areas of interest from a safe distance. In military contexts, drones are utilized for tactical surveillance, target identification, and reconnaissance missions, reducing risks to personnel and improving strategic decision-making. Their ability to cover vast areas and operate in challenging environments has revolutionized modern warfare and security operations (A. Kumar et al., 2022). Additionally, drones' real-time data transmission and high-resolution imaging contribute to improved surveillance and border security, allowing for rapid response to emerging threats and enhanced monitoring of sensitive locations.

Disaster Management

Effective communication is essential to disaster management in order to plan actions, evaluate damage, and allocate resources effectively. Drones have emerged as critical disaster management tools because of their capacity to examine dangerous conditions swiftly and safely (Dambal et al., 2019), (Jabbar et al., 2024), (Khun et al., 2023), (Pereira et al., 2020). Naturally, drones can be utilized in isolated or underdeveloped regions with spotty or non-existent internet connectivity (Francis, Brandis, & McCann, 2022). They provide a wide range of capabilities, from real-time monitoring and data collecting to search and rescue operations, making them invaluable in reacting to natural and man-made disasters.

Quantum computing can revolutionize disaster management through advanced data analysis and simulation capabilities. They can process and analyze large volumes of data from various sources, such as satellite imagery, sensor networks, and historical records, with unprecedented speed and accuracy. This capability enables more accurate modeling and simulation of disaster scenarios, such as hurricanes, floods, and wildfires (A. Kumar et al., 2022). Enabling rapid analysis and decision-making capability allows drones to provide timely and actionable insights during disaster response, such as identifying damage patterns, assessing risk areas, or guiding search and rescue efforts. Disaster management involves complex decision-making processes, including resource allocation, logistics, and evacuation planning. Quantum algorithms can optimize these processes by solving complex combinatorial problems more efficiently than classical algorithms. QKD provides a level of security, making it resistant to eavesdropping and interception (Conrad et al., 2023). This ensures that communication among disaster response teams remains confidential and secure, even in challenging and high-risk environments. Further, quantum accelerometers and gyroscopes offer precise measurements of acceleration and rotational rates, which can improve the accuracy of inertial navigation systems (Al-Hraishawi et al., 2024). This is particularly valuable in environments where

GPS signals may be weak or unavailable, such as urban canyons or remote areas affected by disasters (Ralegankar et al., 2022).

According to (A. Kumar et al., 2022), quantum computing is a major contender for addressing climate change and related events like hurricanes, tsunamis, earthquakes, and heat waves because of its rapid computation, unmatched security, high-reliability potential and data processing capabilities. They further emphasized that more dependable data sensing and long-distance data sharing will be facilitated by an IoQD constellation (Al-Hraishawi et al., 2024). The capacity to fly drones at both low and high altitudes makes IoQD-based climatic data readily available, which is crucial for disaster management. Overall, the potential advancements offered by quantum technologies have the capacity to transform disaster management and drone systems, providing new tools and solutions to address complex challenges and improve response effectiveness in critical situations.

Agriculture and Environmental Monitoring

Drones have revolutionized environmental monitoring and agriculture by offering accurate, real-time data that improves operational efficiency and decision-making (Jabbar et al., 2024), (Pereira et al., 2020). Drones for agriculture have shown to guarantee the growth of sustainable agriculture when used and executed well. Over agricultural areas, drones offer superior quality and real-time aerial imagery compared to satellite imagery (Gopal Dutta, 2020). Additionally, drones can currently be used for applications such as weed and disease localization (Barnas et al., 2020), mapping, seed planting, spraying, soil property determination, vegetation detection, and the creation of precise elevation models (Gopal Dutta, 2020), (Yashika Gupta, 2022).

The combination of quantum sensors, real-time data processing, and sophisticated predictive analytics presents a huge opportunity to transform environmental monitoring and agriculture. According to the authors of (A. Kumar et al., 2021), quantum technologies can improve pollution detection by providing more sensitive and accurate measurements of pollutants in air, water, and soil. Quantum sensors could detect trace amounts of pollutants, such as particulate matter or chemical contaminants, with higher precision than classical sensors (Budker et al., 2022), (Crawford et al., 2021), (Zaiser et al., 2016). This capability enables more effective monitoring and management of environmental pollution. Moreover, they stated that quantum technologies can enhance biodiversity and ecosystem monitoring by providing detailed and precise measurements of environmental parameters that affect ecosystems (A. Kumar et al., 2022). They further emphasized quantum-enhanced sensors which could track changes in habitat conditions, species distributions, and ecosystem health with greater accuracy. This information is vital for conservation efforts and understanding the impacts of environmental changes on biodiversity.

Furthermore, quantum sensors could revolutionize precision farming by providing highly accurate measurements of soil conditions, moisture levels, and nutrient content (A. Kumar et al., 2021). Additionally, they can play a crucial role in monitoring climate change by providing highly accurate measurements of atmospheric and environmental parameters. Quantum-enhanced sensors, such as quantum gravimeters and magnetometers, can deliver unprecedented sensitivity in detecting soil changes and variations (Al-Hraishawi et al., 2024). This increased precision allows for more accurate and timely applications of fertilizers, pesticides, and water, leading to optimized crop yields and reduced resource wastage. Quantum imaging and spectroscopy can improve crop health monitoring by detecting subtle changes in plant health and identifying diseases or nutrient deficiencies early (Eckert et al., 2006). Quantum-enhanced imaging techniques could offer higher resolution and sensitivity, enabling farmers to monitor crop conditions with greater detail and accuracy (Al-Hraishawi et al., 2024). This capability can lead to more targeted interventions and better overall crop management. Quantum sensors could enhance the detection of pests and diseases by providing more sensitive and accurate measurements of environmental changes that indicate pest activity or disease outbreaks. For instance, they could detect minute changes in plant emissions or soil composition that signal the presence of pests or pathogens, allowing for earlier and more effective treatment.

Quantum technologies can contribute to more efficient water management by providing precise measurements of water usage and distribution. Quantum sensors could monitor soil moisture levels and water flow with high accuracy, enabling more efficient irrigation practices and reducing water wastage (Al-Hraishawi et al., 2024). This is particularly important in regions facing water scarcity and the need for sustainable agricultural practices. Although the integration of quantum technologies into current systems offers considerable advances in efficiency, sustainability, and overall effectiveness, adoption of these technologies is hampered by issues related to cost and technical competence (Chiti et al., 2022).

ADVANCEMENTS IN IOQDS

IoQDs with Satellite Integration

The fundamental idea behind advanced quantum-based satellite communication with Earth is quantum entanglement, which is the process by which subatomic particles become entangled or inseparably linked (A. Kumar et al., 2021). This proves that even if entangled particles are attempted, hacking will be impossible. However, there are some limitations in these techniques that could be reduced through the integration of modern IoQDs with satellite. This integration could provide unprecedented prospects for expanding quantum-enabled drone networks across large geographic areas, allowing secure and effective communication, navigation, and data processing (A. Kumar et al., 2022).

Satellites enable secure quantum communication with QKD via protocols such as BB84 and entanglement distribution, resulting in secure encryption and high-speed data sharing while maintaining privacy across immense distances (A. Kumar et al., 2022). Quantum-enhanced GPS, accelerometers, gyroscopes, and other satellite-based sensors enable ultraprecise navigation and localization even in GPS-denied situations, hence improving IoQD operations such as weed mapping and disaster response (Ralegankar et al., 2022). Satellites offer real-time data transfer that ensures smooth, low-latency operations with redundancy for mission-critical applications, serves as a reliable backup during network outages, provides global coverage, and manages massive IoQD swarms. Quantum enhanced satellite imaging supports precision agriculture, environmental monitoring, and disaster management by integrating with drone-sensed data (Al-Hraishawi et al., 2024), (Ralegankar et al., 2022). They also offer quantum-enhanced remote sensing and imaging, which complement IoQD sensing for precision applications.

As satellite-IoQD integration becomes a key component of global quantum networks, future developments such as satellite swarms, inter-satellite quantum communication, and hybrid quantum-classical networks will further improve scalability, lower costs, and optimize operations with AI-driven solutions for disaster monitoring, defense, and agriculture.

IoQDs with 6G Integration

6G networks are expected to provide both evolutionary and revolutionary improvements over 5G, offering unparalleled capabilities for future technologies (C. Wang & Rahman, 2022), (Prateek, Ojha, Altaf, & Maity, 2023). With the integration of 6G networks, IoQDs have the potential to transform standards for drone connectivity, communication, navigation, and data processing. These breakthroughs will enable unprecedented applications in smart agriculture, disaster response, environmental monitoring, and urban planning. 6G networks may deliver the secure infrastructure needed for IoQDs, with ultra-low latency, high bandwidth, and extensive device connectivity (Maher et al., 2020), (N. Sharma et al., 2021). This allows flawless real-time communication, which is required for coordinated drone swarms and high-resolution data transfer (C. Wang & Rahman, 2022). 6G enables edge and quantum cloud integration, which supports computationally intense tasks and improves decision-making capabilities during complex drone operations. Furthermore, satellite integration with 6G assures that IoQDs remain operational in remote or signal-deprived areas, enabling global-scale applications.

By integrating with quantum systems, IoQDs achieve real-time swarm coordination, sharing geolocation, environmental, and task data to perform missions with extreme precision (C. Wang & Rahman, 2022). Furthermore, 6G offers quantum-enhanced AI models that run on quantum cloud servers to optimize flight paths, collision avoidance, and resource allocation (Prateek et al., 2023). These developments allow IoQDs to scale globally while maintaining energy-saving and affordable operations.

Despite these advances, limitations such as hardware scalability, quantum stability and implementation of 6G infrastructure persist (Cheng et al., 2024). Collaborative researches will be essential in advancing the deployment of 6G for IoQDs. The combination of 6G and quantum technology provides a solid platform for IoQDs, enabling intelligent, autonomous, and secure drone operations across multiple industries.

Quantum Machine Learning Assisted IoQDs

The combination of Quantum Machine Learning (QML) with the IoQDs is a groundbreaking invention that combines the computing capacity of quantum systems with the connectivity and efficiency of IoQDs. QML optimizes the decision-making, data processing, and autonomous capabilities of drone networks, making them smarter, quicker, and more responsive to complicated missions in different domains such as smart agriculture, environmental monitoring, and disaster management.

QML employs quantum principles like superposition and entanglement to improve traditional machine learning tasks (Nawaz, Sharma, Wyne, Patwary, & Asaduzzaman, 2019). In IoQDs, QML outperforms classical algorithms in processing large-scale and complex datasets, allowing for real-time flight route optimization, resource allocation, and autonomous decision making. QML's ability to analyze data rapidly and accurately, including as high-resolution imagery and geospatial data, puts it as an essential component for IoQDs (Nawaz et al., 2019). Additionally, 6G networks enhance the potential of QML-assisted IoQDs by offering ultra-low latency, high bandwidth, and uninterrupted connectivity (Maher et al., 2020), (N. Sharma et al., 2021).

Although its potential, QML-assisted IoQDs face several obstacles, including the need for more robust and scalable quantum technology, integration with IoQD systems and 6G infrastructure, and power limitations in drones (Prateek et al., 2023). Future efforts will concentrate on creating lightweight QML models, hybrid quantum-classical frameworks, and energy-efficient quantum processors specifically designed for drones. QML is a key driver in the evolution of IoQDs, making them into intelligent, adaptable systems capable of handling different applications with outstanding performance (Nawaz et al., 2019). QML, when combined with IoQD connection and 6G strength, will drive industry-wide innovation, setting new standards for performance, security, and cognition in drone networks (Prateek et al., 2023).

PROPOSED IOQD BASED WEED DETECTION

The single quantum-enabled drone transforms sustainable farming by combining several sophisticated features into a single unit, rather than employing multiple drones to accomplish various tasks. Figure 5 depicts a futuristic quantum-enabled drone system designed for precise weed detection and management in agricultural instances. The procedure uses cutting-edge quantum technology to improve efficiency, security, and accuracy. Below is a full explanation of how this drone operates and how it will revolutionize agriculture:

Figure 5. IoQD based weed detection framework for precise farming

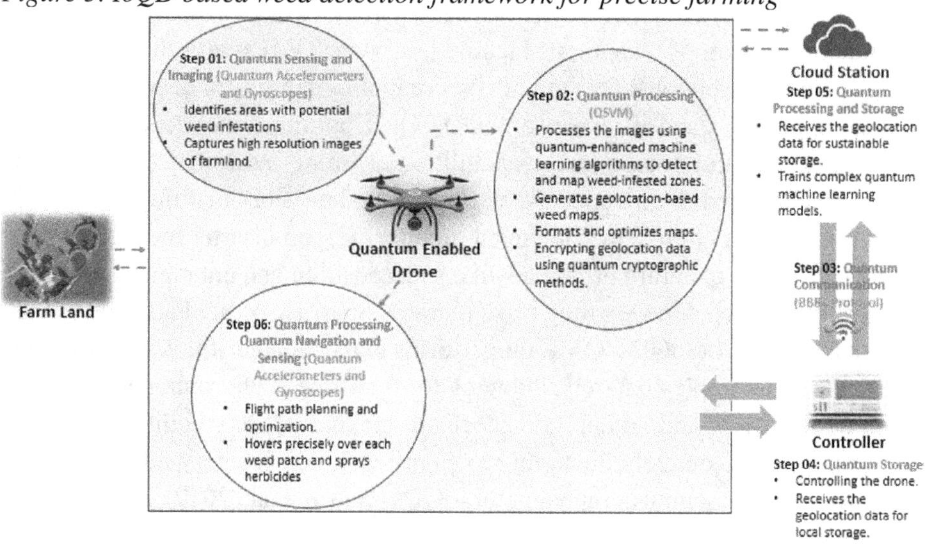

The drone begins by scanning farmland using powerful quantum sensors, such as quantum accelerometers and gyroscopes, to capture images of excellent quality with extreme precision. It uses quantum-enhanced imagery to identify possible weed-infested areas, assuring full coverage of the land. Subsequently, quantum computing examines the collected data with Quantum Support Vector Machines (QSVM) and other quantum-enhanced machine learning methods. This system categorizes and maps weed-infested areas, producing geolocation-based weed maps to guide subsequent activities. To protect the data, crucial geolocation information is encrypted with strong quantum cryptography techniques.

Next, geolocation data is exchanged securely between the drone and ground-based systems via quantum communication protocols such as BB84 QKD, which create secure channels to prevent unauthorized access or tampering. When the data reaches the local controller, it acts as a temporary storage unit and controls the drone's real-time activities. The controller stores geolocation and weed map data locally for quick use before sending it to the cloud station for advanced processing and long-term storage. Following that, the cloud station is critical for evaluating and storing data indefinitely. It uses advanced quantum computing models to forecast weed growth trends, optimize herbicide application, and improve detection procedures, resulting in more successful long-term weed management strategies. The drone returns to the field with the processed data, where it uses real-time quantum data to manage weeds precisely. Quantum technologies optimize its flying route, allowing it to hover directly above weed patches and deliver pesticides with great precision. This focused method eliminates chemical waste while minimizing environmental effect.

The system functions smoothly through its successive processes. Initially, the drone explores the farmland to gather precise information. This data is then analyzed to generate weed maps, which are securely sent to the controller. The controller manages operations and temporarily stores data, whereas the cloud station does sophisticated analysis and saves data for later use. Finally, the drone returns to the indicated weed patches to spray precisely, ensuring effective and sustainable weed management.

CHALLENGES AND FUTURE DIRECTIONS OF IOQDS

Challenges

Drone communication using quantum technology holds transformative potential across various industries. Its application in drone networks is poised to offer more inventive, durable, and efficient solutions to existing problems while unlocking new opportunities for the future (Al-Hraishawi et al., 2024). However, the integration of quantum technologies into drones faces several challenges and necessitates further exploration in key areas, including technological maturity or knowledge, cost and scalability, regulatory considerations, and research needs.

Quantum technologies often require specialized equipment that is large, delicate, and sensitive (Cheng et al., 2024), (Tu, Shen, Dai, Zhang, & Wang, 2022). Miniaturizing quantum components—such as quantum sensors, communication devices, and processing units—to fit within the compact and robust form factors of drones is a substantial challenge. Achieving the necessary size and weight constraints while maintaining performance is crucial for practical deployment (Cheng et al., 2024). Integrating quantum technologies with existing drone systems involves addressing compatibility issues.

Quantum sensors and communication devices need to interface with classical drone components, such as navigation controllers, power systems, and data processors (Ralegankar et al., 2022). Developing effective interfaces and ensuring seamless communication between quantum and classical systems are essential for functional integration. Quantum technologies introduce new complexities that must be managed alongside traditional drone systems. For instance, quantum sensors might require specific operating conditions, such as low temperatures or controlled or cold environments, which could complicate their integration and operation within drones (A. Kumar et al., 2022). The temperature rises as the number of qubits grows along with power consumption. For the device to function and cool down, a very cold atmosphere is required. Hence, managing these complexities while ensuring reliable performance is a major challenge.

Quantum components are highly sensitive to environmental conditions, including temperature fluctuations, electromagnetic interference, and vibrations (Cheng et al., 2024). Drones operate in dynamic and often harsh environments, which can impact the stability and performance of quantum sensors and communication devices. Ensuring that quantum technologies remain stable and functional under such conditions is a significant challenge. Quantum systems are susceptible to errors and noise, which can affect measurement accuracy and communication reliability. Addressing these issues requires the development of advanced error correction techniques and robust shielding methods to protect quantum devices from environmental disturbances (Sidhu et al., 2021). Ensuring high performance and reliability in the challenging conditions experienced by drones is critical.

Future Directions

While quantum communication provides unmatched reliability and safety for data transfer, quantum computing shows promise for tackling challenging optimization issues and improving real-time data processing in drone technology (Chen, 2021), (Al-Hraishawi et al., 2024), (Hill et al., 2017). Drones can respond more quickly and efficiently in disaster management and other vital applications when quantum technologies advance (Dambal et al., 2019), (Jabbar et al., 2024), (Khun et al., 2023), (Pereira et al., 2020), (A. Kumar et al., 2022). The potential advantages of quantum computing and communication in drone technology are significant and might lead to more robust and adaptable systems with continued investment and advancement. However, their technological maturity is still restricted, and there are still a lot of obstacles to be solved before these technologies are completely incorporated into useful applications.

- Advancements in Quantum Processors:

In the case of current quantum processors, their scalability and error rates are impacted by limitations in qubit count and coherence times (Sidhu et al., 2021). Optimization issues like resource allocation in disaster management and flight route optimization may be revolutionized by quantum computing. However, large-scale, real-time optimization tasks necessary for drone operations are beyond the capabilities of present quantum processors.

- Optimization Algorithms for Drone Operations:

Although IoQD can improve data processing capabilities, high error rates and technological constraints still limit practical applications. On the other hand, although quantum algorithms have been demonstrated theoretically, their practical application demands more robust hardware.

- Enhanced Quantum Communication Systems:

Drones may communicate securely with QKD, shielding data from manipulation and interception (Ghamari et al., 2022), (Hill et al., 2017). According to (Michael, 2021), a drone system's ability to facilitate free-space communication between partners without a direct line of sight is one of its benefits. However, the current implementations need fiber optic or line-of-sight links and are restricted to short distances. It also emphasizes that in order for a multi-drone system to be competitive with other quantum network technologies, photon transmission losses must be decreased (Michael, 2021). In order for quantum computing and communication to become more mature and have a revolutionary impact on drone technology and other fields, more research, cooperation, and funding are needed. To use quantum in real-world scenarios, developments in qubit technology, error correction, and coherence times are required (Sidhu et al., 2021). Extending the range and reliability of quantum communication requires more resilient quantum communication hardware and quantum repeaters (Sidhu et al., 2021).

- Hybrid Systems Development:

The development of hybrid systems that integrate quantum and conventional technology can close the gap between available capabilities and potential uses (A. Kumar et al., 2022). This entails fusing quantum-enhanced methods with classical optimization algorithms and integrating QKD with the current communication infrastructure. In addition, for quantum computing and communication to be widely used and interoperable, standard protocols must be established. Moreover, developing quantum communication networks to support more users and greater distances is an important field of research. Key areas of research include the development of efficient quantum repeaters and the optimization of entanglement distribution mechanisms for robust networks.

- Cost Reduction and Scalability:

Cost and scalability are significant barriers to the widespread adoption of quantum technologies in drone systems (Chiti et al., 2022). The development and production of quantum components such as sensors, communication devices, and processors remain expensive due to the specialized materials and equipment required (Potter et al., 2019). Economies of scale, technological breakthroughs, and strategic investments in research and education are essential to reduce costs and enhance production scalability. Increasing qubit counts while preserving coherence and reducing error rates remains a critical technical challenge (A. Kumar et al., 2022). Scaling up quantum systems also necessitates standard protocols and interfaces to ensure compatibility with existing classical systems (Chiti et al., 2022). Moreover, the environmental sensitivities of quantum components pose scalability issues for drones operating in diverse and potentially harsh conditions.

- Quantum Sensor Innovation:

Improving quantum sensor sensitivity, stability, and scalability while investigating new sensing modalities will broaden the range of applications for drones equipped with quantum technologies. Quantum-enabled drones can navigate difficult situations like fog or smog, eliminating the limits of standard fixed systems. Quantum sensors, when combined with advances in materials and emerging technologies, can allow drones to establish mobile quantum networks anywhere and at any time. This versatility increases their usefulness in real-time applications like obstacle avoidance and precision navigation.

- Commercialization and Field Testing:

The commercialization of quantum technologies is gaining prominence, with practical applications appearing in finance, healthcare, and national security (Ghamari et al., 2022). Field testing in real-world scenarios is critical for evaluating the performance and dependability of these systems. Quantum algorithms suited to drone-specific needs, such as optimizing resource allocation and flight routes, will supplement advances in drone data processing via machine learning and hybrid communications systems (A. Kumar et al., 2022).

- Ethical and Governance Considerations:

The use of quantum-enabled drones demands strong regulatory frameworks to assure safety and compliance with airspace management and collision avoidance regulations (A. Kumar et al., 2022). Ethical problems, such as data privacy and the possible misuse of monitoring capabilities, must also be considered (A. Kumar et al., 2022). Comprehensive governance structures and stakeholder participation are critical for balancing innovation and social responsibility. While quantum-enhanced drones can provide secure communication during disaster relief efforts, their use must respect privacy and prevent escalating disputes (Ghamari et al., 2022).

- Interdisciplinary Collaboration and Investment:

The broad deployment of quantum technology in drones necessitates interdisciplinary collaboration among academics, industry, and governments. Investments in scalable qubit systems, cloud-based quantum computing, and free-space optical communication will promote innovation. Furthermore, combining quantum sensors with new materials and emerging technologies will enable realistic solutions for a wide range of applications, ensuring scalability and effectiveness.

Although incorporating quantum technology into drones presents considerable obstacles, ongoing investment, research, and collaboration are required to realize their full potential. Overcoming technological, economic, regulatory, and ethical constraints will result in robust, secure, and efficient quantum-based drone systems. These improvements will revolutionize industries by allowing for novel applications in agriculture, environmental monitoring, disaster management, national security, and so on.

CONCLUSION

By utilizing quantum innovations to overcome the drawbacks of conventional systems, the IoQDs represents a revolutionary breakthrough in drone technology. IoQDs facilitate ultra-secure communication, rapid data processing, and accurate navigation by combining quantum computing, communication, and sensing capabilities. Critical sectors including smart agriculture, national security, disaster management, and environmental monitoring have evolved as a result of these capabilities.

The case study on IoQD-based weed detection in precision agriculture, in particular, demonstrates how these technologies are used in real-world settings. IoQDs can optimize herbicide usage, enhance farming productivity, and detect weed infestations with remarkable precision by leveraging quantum image processing, machine learning, and communication. This contributes to promote sustainable agricultural practices.

Despite the potential, there are some obstacles to overcome in the development and application of IoQDs, such as scalability, hardware downsizing, and integrating quantum systems with current technology. It involves interdisciplinary cooperation, developments in quantum hardware, and the development of hybrid systems to overcome these obstacles. Furthermore, utilizing quantum-enhanced drone technologies in a secure manner will require close attention to ethical and legal considerations.

In conclusion, IoQDs represent a turning point era in drone technology, with transformative benefits and potential. Continued investing, research, and collaboration will be critical for realizing their full potential and influence across industries. This idea has the potential to reshape the future of drone networks by providing unprecedented efficiency, security, and intelligence. Moreover, IoQDs are a revolutionary development in drone technology that have the ability to entirely transform a variety of industries by improving efficiency, safety, and decision-making. Continued investment, research, and collaboration are essential to fully harness the benefits of IoQDs and unlock their potential in reshaping the future of drone networks.

REFERENCES

Al-Adhami, A., Hasoun, R. K., Gbashi, E. K., & Hassan, S. (2022). A secure communication protocol for civil drones. *Indonesian Journal of Electrical Engineering and Computer Science*, *27*(3). Advance online publication. https://ijeecs.iaescore.com/index.php/IJEECS/article/view/26098. DOI: 10.11591/ijeecs.v27.i3.pp1490-1501

Al-Hraishawi, H., Rehman, J. U., Razavi, M., & Chatzinotas, S. (2024). Characterizing and utilizing the interplay between quantum technologies and non-terrestrial networks. *IEEE Open Journal of the Communications Society*, *5*, 1937–1957. DOI: 10.1109/OJCOMS.2024.3380508

Ali, M. Z., Abohmra, A., Usman, M., Zahid, A., Heidari, H., Imran, M. A., & Abbasi, Q. H. (2023). Quantum for 6G communication: A perspective. *IET Quantum Communication*, *4*(3), 112–124. DOI: 10.1049/qtc2.12060

Alsamhi, S. H., Ma, O., Ansari, M. S., & Gupta, S. K. (2019). Collaboration of drone and Internet of Public Safety Things in smart cities: An overview of QoS and network performance optimization. *Drones (Basel)*, *3*(1), 13. Advance online publication. DOI: 10.3390/drones3010013

Arya, S., Yang, J., & Wang, Y. (2023). Towards the designing of low-latency SAGIN: Ground-to-UAV communications over interference channel. *Drones (Basel)*, *7*(7), 479. Advance online publication. DOI: 10.3390/drones7070479

Barnas, A. F., Chabot, D., Hodgson, A. J., Johnston, D. W., Bird, D. M., & Ellis-Felege, S. N. (2020). A standardized protocol for reporting methods when using drones for wildlife research. *Journal of Unmanned Vehicle Systems*, *8*(2), 89–98. DOI: 10.1139/juvs-2019-0011

Behjati, M., Mohd Noh, A. B., Alobaidy, H. A. H., Zulkifley, M. A., Nordin, R., & Abdullah, N. F. (2021). LoRa communications as an enabler for internet of drones towards large-scale livestock monitoring in rural farms. *Sensors (Basel)*, *21*(15), 5044. Advance online publication. DOI: 10.3390/s21155044 PMID: 34372281

Bova, F., Goldfarb, A., & Melko, R. G. (2021). Commercial applications of quantum computing. *EPJ Quantum Technology*, *8*(1), 2. DOI: 10.1140/epjqt/s40507-021-00091-1 PMID: 33569545

Budi Rahmadya. (2021). *Drone-based wireless sensor systems using Zigbee and RFID technology* (Ph.D. dissertation). Ibaraki University.

Budker, D., Cecil, T., Chupp, T. E., Geraci, A. A., Kimball, D. F. J., Kolkowitz, S., Rajendran, S., Singh, J. T., & Sushkov, A. O. (2022). Quantum sensors for high precision measurements of spin-dependent interactions. *arXiv preprint arXiv:2203.09488*. https://doi.org//arXiv.2203.09488DOI: 10.48550

Caruso, A., Chessa, S., Escolar, S., Barba, J., & López, J. C. (2021). Collection of data with drones in precision agriculture: Analytical model and LoRa case study. *IEEE Internet of Things Journal*, *8*(22), 16692–16704. DOI: 10.1109/JIOT.2021.3075561

Chen, J. (2021). Review on quantum communication and quantum computation. *Journal of Physics: Conference Series*, *1865*(2), 022008. DOI: 10.1088/1742-6596/1865/2/022008

Cheng, S., Qi, H., Sun, Y., Liu, L., & Xie, S. (2024). Toward quantum unmanned systems. *Science China. Technological Sciences*, •••, 67.

Chiti, F., Picchi, R., & Pierucci, L. (2022). Metropolitan quantum-drone networking and computing: A software-defined perspective. *IEEE Access : Practical Innovations, Open Solutions*, *10*, 126062–126073. Advance online publication. DOI: 10.1109/ACCESS.2022.3226127

Crawford, S. E., Shugayev, R. A., Paudel, H. P., Lu, P., Syamlal, M., Ohodnicki, P. R., Chorpening, B., Gentry, R., & Duan, Y. (2021). Quantum sensing for energy applications: Review and perspective. *Advanced Quantum Technologies*, *4*(8), 2100049. DOI: 10.1002/qute.202100049

Dambal, V. A., Mohadikar, S., Kumbhar, A., & Guvenc, I. (2019). Improving LoRa signal coverage in urban and sub-urban environments with UAVs. *2019 International Workshop on Antenna Technology (iWAT)*, 210–213. https://api.semanticscholar.org/CorpusID:86764193

Dutta, G., & Goswami, P. (2020). Application of drone in agriculture: A review. *International Journal of Chemical Studies*, *8*(8), 181–187. DOI: 10.22271/chemi.2020.v8.i5d.10529

Eckert, K., Hyllus, P., Brüß, D., Poulsen, U. V., Lewenstein, M., Jentsch, C., Müller, T., Rasel, E. M., & Ertmer, W. (2006). Differential atom interferometry beyond the standard quantum limit. *Physical Review A*, *73*(1), 013814. DOI: 10.1103/PhysRevA.73.013814

Fouda, A., Ibrahim, A. S., Güvenç, Í., & Ghosh, M. (2019). Interference management in UAV-assisted integrated access and backhaul cellular networks. *IEEE Access : Practical Innovations, Open Solutions*, *7*, 104553–104566. DOI: 10.1109/ACCESS.2019.2927176

Francis, R. J., Brandis, K. J., & McCann, J. A. (2022). Offline imagery checks for remote drone usage. *Drones (Basel)*, *6*(12), 395. Advance online publication. DOI: 10.3390/drones6120395

Ghamari, M., Rangel, P., Mehrubeoglu, M., Tewolde, G. S., & Sherratt, R. S. (2022). Unmanned aerial vehicle communications for civil applications: A review. *IEEE Access : Practical Innovations, Open Solutions*, *10*, 102492–102531. DOI: 10.1109/ACCESS.2022.3208571

Ghazali, M. H. M., Teoh, K., & Rahiman, W. (2021). A systematic review of real-time deployments of UAV-based LoRa communication network. *IEEE Access : Practical Innovations, Open Solutions*, *9*, 124817–124830. DOI: 10.1109/ACCESS.2021.3110872

Gupta, Y., Mudgil, P., Sharma, V., & Hussain, Z. (2022). The smart technology in modern agriculture. *Proceedings of the International Conference on Innovative Computing & Communication (ICICC) 2022.*https://doi.org/DOI: 10.22271/chemi.2020.v8.i5d.10529

Haider, S. K., Nauman, A., Jamshed, M. A., Jiang, A., Batool, S., & Kim, S. W. (2022). Internet of drones: Routing algorithms, techniques and challenges. *Mathematics*, *10*(9), 1488. Advance online publication. DOI: 10.3390/math10091488

Hasan, S. R., Chowdhury, M. Z., Saiam, M., & Jang, Y. M. (2023). Quantum communication systems: Vision, protocols, applications, and challenges. *IEEE Access : Practical Innovations, Open Solutions*, *11*, 15855–15877. DOI: 10.1109/ACCESS.2023.3244395

Hemmati, A., Zarei, M., & Souri, A. (2023). UAV-based Internet of Vehicles: A systematic literature review. *Intelligent Systems with Applications*, *18*, 200226. Advance online publication. DOI: 10.1016/j.iswa.2023.200226

Hill, A. D., Chapman, J., Herndon, K., Chopp, C., Gauthier, D. J., & Kwiat, P. (2017). Drone-based quantum key distribution. *Urbana (Caracas, Venezuela)*, *51*, 61801–63003.

Jabbar, W. A., Tan, M. T., Hamidun, M. F. I., Che Kamarudin, A. H., Wu, W., Sultan, J., Alsewari, A. A., & Ali, M. A. H. (2024). Development of LoRaWAN-based IoT system for water quality monitoring in rural areas. *Expert Systems with Applications*, *242*, 122862. Advance online publication. DOI: 10.1016/j.eswa.2023.122862

Kantsepolsky, B., Aviv, I., Weitzfeld, R., & Bordo, E. (2023). Exploring quantum sensing potential for systems applications. *IEEE Access : Practical Innovations, Open Solutions*, *11*, 31569–31582. DOI: 10.1109/ACCESS.2023.3262506

Khan, N. A., Jhanjhi, N. Z., Brohi, S. N., & Nayyar, A. (2020). Chapter three - Emerging use of UAVs: Secure communication protocol issues and challenges. In Al-Turjman, F. (Ed.), *Drones in smart-cities* (pp. 37–55). Elsevier., DOI: 10.1016/B978-0-12-819972-5.00003-3

Khoshnoud, F., Esat, I. I., de Silva, C. W., & Quadrelli, M. B. (2019). Quantum network of cooperative unmanned autonomous systems. *Unmanned Systems*, *7*(02), 137–145. DOI: 10.1142/S2301385019500055

Khun, A. T. P., Shan, L., Lim, Y., & Tan, Y. (2023). MCST scheme for UAV systems over LoRa networks. *Drones (Basel)*, *7*(6), 371. DOI: 10.3390/drones7060371

Ko, Y., Kim, J., Duguma, D. G., Astillo, P. V., You, I., & Pau, G. (2021). Drone secure communication protocol for future sensitive applications in military zone. *Sensors (Basel)*, *21*(6), 2057. Advance online publication. DOI: 10.3390/s21062057 PMID: 33804023

Kumar, A., & Ambika, R. (2020). Drone integrated weather sensors for agriculture purpose. *International Journal of Electrical Engineering & Technology*, *11*(5), 83–90.

Kumar, A., Bhatia, S., Kaushik, K., Gandhi, S. M., Devi, S. G., De, J., Pacheco, D. A., & Mashat, A. (2021). Survey of promising technologies for quantum drones and networks. *IEEE Access : Practical Innovations, Open Solutions*, *9*, 125868–125911. DOI: 10.1109/ACCESS.2021.3109816

Kumar, A., de Jesus Pacheco, D. A., Kaushik, K., & Rodrigues, J. J. P. C. (2022). Futuristic view of the Internet of quantum drones: Review, challenges and research agenda. *Vehicular Communications*, *36*, 100487. DOI: 10.1016/j.vehcom.2022.100487

Kumar, P., Ojha, N. K., Altaf, F., & Maity, S. (2023). Quantum secured 6G technology-based applications in Internet of Everything. *Telecommunication Systems*, *82*(2), 315–344. DOI: 10.1007/s11235-022-00979-y

Kumar, P. P., Kumar, T. A., Muthu, P., Rajendirane, R., & Samuel, R. D. J. (2022). Security analysis of UAV communication protocols: Solutions, prospects, and encounters. In *The Internet of Drones* (pp. 167–190). Apple Academic Press. DOI: 10.1201/9781003277491-9

Liu, H.-Y., Tian, X.-H., Gu, C., Fan, P., Ni, X., Yang, R., Zhang, J.-N., Hu, M., Guo, J., Cao, X., Hu, X., Zhao, G., Lu, Y.-Q., Gong, Y.-X., Xie, Z., & Zhu, S.-N. (2020). Drone-based entanglement distribution towards mobile quantum networks. *National Science Review*, *7*(5), 921–928. DOI: 10.1093/nsr/nwz227 PMID: 34691535

Maher, A., Inoue, M., Watanbe, A., Yokemura, T., Ogyu, F., & Iida, H. (2020). UAV communication system integrated into network traversal with mobility. *SN Applied Sciences*, *2*(6), 1057. Advance online publication. DOI: 10.1007/s42452-020-2749-5

Maraveas, C., Konar, D., Michopoulos, D. K., Arvanitis, K. G., & Peppas, K. P. (2024). Harnessing quantum computing for smart agriculture: Empowering sustainable crop management and yield optimization. *Computers and Electronics in Agriculture*, *218*, 108680. DOI: 10.1016/j.compag.2024.108680

Millner, N., Cunliffe, A. M., Mulero-Pázmány, M., Newport, B., Sandbrook, C., & Wich, S. A. (2023). Exploring the opportunities and risks of aerial monitoring for biodiversity conservation. *Global Social Challenges Journal*.https://api.semanticscholar.org/CorpusID:259430259

Mozaffari, M. (2018). *Wireless communications and networking with unmanned aerial vehicles: Fundamentals, deployment, and optimization*. Virginia Tech.

Mustafa, N. E., & Alizadeh, F. (2024). Unmanned aerial vehicle (UAV) images of road vehicles dataset. *Data in Brief*, *54*, 110264. Advance online publication. DOI: 10.1016/j.dib.2024.110264 PMID: 38516279

NASA. (2020). Quantum technologies are changing the face of unmanned aircraft communications. https://www.nasa.gov/aeronautics/quantum-technologies-are-changing-the-face-of-unmanned-aircraft-communications/ (Accessed: 2024-08-21).

Nawaz, S. J., Sharma, S. K., Wyne, S., Patwary, M. N., & Asaduzzaman, M. (2019). Quantum machine learning for 6G communication networks: State-of-the-art and vision for the future. *IEEE Access : Practical Innovations, Open Solutions*, *7*, 46317–46350. DOI: 10.1109/ACCESS.2019.2909490

Pereira, D. S., De Morais, M. R., Nascimento, L. B. P., Alsina, P. J., Santos, V. G., Fernandes, D. H. S., & Silva, M. R. (2020). Zigbee protocol-based communication network for multi-unmanned aerial vehicle networks. *IEEE Access : Practical Innovations, Open Solutions*, *8*, 57762–57771. DOI: 10.1109/ACCESS.2020.2982402

Potter, B., Valentino, G., Yates, L., Benzing, T., & Salman, A. (2019). Environmental monitoring using a drone-enabled wireless sensor network. In *2019 Systems and Information Engineering Design Symposium (SIEDS)* (pp. 1-6). IEEE. DOI: 10.1109/SIEDS.2019.8735615

Ralegankar, V. K., Bagul, J., Thakkar, B., Gupta, R., Tanwar, S., Sharma, G., & Davidson, I. E. (2022). Quantum cryptography-as-a-service for secure UAV communication: Applications, challenges, and case study. *IEEE Access : Practical Innovations, Open Solutions*, *10*, 1475–1492. DOI: 10.1109/ACCESS.2021.3138753

Sakib, S. M. (2022). LiDAR technology—An overview. *IUP Journal of Electrical & Electronics Engineering*, *15*(1), 10–20.

Schirber, M. (2021). Quantum drones take flight. *Physics*. https://physics.aps.org/articles/v14/7

Shafiee, M., Zhou, Z., Mei, L., Dinmohammadi, F., Karama, J., & Flynn, D. (2021). Unmanned aerial drones for inspection of offshore wind turbines: A mission-critical failure analysis. *Robotics (Basel, Switzerland)*, *10*(1), 26. Advance online publication. DOI: 10.3390/robotics10010026

Sharma, A., Vanjani, P., Paliwal, N., Basnayaka, C. M. W., Jayakody, D. N. K., Wang, H.-C., & Muthuchidambaranathan, P. (2020). Communication and networking technologies for UAVs: A survey. *Journal of Network and Computer Applications*, *168*, 102739. DOI: 10.1016/j.jnca.2020.102739

Sharma, N., Kumar, A., Pervaiz, H., Magarini, M., Musavian, L., Alam, M., Jindal, A., & Imran, M. (2021). Aerial base station assisted cellular communication: Performance and trade-off. *IEEE Transactions on Network Science and Engineering*, *8*(4), 1. DOI: 10.1109/TNSE.2021.3052984

Sidhu, J. S., Joshi, S. K., Gündoğan, M., Brougham, T., Lowndes, D., Mazzarella, L., Krutzik, M., Mohapatra, S., Dequal, D., Vallone, G., Villoresi, P., Ling, A., Jennewein, T., Mohageg, M., Rarity, J. G., Fuentes, I., Pirandola, S., & Oi, D. K. L. (2021). Advances in space quantum communications. *IET Quantum Communication*, *2*(4), 182–217. DOI: 10.1049/qtc2.12015

Singh, I. (2022). US is developing drones with 'unhackable' quantum communication technology. https://dronedj.com/2022/05/26/quantum-drones/

Tiurlikova, A., Stepanov, N., & Mikhaylov, K. (2019). Wireless power transfer from unmanned aerial vehicle to low-power wide area network nodes: Performance and business prospects for LoRaWAN. *International Journal of Distributed Sensor Networks*, *15*(11), 1550147719888165. Advance online publication. DOI: 10.1177/1550147719888165

Tu, C., Shen, J., Dai, J., Zhang, L., & Wang, J. (2022). A lower size, weight acquisition and tracking system for airborne quantum communication. *IEEE Photonics Journal*, *14*(6), 1–8. DOI: 10.1109/JPHOT.2022.3212590

Vallance, C. (2024). Google unveils 'mind-boggling' quantum computing chip. https://www.bbc.com/news/articles/c791ng0zvl3o

Wadod, M., & Mohammed, F. (2023, July). Review on drone application methodologies in agriculture precision. *IOP Conference Series. Earth and Environmental Science, 1202*(1), 012001. Advance online publication. DOI: 10.1088/1755-1315/1202/1/012001

Wang, B., Xie, J., Li, S., Wan, Y., Fu, S., & Lu, K. (2018). Enabling high-performance onboard computing with virtualization for unmanned aerial systems. In *2018 International Conference on Unmanned Aircraft Systems (ICUAS)* (pp. 202-211). IEEE. DOI: 10.1109/ICUAS.2018.8453368

Wang, C., & Rahman, A. (2022). Quantum-enabled 6G wireless networks: Opportunities and challenges. *IEEE Wireless Communications, 29*(1), 58–69. DOI: 10.1109/MWC.006.00340

Yaacoub, J. P., Noura, H., Salman, O., & Chehab, A. (2020). Security analysis of drones systems: Attacks, limitations, and recommendations. *Internet of Things : Engineering Cyber Physical Human Systems, 11*, 100218. DOI: 10.1016/j.iot.2020.100218 PMID: 38620271

Zaiser, S., Rendler, T., Jakobi, I., Wolf, T., Lee, S.-Y., Wagner, S., Bergholm, V., Schulte-Herbrüggen, T., Neumann, P., & Wrachtrup, J. (2016). Enhancing quantum sensing sensitivity by a quantum memory. *Nature Communications, 7*(1), 12279. DOI: 10.1038/ncomms12279 PMID: 27506596

Zhang, J., Yu, F., Zhang, Q., Wang, M., Yu, J., & Tan, Y. (2024). Advancements of UAV and Deep Learning Technologies for Weed Management in Farmland. *Agronomy (Basel), 14*(3), 494. DOI: 10.3390/agronomy14030494

Zolanvari, M., Jain, R., & Salman, T. (2020). Potential data link candidates for civilian unmanned aircraft systems: A survey. *IEEE Communications Surveys and Tutorials, 22*(1), 292–319. DOI: 10.1109/COMST.2019.2960366

Chapter 9
A New Way to Represent an Image in the Quantum Circuit for Compression and Representation Perspective

Ershadul Haque
https://orcid.org/0000-0003-4322-4533
Charles Sturt University, Australia

Faranak Tohidi
https://orcid.org/0000-0003-3136-2584
Charles Sturt University, Australia

Anwaar Ulhaq
https://orcid.org/0000-0002-5145-7276
CQUniversity, Australia

ABSTRACT

Quantum image processing has recently received much attention due to the opportunity to place image data in quantum mechanics systems. The literature survey has shown that the complexity of preparing image data raises the question of how and in what efficient way a medium- and high-size image is represented inside the quantum domain, which is still a big question. Quantum properties such as entanglement, superposition, and parallelism make the quantum system faster and more robust. Qubit is the central basic unit that maps the image data in the quantum processor, which propagates from left to right. The number of qubits and their connection

DOI: 10.4018/979-8-3373-0020-7.ch009

Copyright © 2025, IGI Global Scientific Publishing. Copying or distributing in print or electronic forms without written permission of IGI Global Scientific Publishing is prohibited.

determine the efficiency of the circuit architecture. This chapter aims to investigate the performance of the transformed coefficient-based quantum image representation and compression approaches in terms of the required number of gates and PSNR values. Moreover, the challenges and prospects of quantum image compression and processing will be discussed as a future research direction.

1. INTRODUCTION

In the quantum computer discipline, physics, mathematics, and computer science play an important role in concreting the quantum information processing (QIP) field (Khan et al. 2019). In QIP, superposition, entanglement, and parallelism are the three main properties that make it unique (Jacobs 1963; Rehman 2010; Aaronan 2008). Meanwhile, the computing power of the classical computer has not increased significantly in the past decade (Wang and Zhang 2021) . Feynman et al., introduced quantum computers as another way to increase computing power (Feyman). Quantum-based factorial calculations of integer and database search algorithms were proposed in Shor (1194) and Grover (1996). The various applications of quantum computers in image computing are found in (Rehman 2010; Kalonia and Gupta 2021). The number of required operational gates in quantum computers determines the system's complexity (Illiyasu et al. 2016).

In 2003, Venegas-Andraca proposed a qubit lattice approach for storing, processing, and retrieving images using a multi-particle quantum mechanics system. It could store only four-pixel values of the image randomly using an electromagnetic frequency that generates pixel and state output via qubits (Venegas-Andraca and Bose 2003). How and in what way to map the pixel value beyond four is still unclear. Following this approach, various representation approaches have been investigated for both gray and color images, primarily focusing on pixel storing and reproducing. After that, an FRQI (efficient flexible representation of the quantum image) was proposed to store pixels of the grayscale image using a quantum rotation gate and the corresponding position in the qubit's basic state (Dong et al. 2011). Qubit in the FRQI is the entangled qubit and stores the pixel value in a normalized way. Due to the probabilistic-based outcome, the reconstruction of the image is still a gap.

The novel enhanced quantum representation (NEQR) demonstrates the pixel-based mapping of the image in the quantum circuit encoding each number of ones in the pixel and its corresponding position (Gao et al. 2013). The frequent number of ones in pixel value leads to complexity of storing, including processing because of considering each one's positional value separately. Also, the representation of the rectangular size of the image is another gap. To solve the rectangular issue of the NEQR approach, the INEQR approach was proposed, which represents the color image (Jiang and Wang 2015). It requires huge quantum resources due to the complexity of the circuit diagram. The multichannel representation of the color image was developed by the novel quantum representation of the color digital images (NCQI) by encoding the information using the NEQR circuit diagram (Sang et al. 2017). It still has measurement limitations due to probabilistic outcomes, and the quality measure of the reconstruction image is still an issue. The transform coefficient-based image representation and compression were proposed in Haque et al. 2023. This approach demonstrates high efficiency for both compression and representation between DCT-EFRQI, SCMNEQR, and ZSCNEQR approaches, respectively. The quantum color digital image has been proposed by Sang et al. (2017) and uses the basis sequence of the qubit to store the color value (R, G, B) of the pixel, known as the NCQI approach (Sang et al. 2017). For pixel preparation, all the values are normalized and create the superposition, which can operate simultaneously. In the quantum circuit, it maps the corresponding frequent number of one and its corresponding state position. As a result, the state label circuit complexity becomes very high and requires a higher number of gates. Transparency and transformations based on quantum color image processing have been proposed by Jie et al. (2021) to address the vertical and horizontal inconsistency size of the NCQI approach (Liu et al. 2021). The encoding method of this approach is the same as the NCQI approach. Besides, it also uses the accessory quantum bit to store the intermediate value of the pixel. As a result, rather than decreasing the complexity, it increases. Moreover, to map the value in the quantum circuit, higher quantum resources are required due to the mapping procedure.

In this work, a transformed coefficient-based approach has been proposed to compress quantum grayscale images. It encodes the grayscale image's intensity and state label information using transformed coefficient and quantization factors. It also represents the state label information after compression by the superposition and entanglement of the qubits. The goal of this work is to store, compress, and reproduce the image using quantum mechanics properties. The remaining paper is organized as follows. Section 2 outlines the related works. Section 3 discusses the proposed approach. Section 4 analyses the computational result and its discussion. Section 5 concludes the findings of the proposed work.

2. RELATED WORK

Recently, quantum computers have attracted a lot of attention in quantum image processing due to their huge data storage facility with faster computation than classical computers. In addition, the computational complexity of quantum computers is quadratically increased compared to classical computers (Feyman 2018). As a result, it is gaining popularity in research communication. The effective representation and compression of the quantum image make the system more efficient and provide a better reproduction quality by saving storage. Qubit lattice is the first quantum approach that demonstrates how to represent and restore a quantum image (Gao et al. 2013). Due to the use of one qubit, representing a real-size image is challenging. Latorre proposed a real-ket image representation algorithm in (Latorre 2005). In the real-ket approach, the outcome is probabilistic, and the reconstruction of the image is still a gap.

In Dong et al. 2011, an FRQI algorithm was developed that converts four random pixels into its equivalent quantum system using a single ancillary qubit as an angle. The image representation based on entanglement was proposed in 2010 (Venegas-Andraca and Bell 2010) . JPEG-based image compression was proposed by Jiang et al. in 2017 (Hu et al. 2018) and uses GQIR (general quantum image representation) and DCT approaches together. An equivalent bit-pixel image is investigated by Laurel et al. in Hu et al. 2018. An NEQR was proposed to represent a square grayscale image and resolve the FRQI issue (Sang 2017). An INEQR (improved novel enhanced quantum representation) approach was proposed (Liu et al. 2021) that represents tiny rectangular grayscale images rather than a color image. A NASS (normal arbitrary superposition state) was proposed in Li et al. 2014.

Sun et al. proposed a multi-channel representation for quantum image (MCQI) that captures RGB information of the color image and its corresponding position (Yan et al. 2012) . This method is limited to a 64x64 image size, and it requires 15 qubits. The required number of qubits makes it expensive, and the compression study was not investigated. The multi-object-based quantum neural classification was proposed in Boev et al. 2022. The tensor flow quantum library was used to classify the MNIST dataset that uses 12 qubits; among those, 8 qubits were used for data encoding and 4 for ancillary qubits to address the 3-class classification issue. In Olson et al. 2017, a quantum-based auto-encoder was proposed that compresses the quantum data by representing the lower dimension of the image compared to the given input. An EFRQI algorithm was proposed in 2021 to minimize the NEQR complexity. Integration of the DCT approach with EFRQI for representation and compression of the quantum image was proposed in Haque et al. 2023.

2.1 Quantum Image Computing Applications

There are a lot of areas of application and development of the quantum computer. Applications of a quantum computer in terms of image processing are increasing day by day. Some of those applications are image representation, image compression, watermarking, information security, and high privacy, which still suffer the same algorithmic limitation as today's classical computer (Guerreschi et al. 2016; Aaronson 2004; Kalonia and Gupta 2021; Dendukuri and Luu 2018).

2.2 Quantum Image Representation and Compression

In classical image computing, images are metrics numbers representing the color or intensity of every pixel in a Cartesian coordinate system with its location. In classical image processing, spatial correlation is normally exploited within explicit consideration of the location of the pixel as those are always associated with pixel intensities. However, in quantum image computing, we need to explicitly encode the pixel intensities and the corresponding locations using qubits. In the pixel intensities, we only consider the '1' position of the binary representation. The traditional image-processing approaches require higher computational resources to process and store the images.

On the other hand, quantum computing (QC) in image processing provides a quadratic speedup compared to classical computation (Dendukuri and Luu 2018). In quantum computing, a qubit or quantum bit is the basic unit of quantum information. The quantum version of the classical binary bit is physically realized with a two-state device. Figure 1 shows the frequency of publications and citations on quantum image representation published between 1991 and 2022. Interest in this field is gradually increasing, and the publication trends predict that interest in this field will be higher in 2022.

Figure 1. Frequent publications and citations for quantum image representation.

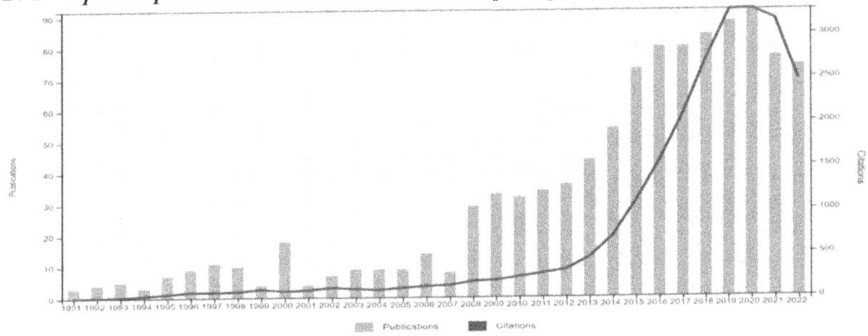

The quantum image compression approach is used to reduce the number of required resources inside the quantum processor (Dong et al. 2011). The main resource is the number of the required operational gates, not the qubits, because all the operational gates are participating resources to perform the image operation. In reference (Hu et al. 2018), the BEC (Boolean expression compression) approach is used to compress the image. If x is the Boolean variable, and 1 is its value, then the lateral x is used; otherwise, otherwise $\overline{x}$ is used as lateral. Therefore, a Boolean expression can be written as

$$e = x_0 x_1 \overline{x_2} + x_0 x_1 x_2$$

is compressed as

$$e = x_0 x_1$$

If two c-not gates operate on the same qubits, which means the same color and their control qubits are different, then the two c-nots can be combined into one. Figure 2 shows an example of the BEC approach for quantum image compression.

Figure 2. Quantum image compression using BEC approach (Hu et al. 2018)

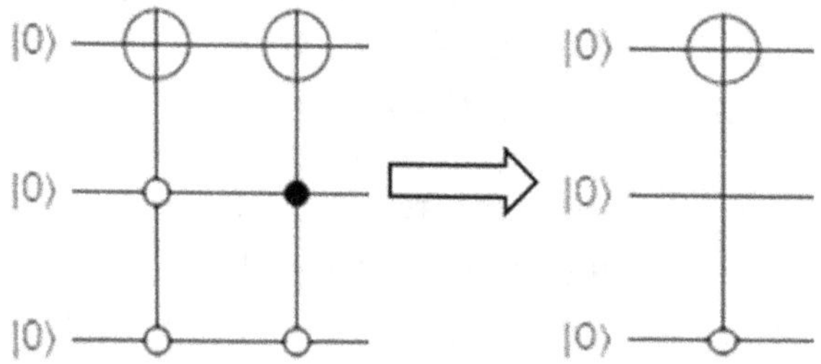

However, it has the following defects. In a pre-processing step, it's always required to identify which CNOT gate can be combined. It's time-consuming since every time it's needed to look over pair by pair before compression happens. Its complexity is $O(q.2^{4n})$ without compression. Figure 3 shows the publications and citations trend for more than three decades for image representation and compression. The graph shows that the initial work of quantum representation and its compression was seen in 1998. After that, no significant work was done until 2011.

Figure 3. Frequent publications and citations for quantum image representation and compression.

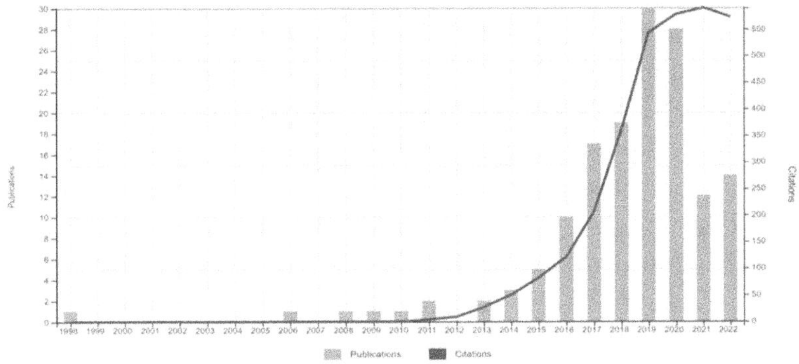

The publications report increased exponentially till 2019. Between 2019 and 2021, the number of publications changed randomly. The solid line in Figure 3 also shows the citation report for image representation and compression. The citation report started after 2011, and thereafter, it grew tremendously. The citation report indicates that quantum computing is a growing field in the research community. However, there are a lot of areas of development issues that are the main motivation of this research.

2.3 Quantum Image Encryption and Watermarking

In classical image processing, image protection is a very mature field. Image protection is deemed necessary because the attacker or unauthorized person always tries to attack in many ways. Therefore, it's necessary to protect the quantum image. Quantum cryptography and encryption play the main role in protecting the image against malware and other unethical hacking (Zhou et al. 2018). Quantum watermarking is another field of quantum image application. An FRQI-based image is watermarked, which is embedded $2^{n1} \times 2^{n2}$ binary image into a grayscale image. Quantum wavelet transform is applied to images for decomposition purposes. Different carriers and water in images are used to demonstrate the system's performance (Hu et al. 2019).

2.4 Essential Need for Representation and Compression of the Quantum Image

Quantum computing is a combined application of quantum mechanics, computer science, and mathematics and provides a new para-diagram to the probable solution of Moore's law failure to increase the computational power (Relese 2023). In quantum image processing, image presentation and image retrieval still are the main concerns. In quantum computing, a qubit or quantum bit is the basic unit of quantum information. In 2003, Venegas-Andraca and Bose proposed a method called Qubit lattice to map the four different random pixel values into single qubits. Storing, processing, and retrieving an image using quantum mechanics. Resolving the presentation and processing issue of classical images on the quantum computer is gradually accelerating the combined applications of a quantum computer and classical computer (Grover 1996). In many applications, image processing plays an important role in image operation (Haque et al. 2023). In classical computation, processing huge data images requires a lot of memory and hardware (Tohidi et al. 2023). Therefore, to perform a high amount of image data, it is necessary to provide computational support and compress the image in a compact way to save memory and ultimately minimize the system cost (Latorre 2005).

2.5 Existing Approach for Quantum Image Representation and Compression

In 2003, Qubit lattice was proposed by Venegas-Andrac et al., which is like the classical approach where each qubit is capable of storing one pixel (Venegas-Andraca and Bose 2003) . It is the first quantum image representation approach. After that, a real-ket approach was proposed (Latorre 2005) in 2005. Entanglement-based image representation was proposed in 2010 and uses the quantum entanglement property (Venegas-Andraca and Ball 2010). Inspired by the pixel-wise representation of the image, FRQI (Flexible Representation of Quantum Image) was developed to represent and encode the binary classical image into the quantum domain based on the probabilistic representation of amplitude (Dong et al. 2011). Figure 4 shows the quantum image, including its state for the 2×2 image size.

Figure 4. A 2 × 2 FRQI quantum image and quantum state (Dong 2011)

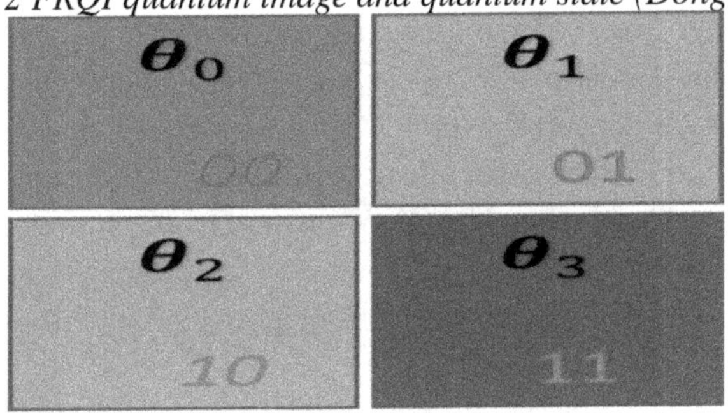

An FRQI image can be expressed as

$$|I\rangle = \frac{1}{2}\left[(\cos\theta_0|0\rangle + \text{sine}\theta_0|1\rangle)\otimes|00\rangle + (\cos\theta_1|0\rangle + \text{sine}\theta_1|1\rangle)\otimes|01\rangle + (\cos\theta_2|0\rangle + \text{sine}\theta_2|1\rangle)\otimes|10\rangle\right] + \left(\cos\theta_3|0\rangle + \text{sine}\theta_3|1\rangle\right)\otimes|11\rangle$$
(2.1)

Figure 5 shows the FRQI representation circuit diagram of the $|I\rangle$,

Figure 5. FRQI circuit diagram for image $|I\rangle$ (Dong et al. 2011).

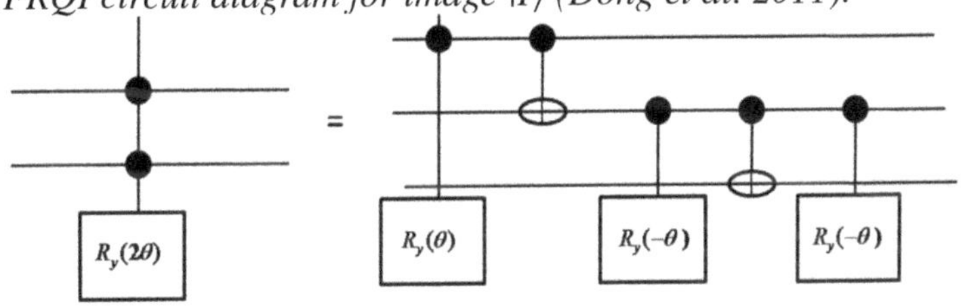

The standard rotation matrix $R_y(2\theta)$ is given as,

$$R_y(2\theta) = \begin{pmatrix} \cos\theta_1 & -\sin\theta_1 \\ \sin\theta_1 & \cos\theta_1 \end{pmatrix}$$
(2.2)

It represents the image pixel values in the quantum domain using the control rotation matrices. The control rotational matrices can be implemented using a c-not gate and standard rotation. An FRQI was unable to represent pixel-wise grayscale complex operation since it used only one qubit. It represents the grayscale image information as an angle in the quantum system. The main drawback of the FRQI cannot represent pixel-wise grayscale complex operation since it uses only one qubit.

It represents the color information as an angle and position in a qubit sequence. It represents the image pixel values in the quantum domain using the control rotation matrices. The control rotational matrices can be implemented using a c-not gate and standard rotation. Generally, the FRQI approach uses rotating gates to store grayscale pixel values in the probabilistic-based amplitude mapping in the Bloch sphere. Sun et al., proposed the MCRQI (Multichannel Representation of Quantum Image) approach (Sun et al. 2011) that represents color images and extends to the RGB space. To store RGB and opacity information, three qubits were used.

$$|I_{RGB\alpha}{}^{i}\rangle = \cos\theta_{Ri}|000\rangle + \cos\theta_{Gi}|001\rangle + \cos\theta_{Bi}|010\rangle + \cos\theta_{\alpha i}|011\rangle$$

$$+\sin\theta_{Ri}|100\rangle + \sin\theta_{Gi}|101\rangle + \sin\theta_{Bi}|110\rangle + \sin\theta_{\alpha i}|111\rangle \quad (2.3)$$

Where θ_{Ri}, θ_{Gi}, θ_{Bi} and $\theta_{\alpha i}$ represent the vectors encoding of R, G, B colors and α channels respectively. It encoded the color information of the $2^n \times 2^n$ image with the probabilistic amplitude of the considered qubits and position with $2n$ qubits. It is only able to represent square images with eight-pixel values. In 2013, the NEQR (Novel Quantum Representation of Color Digital Images) was proposed to represent the grayscale image (Gao et al. 2013). In NEQR, after converting the pixel values into a binary system, only the frequent number of ones is considered to map the pixel value in a quantum system. The pixel and state (position) preparation qubits are used to map pixel values and their corresponding state position of the pixel. For example, an image whose pixel values are 0 (Y=0, X=0), 100 (Y=0, X=1), 200 (Y=1, X=0), and 255 (Y=1, X=1). The quantum representation of those pixels is known as the NEQR representation approach,

$$|I\rangle_{NEQR} = \frac{1}{2}\left[|0\rangle\otimes|00\rangle + |100\rangle\otimes|01\rangle + |200\rangle\otimes|10\rangle + |255\rangle\otimes|11\rangle\right]$$

$$|I\rangle_{NEQR} = \frac{1}{2}\left[|00000000\rangle\otimes|00\rangle + |01100100\rangle\otimes|01\rangle + |11001000\rangle\otimes|10\rangle + |11111111\rangle\otimes|11\rangle \quad (2.4)$$

Figure 6 shows the NEQR circuit diagram for the image $|I\rangle_{NEQR}$. It resolves the FRQI issues because it provides several qubits to represent the pixel of the grayscale image.

Figure 6. Quantum circuit for NEQR preparation for the image $|I\rangle_{NEQR}$ (Gao et al. 2013)

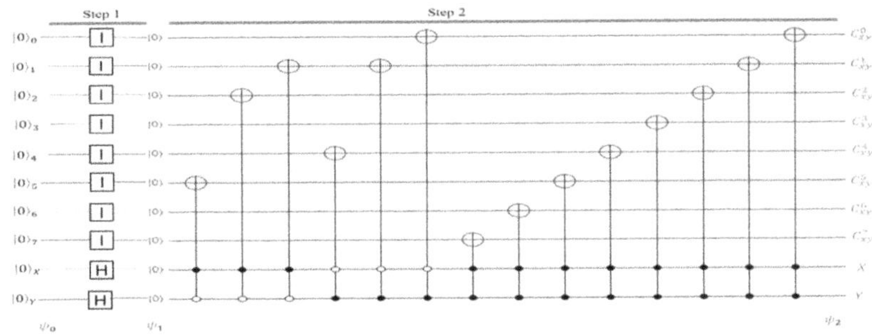

It resolves the FRQI issues because it provides several qubits to represent the grayscale image. In NEQR, after converting the pixel values into a binary system, only the frequent numbers of ones are considered to map the pixel value. It's limited to the tiny size of a grayscale square image. To resolve the color issue of the NEQR approach, INEQR was proposed (Jiang et al. 2015). In the INEQR approach, how to represent a medium or big-size color image is still unclear. A GQIR (General Quantum Image Representation) uses a logarithmic scale to represent the rectangular image, but it generates a lot of redundant bits (Jiang et al. 2015). For example, for a 512 × 512 image, the number of required qubits for encoding Y and X positions are 9 and 9 qubits, respectively. In practical application, Figure 7 shows a 16 × 16 deer image quantum circuit as an example after performing the DCT preparation approach and 70 quantization factors using the Quirk simulation tool (Nielson et al. 2015). The coefficient values of a 16 × 16 deer image are 126(X = 1,Y = 1), 1(X = 1,Y = 9), 4(X = 2,Y = 1), 4(X = 2,Y = 9), 1(X = 2,Y = 10), 1(X = 4,Y = 1), 1(X = 6,Y = 9), 138 (X = 9,Y = 1), 140(X = 9,Y = 9), 2(X = 10,Y = 1), 2(X = 10,Y = 9) and 1(X = 11,Y = 9) respectively. To prepare the coefficient value in the quantum domain, first, convert the coefficient of nonzero values into binary and then map the corresponding ones only using 8 numbers of qubits shown at the top of the quantum circuit in Figure 7. For state preparation, the corresponding coefficient nonzero position values are also recorded and located in the two-dimensional YX -position using the remaining 8 numbers of qubits shown at the bottom of Figure 7. After that, convert them into a binary system and count how many times one's happened in one coefficient. The quantum circuits only deal with ones and zeros that are governed by the qubits. The corresponding position value of each coefficient is used to prepare the quantum state. There were two kinds of bit rates; one is for the coefficient and another for its position.

Figure 7. A 16 × 16 DCT-GQIR quantum image including state preparation (Venegas-Andraca and Ball 2010)

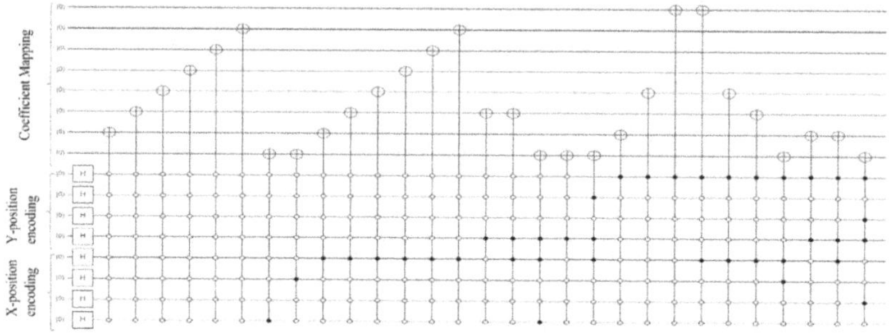

Polar coordinate-based quantum is known as QUALPI (Sang et al. 2017). In 2014, Li et al., proposed a normal arbitrary superposition state (NASS) approach for representing multidimensional color images. It stores pixel values in angles and positions in states. It is also applicable for color image compression on a quantum computer (Li et al. 2014). This method is suitable for well-image segmentation. In 2014, Yuan et al. proposed a simple quantum representation (SQR) approach (Yuan et al. 2014) based on the infrared image characteristics. Rather than use the angle parameter of the qubit to store a color, it uses the radiation energy of the pixel. It's limited to the infrared image. In terms of reconstruction image, the quality measurement is still a gap. Li et al., proposed an improved FRQI approach known as FRQCI (Flexible Representation of Quantum Color Image) (Li et al. 2016) in 2016. This method uses the probabilistic approach to store the RGB values of pixels. Sang et al. proposed a novel quantum representation of color digital images (NCQI) approach (Sang et al. 2017) that improved the MCQRI and NEQR approaches. It uses $3q$ qubits rather than q qubits to represent grayscale images. Although the time complexity of this method decreases quadratically compared to MCRQI, it still needs too many qubits. Also, this method did not measure the reconstruction images' quality. In 2017, Abdolmaleky et al. proposed a quantum multichannel Red-Green-Blue (RGB) representation (QMCR) approach (Abdolmaleky et al. 2017) for color image, which is the NEQR approach. It reduces the complexity preparation but requires more qubits compared to the MCRQI approach. In the same year, Jiang et al. proposed a quantum point cloud (Jiang et al. 2017) for color digital images. This method usually contains millions or hundreds of millions of points to record the surface. Therefore, the system becomes more complex to process this bulk amount of data. Li et al., proposed a bitplane representation of quantum images (BRQI) (Li et al. 2018) in 2018. It improves the NEQR approach by using binary bits and binary images. On accounting of all qubits connection including bit planes, the system becomes more complex and ultimately requires more qubits to arbitrary size of images. It requires (2n+4) or (2n+6) qubits to represent a grayscale or RGB color image of $2^n \times 2^n$.

In 2018, Li et al. proposed an optimized quantum representation for color images (OCQR) approach (Liu et al. 2018) that represents the pixels of the color image. Although its uses qubits nearly one-third of NCQI approach but still it requires more qubits compare to NEQR approach since it consider indices values. Besides, it's also limited to a 2×2 image size. In the same year, Sahin et al. proposed a quantum representation of multiwavelength images (QRMW) (Şahin et al. 2018) to address the representation issue of multi-spectral issues. In the same year, Nanrun et al., proposed a quantum representation model for multiple images (QRMMI) approach (Boev et al. 2022) to represent multiple images that could save hardware space for image representation. For state label circuit connection, this method uses the NEQR approach, which is why it requires a huge number of bits to complete full circuit connection, which leads to the complexity of preparation.

Figure 8. Example of QRMMI approach for multi-image representation

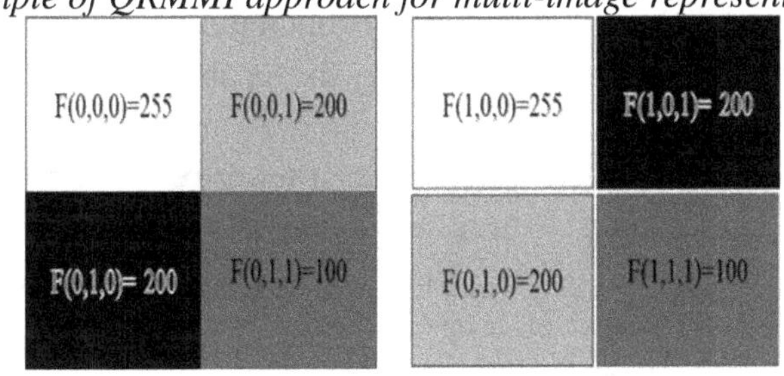

The QRMMI representation of the two 2×2 images shown in Figure 8 are expressed as

$$|I_{QRMMI}\rangle = \frac{1}{(\sqrt{2})^3}(|255\rangle|000\rangle + |200\rangle|001\rangle + |0\rangle|010\rangle + |100\rangle|011\rangle + |255\rangle|100\rangle + |0\rangle|101\rangle + |200\rangle|110\rangle + |100\rangle|111\rangle) \quad (2.5)$$

In 2019, Wang et al., also proposed a biplane-based quantum image representation of color images named quantum representation of color image (QRCI) (Wang et al. 2019). This method requires $(2n + 6)$ ubits to store the color of the digital images. Compared to the NCQI, the storage capability of the QRCI improved 2^{18}. In the same year, Khan proposed an improved flexible representation of the quantum image (IFRQI) approach (Khan 2019), which represents the color of a grayscale image based on angle. This method is probabilistic and limited to 4×4 or 8×8 images. Due to the probabilistic-based measurement, the accurate reconstruction of the image with measuring quality it is still challenging.

Wang et al., proposed quantum indexed image representation (QIIR) approach (Wang et al. 2020) that consist of two quantum matrices: one is data matrix, and another is palette matrix. In the data matrix, pixel-related information, such as pixel values with their positions, is considered. On the other hand, in the palette matrix, indexed and color values are considered. This system becomes more complex and needs plenty of qubits to represent the medium or big size. Xe et al., proposed an order-encoded quantum image model (OQIM) approach (Xu et al. 2019) that uses an ascending order of pixel values for real, complex, and integer signals. This complex method is only applicable for 2×2 images. In the case of images of other sizes, how this system works is not mentioned. Also, the performance measure of the proposed system on reconstruction images is not mentioned. In 2020, Grigoryan et al., proposed a Fourier transform representation approach (Grigoryan et al. 2020) that represents the quantum image in a new way. This method is based on the 3-point DFT (discrete Fourier transform) approach. In this method, image reconstruction with measuring quality is still lacking. Wang et al., proposed a double quantum color images representation model (DQRCI) (Wang 2020) that uses quantum superposition state to store two colors of digital image simultaneously. This method uses the projective measurement technique to measure the reconstruction image. Due to two color images separately along with bit-plane and position information, it requires a huge amount of qubit and finally generates a lot of required bits. Hai-Sheng et al. proposed $(n+1)$-qubit normal arbitrary quantum superposition state (NAQSS) approach (Li et al. 2014) that resolves the three-dimensional representational issue of (1024×1024×1024). Moreover, it has also resolved the poor segmentation issue based on content search. It requires higher qubits, such as 31; among those, 30 qubits represent the color and coordinates of two pixels, and the remaining qubits carry the segmentation information. In 2013, Li et al. proposed a quantum state for M- colors (QSMC) and quantum states for N- coordinates (QSNC) of a square image pixel known as QSMC&QSNC (Li et al. 2013) [51]. Sun et al., proposed a multi-channel representation for quantum image (MCQI) [26] that captures RGB information of the color image and its corresponding position. This method is limited to a 64×64 image size, and it requires 15 qubits.

2.6 Taxonomy for Quantum Image Representation and Compression

Figure 9 shows the taxonomy of the quantum image representation and compression scheme. Generally, it represents pixel-based, angle-based, coefficient transfer-based, and HSI (Hue Saturation Intensity) based. It also shows the taxonomy of quantum image representation for binary, gray, and color images. Although various kinds of representation are already developed, among those angles and pixel-based representation approaches got more popular because of operational advantages and easy to integrate others system. Coefficient-based representation is more capable of representing images inside the quantum domain due to the mapping of coefficients rather than pixels. It reduces the state preparation complexity.

Figure 9. Taxonomy to quantum image representation and compression

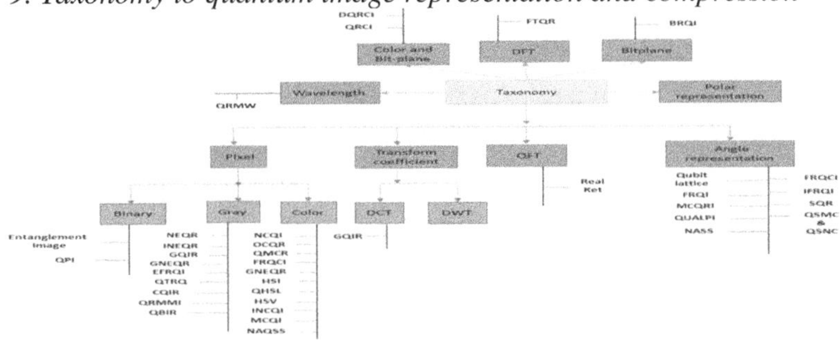

3. PROPOSED APPROACH

Figure 10(a) shows the conventional scheme that converts the pixel value directly into the quantum domain. Figure 10(b) shows the methodology of the proposed approach. In Figure 10, the Orange colored block indicates the classical computation, whereas the green block indicates the quantum. Compared to the traditional scheme, the proposed scheme deals with coefficient transform rather than representing the pixel value directly, which is done via the classical preparation approach. After that, the quantized coefficient value and its corresponding position are prepared to make it suitable for quantum representation. These things happen in quantum computers, but operation calculation is done via classical computers. In a pre-processing step, 8 and 64 block sizes of DWT and 8 block sizes of DCT were considered, including *Q=8, 16, 32, 36, and 70* quantization factors in comparing the performance of our proposed method.

Figure 10. (a) Traditional scheme and (b) Proposed scheme

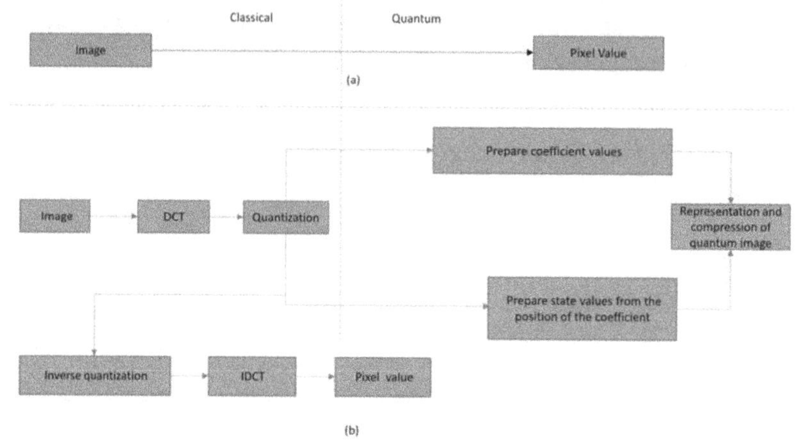

A $2^n \times 2^m$ image size is considered to be compressed in a quantized transform quantum system. Where *m* and *n* are the number of rows and columns of the image. The steps involved in the ZSCNEQR approach are:

Step 1: Divide the transfer coefficient and state position.

Step 2: After performing quantization, pre-process the coefficient and state position, which requires $(q+2n+1)$ qubits. Initially set all qubit's values to $|0\rangle$. Where *q* is the number of required qubits to map transfer coefficient values. On the other hand, $n_p=log_2(S_x)$ and $m_p = log_2(S_y)$ are the state positions representing qubits for representing *X*- and *Y*- positions of non-zero coefficients. Where, S_x and S_y are the block sizes of the non-zero *X*- and *Y*-positional images. An auxiliary qubit is used to make the connection between the transform coefficient and state position. The initial state of qubits can be expressed by the below equation [29].

$$|\Psi_0\rangle = |0\rangle^{\otimes(q+2n+1)} \quad (1)$$

Then, $(q + 1)$ identity gates and $2n$ Hadamard gates are used for coefficient preparation and its state preparation, respectively, and are expressed below.

$$I = \begin{bmatrix} 1 & 0 \\ 0 & 1 \end{bmatrix} \quad (2)$$

$$H = \begin{bmatrix} 1/\sqrt{2} & 1/\sqrt{2} \\ 1/\sqrt{2} & -1/\sqrt{2} \end{bmatrix} \quad (3)$$

In this step, the whole quantum step can be expressed as:

$$U = I^{\otimes q+1} \otimes H^{\otimes 2n} \tag{4}$$

The operator U transforms Ψ_0 from the initial to the intermediate state,

$$\Psi_1 = U(|\Psi_0\rangle) = I|0\rangle^{\otimes q+1} \otimes H^{\otimes 2n} \tag{5}$$

The final preparation step is done using the U_2 quantum operator:

$$\Psi_2 = U_2(|\Psi_1\rangle) = \frac{1}{2^n}\sum_{i=1}\sum^{j=1}|C_{YX}\rangle\left|YX\right\rangle \tag{6}$$

where $|C_{YX}\rangle$ and Y X are coefficients and the positional value of the non-zero values in the block. The quantum transforms operator U_2, which is given below.

$$U_2 = \prod_{X=0,\ldots,2^n-1}\prod_{Y=0,\ldots,2^n-1} U_{YX} \tag{7}$$

The required quantum resources are calculated using the below equation.

$$B_R = q_{\text{ones}} + S_{\text{state}} + S_{\text{bit}} + A_{\text{bit}} + B_e \tag{8}$$

The bit per pixel value is calculated as:

$$bpp = \frac{B_R}{I_s} \tag{9}$$

Where S_{state} is the state preparation bit. q_{ones} is the frequent number of ones from coefficients. The sign of the non-zero coefficient values is represented by S_{bit}. An N_{tcn} is the total number of non-zero coefficient elements. An A_{bit} is the number of bits that come from the auxiliary qubits. B_e is the required bit used to locate the block position errors of each block.

Step 3: After applying the block partitioning, calculate the required bits (BR) considering the block position error.
Step 4: Perform de-quantization.
Step 5: Perform inverse preparation and merging block.
Step 6: Measure rate-distortion (RD) performance metric for reconstruction.

The circuit of the ZSCNEQR is given in Figure 11 below.

Figure 11. Proposed ZSCNEQR circuit diagram

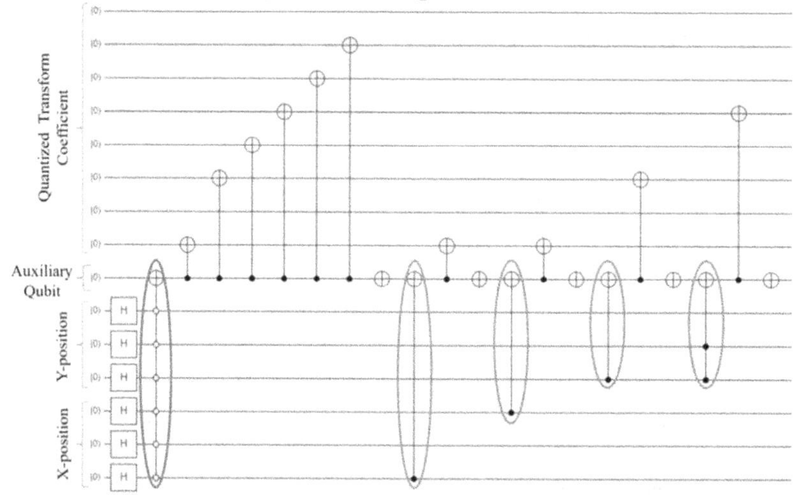

4. RESULTS AND DISCUSSION

In this section, transformed coefficient-based experimental results are analysed using various features and sizes of the image, such as deer(1024×1024), baboons(512×512), scenery (512×512), peppers(512×512), airport(1024×1024), and building (512 × 512) images (USC 2022; UoW et al. 2022; Qing-Bo et al. 2016).

Figure 12 depicts the required number of quantum resources and respective PSNR value of the transformed coefficient-based DCT-EFRQI, SCMNEQR, and ZSCNEQR approaches using different quantization factors for deer images. The comparison results show that for each quantization factor, the ZSCNEQR approach exhibits better performance compared to the SCMNEQR and DCTEFRQI approaches. On the other hand, SCMNEQR demonstrates a better result compared to the DCT-EFRQI approach because of the different state encoding approaches. Inversely, we can say that DCT-EFRQI requires a higher amount of RD values compared to the SCMNEQR approach due to bit values. From this scenario of the result, it is concluded that the proposed ZSCNEQR scheme performs better than DCT-EFRQI, and SCMNEQR approaches.

Figure 12. RD performance of the deer image using quantized transform coefficient approach.

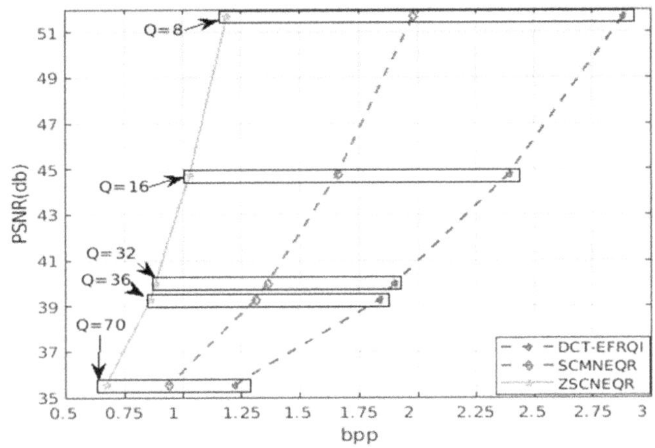

Figure 13 depicts the comparative results of the RD performance metric of the proposed ZSCNEQR approach against the DCTEFRQI and SCMNEQR approaches for airport images. The comparison results show that the proposed approach compresses the gray channel of the deer image efficiently compared to all other considered approaches. On the other hand, the proposed approach requires a lower bpp value, such as operational gates, which means it has more compression ability. For each quantization factor, the proposed approach requires less bpp compared to the SCNEQR and DCT-EFRQI approaches. From this result of the experiment, it is concluded that for deer image, ZSCNEQR performs superior compared to the DCT-EFRQI approach.

Figure 13. RD performance of the airport image using quantized transform coefficient approach

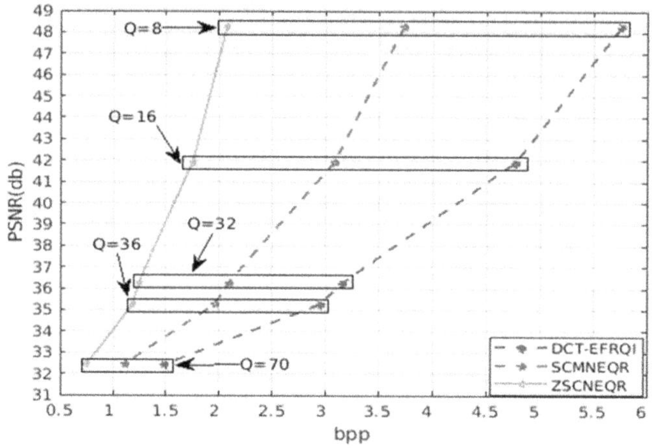

Figure 14 shows the RD computational results of the ZSCNEQR approach for baboon images using 8,16, 32, 16, and 70 quantization factors compared to SCMNEQR and DCTEFRQI approaches. It has been seen that the ZSCNEQR approach has a better ability to compress the baboon image over considered quantization factors compared to the SCMNEQR and DCT-EFRQI approaches. Based on the comparison result of the RD performance metric, the ZSCNEQR approach performs better compared to the DCT-EFRQI approach.

Figure 14. RD performance of the baboon image using quantized transform coefficient approach

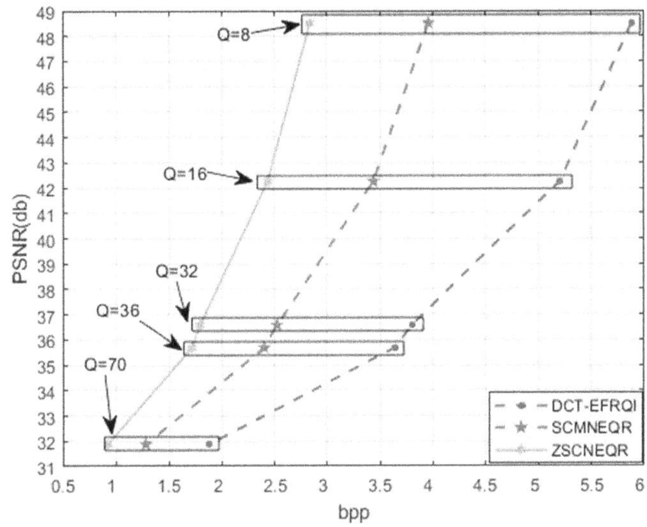

Figure 15 shows the RD performance metric of the ZSCNEQR approach compared to the DCT-EFRQI, and SCMNEQR approaches for the building image using 8,16,32,36, and 70 quantization factors. The comparison result reveals that the ZSCNEQR approach provides a better RD value for each quantization factor compared to the DCTEFRQI, and SCMNEQR approaches. The result is expected because of its efficient circuit encoding technique for each transform coefficient, closing the connection and also dividing the coefficient and state into various blocks after performing quantization.

Figure 15. RD performance of the building image using quantized transform coefficient approach

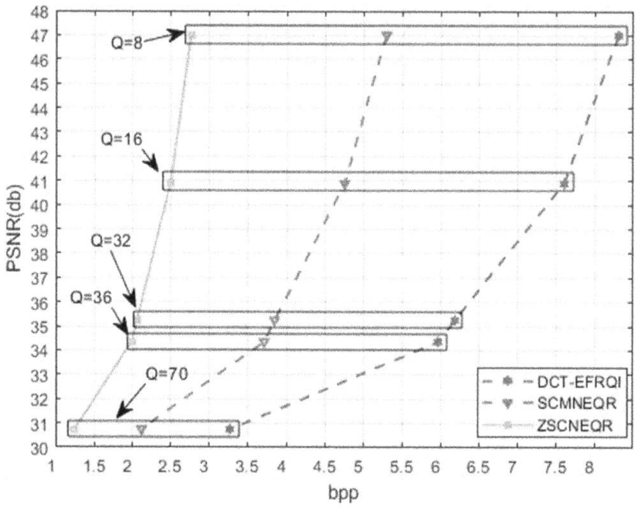

Figure 16 shows the RD performance of the ZSCNEQR approach alongside SCMNEQR and DCT-EFRQI approaches for the peppers image using 8, 16, 32, 36, and 70 quantization factors. Comparison results show that the ZSCNEQR approach exhibits better RD values compared to the DCT-EFRQI approach. Conversely, JPEG performs better RD performance compared to the proposed approach. For each quantization factor, the bpp value of the proposed approach requires fewer values compared to the DCT-EFRQI approach. In the case of PSNR, both approaches exhibit the same number of values. The difference in the bpp value contributes to the performance difference because of the block position encoding alongside universal gates. As a result, the ZSCNEQR approach draws a better RD performance compared to the SCNEQR and DCT-EFRQI approaches.

Figure 16. RD performance of the pepper image using the quantized transform coefficient approach

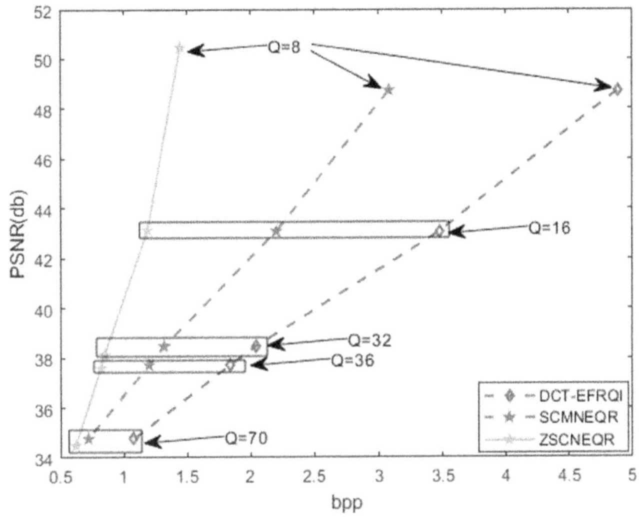

Figure 17 shows the RD computational results of the ZSCNEQR approach for scenery images using 8,16,32,36, and 70 quantization factors compared to the ZSC-NEQR and DCTEFRQI approaches. It has been seen that the proposed approach has a better ability to compress the scenery image over considered quantization factors compared to the SCMNEQR and DCT-EFRQI approaches. Based on the comparison result of the RD performance metric, the ZSCNEQR approach performs better compared to the DCT-EFRQI approach.

Figure 17. RD performance of the scenery image using quantized transform coefficient approach

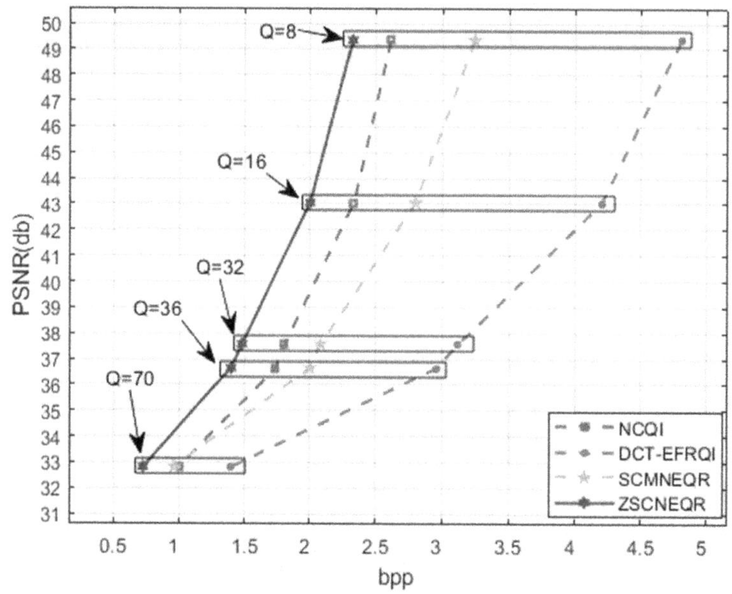

5. OPPORTUNITIES AND CHALLENGES

Quantum computing improves computational efficiency due to the properties of quantum mechanics. The pre-processing step of quantum computation is complex since it needs to create the superposition, entanglement, and parallelism properties of quantum mechanics (Ruan et al. 2021). Generally, at the end of the quantum algorithm, the superposition and entanglement degenerated using the probabilistic-based amplitude of the qubit connection. Image representation and compression are becoming major fields in quantum computing. The existing algorithm involves the FRQI and NEQR quantum image representation. For digital image processing, the FRQI and NEQR approach has a lot of limitations, such as higher complexity of preparation. Moreover, how and in what way the color and its corresponding position are encoded in the quantum circuit is still a big challenge for grayscale images. Since it stores the probability of the grayscale information using a single qubit, it is challenging to retrieve the original information.

A quantum state is a set of a wide range of vectors performed by metrics operation in high-dimensional space. Furthermore, machine learning protocols operate using the matrix operation. Adopting machine learning tasks with quantum circuits would be incredibly beneficial to simulate the natural behaviour of the word. In quantum computing, complexity happens with the state preparation of pixels. In this work, we have introduced a quantized transfer coefficient block partitioning circuit and zero discards approach to compress and represent the grayscale image. Besides, measuring the performance metrics to demonstrate its image compression capacity using the required number of gates per pixel is introduced first. Rather than encoding pixels directly into the quantum circuit, a quantized transform version of pixels is encoded.

Based on the above discussion, a quantum circuit is the right candidate to address the higher requirement of computational resources for storing, processing, and retrieving image data for representation and compression purposes. By processing and mapping classical image data into quantum circuits, superposition, entanglement, and parallelism of a quantum state can be the advantages of data processing (Giovannetti et al. 2008; Giovannetti et al. 2008a) .

QRAM (Quantum Random Access Memory) prepares the superposition state. It distributes the *Nd*-dimensional vectors into *Nd* superposition qubit. The time complexity of the QRAM approach is *O(log(Nd))*. It needs more physical resources such as *O(Nd)*. For the required number of qubits, it increases exponentially. As a result of experimental verification, it is a question of providing the advantages or not (Aaronson 2015; Jeremy et al. 2015).

Noise reduction is another challenging factor for quantum image computation. In the 1990s, quantum error correction codes (Knill et al. 2000) were studied to improve the fault-tolerant quantum computing system (Steane 1999). Nowadays, noise elimination is one of the fascinating research fields where topology quantum computing is the most concerning issue (David et al. 2016). Disjunctive Normal Form (DNF) (Bshouty et al. 1995) s an exciting function that can learn about noisy environments where classical cannot. Researchers face challenges integrating even and odd functions with DNF (Grilo et al. 2019). Moreover, linearity is the nature of quantum computing. Further investigation is required to induce the non-linearity properties of quantum computing, which is another big challenge.

6. CONCLUSION

This article introduces an efficient state modification approach compressing quantum grayscale images using the transformed coefficient-based quantized approach. Employing a lossy preparation method enhances compression as well as representation. One of its key advantages lies in utilizing the quantization factor, separating the state connection for further processing. In addition, rather than encoding the pixel value in the quantum processor, the transformed coefficient and its corresponding state connection circuit diagram demonstrate better performance. Furthermore, only six qubits are required for state connection, enabling the compression of grayscale images of any size. To optimize the circuit, the ZSCNEQR approach incorporates the universal quantum Toffoli gate and block-wise division. Another benefit is that it does not depend on a lookup table for operation. Moreover, the outcome of this approach is deterministic. In conclusion, our proposed ZSCNEQR scheme outperforms the DCT-EFRQI and SCMNEQR approaches in terms of RD performance.

REFERENCES

Aaronson, S. (2008). The limits of quantum. *Scientific American, 298*(3), 62–69. DOI: 10.1038/scientificamerican0308-62 PMID: 18357822

Aaronson, S. (2015). Read the fine print. *Nature Physics, 11*(4), 291–293. DOI: 10.1038/nphys3272

Abdolmaleky, M., Naseri, M., Batle, J., Farouk, A., & Gong, L.-H. (2017). Red-Green-Blue multi-channel quantum representation of digital images. *Optik (Stuttgart), 128*, 121–132. DOI: 10.1016/j.ijleo.2016.09.123

(2013).. . *Adv. Comput. Intell. Intell. Inform, 17*(3).

Avaliani, A., "Quantum computers," arXiv preprint cs/0405004, 2004.

Boev, A. S. (2022). D. N. Trubnikov D. Bokhan, A. S. Mastiukova and A. K. Fedorov, "Multiclass classification using quantum convolutional neural networks with hybrid quantum-classical learning,". *Frontiers in Physics, 10*, 1173.

Bshouty, H. N., & Jackson, J. C. "Learning DNF over the uniform distribution using a quantum example oracle", Proceedings of the eighth annual conference on Computational learning theory, vol., no., pp. 118-127, 1995. DOI: 10.1145/225298.225312

David, . (2016). Milestones toward Majorana-based quantum computing. *Physical Review X, 6*(3), 031016. DOI: 10.1103/PhysRevX.6.031016

Dendukuri, A., & Luu, K. Image processing in quantum computers. arXiv preprint arXiv:1812.11042, 2018.

Dong, P. Q. (2011). Le and Hirota, K., "A flexible representation of quantum images for polynomial preparation, image compression, and processing operations,". *Quantum Information Processing, 10*(1), 63–84. DOI: 10.1007/s11128-010-0177-y

Feynman, R. P. (2018). *"Simulating physics with computers," in Feynman and computation*. CRC Press.

Gao, Y., Zhang, Y., Lu, K., & Wang, M. (2013). NEQR: A novel enhanced quantum representation of digital images. *Quantum Information Processing, 12*(8), 2833–2860. DOI: 10.1007/s11128-013-0567-z

Giovannetti, V., Lloyd, S., & Maccone, L. (2008). Architectures for a quantum random access memory. *Physical Review A, 78*(5), 052310. DOI: 10.1103/PhysRevA.78.052310

Giovannetti, V., Lloyd, S., & Maccone, L. (2008). Quantum random access memory. *Physical Review Letters*, *100*(16), 160501. DOI: 10.1103/PhysRevLett.100.160501 PMID: 18518173

Grigoryan, A. M., & Agaian, S. S. (2020). New look on quantum representation of images: Fourier transform representation. *Quantum Information Processing*, *19*(5), 148. DOI: 10.1007/s11128-020-02643-3

Grilo, A. B., Kerenidis, I., & Zijlstra, T. (2019). Learning-with- errors problem is easy with quantum samples. *Physical Review. A*, *99*(3), 032314. DOI: 10.1103/PhysRevA.99.032314

Grover, L. K. "A fast quantum mechanical algorithm for database search," in *Proceedings of the twenty-eighth annual ACM symposium on Theory of computing*, 1996, pp. 212–219. DOI: 10.1145/237814.237866

Guerreschi, G. G., Mandra, S., & Aspuru-Guzik, A. (2016). Faster than classical quantum algorithm for dense formulas of exact satisfiability and occupation problems. *New Journal of Physics*, *18*(7), 073003. DOI: 10.1088/1367-2630/18/7/073003

Haque, E., Paul, M., Ulhaq, A., & Debnath, T. "Advance quantum image representation and compression using DCT-EFRQI approach," Scientific Report, vol. 13, 2023.

Hu, H. Y. (2018). Dang, N. Jiang, X. Lu and Y. Cai, "A novel quantum image compression method based on jpeg,". *International Journal of Theoretical Physics*, *57*(3), 611–636. DOI: 10.1007/s10773-017-3593-2

Hu, W. W., Zhou, R. G., El-Rafei, A., & Jiang, S. X. (2019). Quantum image watermarking algorithm based on Haar wavelet transform. IEEE Access, 7, 121303 121320.

Iliyasu, F. (2016). Yan and Venegas-Andraca. S. E., "A survey of quantum image representations,". *Quantum Information Processing*, *15*(1), 1–35. DOI: 10.1007/s11128-015-1195-6

Information Processing, 2014. 13: p. 1353-1379.

Jacobs, I. S. (1963). *Fine particles, thin films and exchange anisotropy*. Magnetism.

Jeremy., ... "Advances in quantum machine learning",arXiv preprint arXiv:1512.02900,vol.,no.4,pp., 2015.

Jiang, N., Hu, H., Dang, Y., & Zhang, W. (2017). Quantum point cloud and its compression. *International Journal of Theoretical Physics*, *56*(10), 3147–3163. DOI: 10.1007/s10773-017-3483-7

Jiang, N., & Wang, L. (2015). Quantum image scaling using nearest neighbor interpolation. *Quantum Information Processing, 14*(5), 1559–1571. DOI: 10.1007/s11128-014-0841-8

Kalonia, M., & Gupta, T. "Review on applications of quantum image processing," in Thirteenth International Conference on Digital Image Processing (ICDIP 2021). SPIE, 2021, vol. 11878, pp. 519–528. DOI: 10.1117/12.2601025

Kazuhiro, G. (2015). "Quantum computing: Limits, options and applications," 2020, Forbes Technology Council. Jiang, N., J. Wang, and Y. Mu, Quantum image scaling up based on nearest-neighbor interpolation with integer scaling ratio. *Quantum Information Processing, 14*(11), 4001–4026.

Khan, R. A. (2019). An improved flexible representation of quantum images. *Quantum Information Processing, 18*(7), 1–19. DOI: 10.1007/s11128-019-2306-6

Knill, E., Laflamme, R., & Viola, L. (2000). Theory of quantum error correction for general noise. *Physical Review Letters, 84*(11), 2525–2528. DOI: 10.1103/PhysRevLett.84.2525 PMID: 11018926

Latorre, J. I., "Image compression and entanglement," arXiv preprint quant ph/0510031, 2005.

Laurel, C. O., Dong, S., & Cruz-Irisson, M. (2015). Equivalence of a bit pixel image to a quantum pixel image. *Communications in Theoretical Physics, 64*(5), 501–506. DOI: 10.1088/0253-6102/64/5/501

Li, H.-S., Chen, X., Xia, H., Liang, Y., & Zhou, Z. (2018). A quantum image representation based on bitplanes. *IEEE Access : Practical Innovations, Open Solutions, 6*, 62396–62404. DOI: 10.1109/ACCESS.2018.2871691

Li, H.-S., Qingxin, Z., Lan, S., Shen, C.-Y., Zhou, R., & Mo, J. (2013). Image storage, retrieval, compression and segmentation in a quantum system. *Quantum Information Processing, 12*(6), 2269–2290. DOI: 10.1007/s11128-012-0521-5

Li, H.-S., Zhu, Q., Zhou, R.-G., Song, L., & Yang, X. (2014). Multi-dimensional color image storage and retrieval for a normal arbitrary quantum superposition state. *Quantum Information Processing, 13*(4), 991–1011. DOI: 10.1007/s11128-013-0705-7

Li, M. (2014). *H. Ian H. Li, Q. Zhu et al., "Multidimensional color image storage, retrieval, and compression based on quantum amplitudes and phases* (Vol. 273). Information Sciences.

Li, P., Xiao, H., & Li, B. (2016). Quantum representation and watermark strategy for color images based on the controlled rotation of qubits. *Quantum Information Processing*, *15*(11), 4415–4440. DOI: 10.1007/s11128-016-1413-x

Liu, C., S. Lu J. Su, X. Guo and L. Li, "An improved novel quantum image representation and its experimental test on ibm quantum experience," Scientific

Liu, K., Zhang, Y., Lu, K., Wang, X., & Wang, X. (2018). An optimized quantum representation for color digital images. *International Journal of Theoretical Physics*, *57*(10), 2938–2948. DOI: 10.1007/s10773-018-3813-4

Nielson, M. Quirk: quantum circuit simulator, https://algassert.com/quirk, accessed on 14.03.2025.

Olson, J. P., Romero, J., & Aspuru-Guzik, A. (2017). Quantum autoencoders for efficient compression of quantum data. *Quantum Science and Technology*, *2*(4), 045001. DOI: 10.1088/2058-9565/aa8072

Qing-Bo, Y., Pan-Pan, N., Xiang-Yang, W., & Li-Ran, S. (2016). A new robust color image watermarking method for multimedia technology enhanced learning protection. *Journal of Intelligent & Fuzzy Systems*, *31*(5), 2553–2564. DOI: 10.3233/JIFS-169096

Relese, I. T. Advantages and disadvantages of quantum computers. Accessed on 01, 2023. Rehman, J., "Advantages and disadvantages of quantum computers," Learn about IT beyond distance, https://www.itrelease.com/2020/10/advantages-and-disadvantages-of-quantum-computers/

(2021).. . *Reports*, *11*(1), 1–13.

Ruan, Y., Xue, X., & Shen, Y. (2021). Quantum image processing: Opportunities and challenges. *Mathematical Problems in Engineering*, *2021*(1), 6671613.

Şahín, E., & Yilmaz, İ. (2018). QRMW: Quantum representation of multi wavelength images. *Turkish Journal of Electrical Engineering and Computer Sciences*, *26*(2), 768–779. DOI: 10.3906/elk-1705-396

Şahin, E. and I. Yilmaz, QRMW: quantum representation of multi wavelength images.

Sang, J., & Shen, W. (2017). A novel quantum representation of color digital images. *Quantum Information Processing*, *16*(2), 1–14. DOI: 10.1007/s11128-016-1463-0

Shor, P. W. "Algorithms for quantum computation: discrete logarithms and factoring," in Proceedings 35th annual symposium on foundations of computer science. Ieee, 1994, pp. 124 134. DOI: 10.1109/SFCS.1994.365700

Siomau, M. (2019). Any quantum network is structurally controllable by a single driving signal. *Quantum Information Processing*, *18*(1), 1–26. DOI: 10.1007/s11128-018-2112-6

Steane, A. M. (1999). Efficient fault-tolerant quantum computing. *Nature*, *399*(6732), 124–126. DOI: 10.1038/20127

Sun, B., ... *A multi-channel representation for images on quantum computers using the RGBα color space*. in *2011 IEEE 7th International Symposium on Intelligent Signal Processing*. 2011. IEEE.

Sun, B., et al., An RGB multi-channel representation for images on quantum computers. J.

Tohidi, F., Ul-Haq, A., Haque, E., Paul, M., & Debnath, T. "Enhancing image representation and compression: An innovative nz-nqer framework with block truncation quantum coding," in 2023 The International Conference on Digital Image Computing: Techniques and Applications (DICTA). IEEE Xplore, 2023.

University of Washington. "Object and concept recognition for content-based image retrieval," Accessed on 12, Dec, 2022, http://imagedatabase.cs.washington.edu/

University Southern California School of Engineering. "The usc-sipi image database," Accessed on 12, Dec, 2022, https://sipi.usc.edu/database/

Venegas-Andraca, S. E., & Ball, J. L. (2010). Processing images in entangled quantum systems. *Quantum Information Processing*, *9*(1), 1–11. DOI: 10.1007/s11128-009-0123-z

Venegas-Andraca, S. E., & Bose, S. (2003). "Storing, processing, and retrieving an image using quantum mechanics," in Quantum information and computation. *SPIE*, *5105*, 137–147.

Wang, B., Hao, M., Li, P., & Liu, Z. (2020). Quantum Representation of Indexed Images and its Applications. *International Journal of Theoretical Physics*, *59*(2), 374–402. DOI: 10.1007/s10773-019-04331-0

Wang, L., Ran, Q., & Ma, J. (2020). Double quantum color images encryption scheme based on DQRCI. *Multimedia Tools and Applications*, *79*(9-10), 6661–6687. DOI: 10.1007/s11042-019-08514-z

Wang, L., Ran, Q., Ma, J., Yu, S., & Tan, L. (2019). QRCI: A new quantum representation model of color digital images. *Optics Communications*, *438*, 147–158. DOI: 10.1016/j.optcom.2019.01.015

Wang, M. (2021). Xu Z. and Zhang, Y., "Review of quantum image processing,". *Archives of Computational Methods in Engineering*, •••, 1–25.

Wang, B., et al., Quantum representation of indexed images and its applications.

Wang, L., et al., QRCI: A new quantum representation model of color digital images.

Xu, G., et al., Order-encoded quantum image model and parallel histogram specification.

Yan, F. (2013). F. Dong B. Sun, A. Iliyasu and K. Hirota, "An rgb multi-channel representation for images on quantum computers,". *J. Adv. Comput. Intell. Intell. Inform*, *17*(3).

Younes, A., N. Nasr and A. Elsayed, "Efficient representations of digital images on quantum computers," Multimedia Tools and Applications, vol. 80, no. 25, pp. 34019 34034, 2021.

Yuan, S., et al., SQR: a simple quantum representation of infrared images. Quantum

Zhou, N., Yan, X., Liang, H., Tao, X., & Li, G. (2018). Multi-image encryption scheme based on quantum 3D Arnold transform and scaled Zhongtang chaotic system. *Quantum Information Processing*, *17*(12), 1–36. DOI: 10.1007/s11128-018-2104-6

Chapter 10

Cyber Security Risk in Smart Agriculture in Regional Australia

Arjun Neupane
https://orcid.org/0000-0002-1010-7552
CQUniversity, Australia

Tej Bahadur Shahi
CQUniversity, Australia & Queensland University of Technology, Australia

Sameer Sitoula
https://orcid.org/0009-0000-1609-779X
Tribhuvan University, Mechi Multiple Campus, Nepal

Philip Kibet Langat
CQUniversity, Australia

Kerry Walsh
CQUniversity, Australia

ABSTRACT

Agriculture is a fundamental global primary industry, evolving with advances in ICT and automation. The rise in 'smart farming' integrates cutting-edge technologies like IoT, drones, sensors, GPS, big data analytics, and AI to improve efficiency, productivity, and sustainability. These innovations facilitate real-time monitoring, data-driven decisions, and automation in key farming tasks, including soil analysis, irrigation, crop health evaluation, and pest management. However, the adoption of smart technologies introduces cybersecurity risks. This chapter explores cybersecurity threats in smart farming in regional Australia. As farmers increasingly rely on

DOI: 10.4018/979-8-3373-0020-7.ch010

Copyright © 2025, IGI Global Scientific Publishing. Copying or distributing in print or electronic forms without written permission of IGI Global Scientific Publishing is prohibited.

advanced technologies such as IoT, drones and satellites, blockchain, and robotics, implementing cybersecurity measures is vital. Without robust security, farmers may lose trust in these technologies. Farm data and IT assets require raising awareness, promoting best practices and integrating cybersecurity into agricultural systems.

1. INTRODUCTION AND BACKGROUND

The agriculture industry has evolved from Agriculture 1.0 (Manual Era) to Agriculture Industry 5.0 (The Digital Revolution in Farming with Integration of Artificial Intelligence (AI) and Robotics, Bissadu et al., 2024), with the development of farming practices that are more automated and digital. Various terms are used to describe the new farming techniques, e.g., digital farming (DF) refers to the integration of digital technologies into farming practices, while Smart Agriculture (SA) refers to the use of 'smart' technologies to optimise agricultural efficiencies and productivity and promote sustainability (Brisco et al., 1998; Bui et al., 2024; Pierce & Nowak, 1999). Advanced technologies include Artificial Intelligence (AI), Internet of Things (IoT), drones, sensing technologies, robotics, machine and deep learning and advanced data analytics to empower farmers to make data-driven decisions and automate the wide range of farming tasks such as soil analysis (Barathkumar et al., 2024), plant monitoring (Madeira et al., 2024), crop disease detection (Shahi et al. 2023), pest management (Hanif et al., 2022), Irrigation status detection (Shahi et al. 2024), farm management (Mishra & Mishra, 2024) and many more.

In the last two decades, digital agriculture has revolutionised farming practices worldwide, enabling farmers to maximise productivity and sustainability while addressing a country's unique challenges, like Australia's, such as water availability, labour shortages, skills shortages and human capital, and vast geographic diversity (Wu et al., 2019). In particular, regional Australia, smart farming offers transformative solutions for farmers to address their unique challenges where environmental, economic, and logistical factors often pose significant hurdles. Although modern information technology brings new opportunities to the development of agricultural production, it also creates great demands and challenges to security and privacy in the field of smart agriculture.

On the other hand, the adoption of technologies brings the risk of malware, data theft, and software viruses and bugs (Singh & Kaur, 2022, Demestichas et al., 2020). Cybersecurity is the protection of information, services and systems against hackers and disruptive events that can compromise information security and confidentiality (Borchi, et al., 2021, Craigen et al., 2014). Hackers are individuals who attempt to steal, alter, or disrupt information or prevent others from accessing it to operate their businesses. Their motives often include making money or obtaining confidential details about a company's activities, especially related to research or innovations aimed at improving the business (Borchi, et al., 2021).

It is essential to safeguard farmers' data and IT assets by improving awareness, encouraging the adoption of security best practices and standards and integrating security measures into the system (Hazrati et al., 2022). Zanella et al. (2020) focus on the security challenges in SA and identify various security threats in IoT resources, including unauthorised access, Denial of Service (DoS), Distributed Denial of Service (DDoS), and man-in-the-middle (MITM) attacks, among others. A DoS attack renders network and services unavailable to its users by overwhelming them with excessive network traffic. On the other hand, a DDoS attack utilises multiple systems to flood a target system with high traffic, making it inaccessible to legitimate users. Meanwhile, an MITM attack involves an attacker intercepting and potentially altering communication between two parties, enabling them to steal data or cause harm. In this line of research, Alahe et al. (2024) highlighted the overview of security challenges in smart agriculture regarding threat types, current status and future trends.

In recent years, many cybersecurity incidents have been seen in the agriculture sector, with scammers mainly targeting farmers because they are less aware of these incidents. According to reports from the Australian Competition and Consumer Commission (ACCC's) Scam-watch, Australian farm businesses experienced losses exceeding $1.2 million due to scammers between January 1 and August 31, 2022 (ACCC,2022). This represents an increase of over 20% compared to the same timeframe in the previous year. There has been an increase in the reporting of these incidents, and regional Australian farmers are becoming more aware of cybersecurity and its associated risks. However, many still lack comprehensive practices to protect their systems from cyber threats.

This study aims to examine the cybersecurity risks associated with smart farming technologies in regional Australian agriculture, where the reliance on digital solutions is growing rapidly. Given regional Australia's unique geographic and infrastructural challenges, such as limited access to high-speed internet and technical expertise, farmers may face heightened vulnerabilities to cyber threats, including unauthorised access to IoT devices, ransomware attacks on farm management systems, and data breaches. These risks affect the individual farmers' operations and can have broader implications for the agricultural supply chain and food security. In addition, this study assesses the current cybersecurity practices and awareness levels among Australian farmers. It will explore how well farmers understand cyber risks, the measures they adopt to secure their systems, and the gaps in knowledge and infrastructure that may expose them to potential attacks. By investigating these aspects, the study seeks to provide recommendations for improving cybersecurity preparedness, building resilience, and fostering a culture of cyber awareness in regional agriculture. The insights from this research would help Australian farmers continue embracing the benefits of smart agriculture while safeguarding their operations and data against evolving cyber threats.

The rest of the chapter is structured as follows. Section 2 identifies and describes the cybersecurity threat landscape in smart agriculture. Section 3 describes the impact of cybersecurity risks on regional Australian farmers. Section 4 addresses risk mitigation strategies and best practices. Section 5 discusses the challenges in implementing cybersecurity measures in regional Australia. Section 6 provides future directions, emerging trends, and the conclusion of the chapter.

2. CYBERSECURITY THREAT LANDSCAPE IN SA

The digital agriculture system includes four main components: farm data collection, data transfer infrastructure, data analytics and applications (see Figure 1). Building such an ecosystem integrates the various key technologies for data collection (e.g. IoT sensors, drones), data transfers (e.g. Wi-Fi, internet and networking), data analytics (e.g. AI and ML), and applications (e.g. pest management). In fact, the advancement of these technologies has transfigured traditional farming practices and empowered farmers to adapt to the rapidly changing demands of the agriculture industry. The key technologies in SA are reported in Table 1.

Table 1. Key technologies in smart agriculture

Technology	Description in Agriculture	Application	References
Internet of Things (IoT)	Use of sensors, devices and analytics to monitor and manage various aspects of agricultural production	Monitor farming processes such as water management, irrigation management, soil management, precision farming, etc.	(Sinha & Dhanalakshmi, 2022; Verma et al., 2018)
Sensors	Devices that are used to measure and collect data on environmental and farm parameters	Crop quality assessment, crop classification, soil moisture measurement	(Ullo & Sinha, 2021)
Drones	Aerial devices equipped with cameras and sensors to gather high-resolution images and data.	Drones can assist with crop monitoring, irrigation schedule planning, and fertiliser application.	(Kumar et al., 2023; Maddikunta et al., 2021)
Artificial Intelligence (AI)	The use of computational power and algorithms to analyse a variety of data sources facilitates predictive analytics and automation in farming.	Improve the precision of information about the farm, e.g. crop quality, managing pests, monitoring soil, etc.	(Javaid et al., 2023; Smith, 2018)

continued on following page

Table 1. Continued

Technology	Description in Agriculture	Application	References
Big Data Analytics	Use of computational power to analyse large data sets from various sources to identify patterns and trends	Risk assessment of crops, yield estimation, operational planning, resource utilisation, and long-term decision-making capabilities	(Alfred et al., 2021; Wolfert et al., 2017)
Cloud Computing	Enables data storage and the processing of agriculture data to be accessible anywhere and anytime through an internet connection	Farmers keep track of the state of the crops and livestock, real-time processing, and remote monitoring of agricultural resources.	(Ivanochko & Melnyk, 2024; Rathod et al., 2022)
Satellite Imagery	Providing high-resolution, real-time data for monitoring and managing farm activities	Access crop health and soil condition, predict yields and assess drought stress.	(Gore et al., 2024; Virnodkar et al., 2020)
Blockchain	Transforming smart agriculture by providing a secure, transparent and decentralised system for managing data and transactions.	Lower transaction costs improve the efficiency and transparency of agriculture supply chains.	(Kamilaris et al., 2021; Potts, 2019)
Autonomous Machinery	Equipped with advanced sensors, cameras, GPS, AI and Machine Learning technologies	Robotics Harvesters (e.g. Robotic Mango Harvester developed by CQUniversity), Spraying drones, autonomous tractors, intelligent decision support	(Basri et al., 2021; Lochan et al., 2024,Goulart et al. 2023)
Precision Irrigation Systems	Using key technology (IoT, Sensors, data analytics) for farmers to optimise water use.	Remotely monitor and manage water scheduling, drive efficiency, sustainability, and resilience in smart farming practices.	(Adeyemi et al., 2017; Bwambale et al., 2022)
Remote Sensing	Use of satellite / aerial imagery and sensors to collect data from agriculture fields and crops without physical contact	Crop health monitoring, soil analysis, yield estimation and forecasting, water stress, pest and weed detection, mapping and zoning, etc.	(Inoue, 2020; Khanal et al., 2017)

Figure 1. Digital architecture for SA with cybersecurity risk

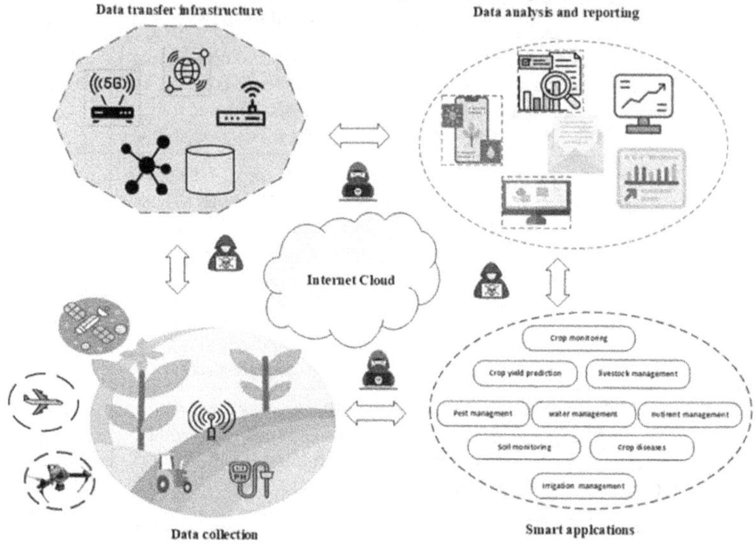

The layered architecture of SA (Figure 2) illustrates the potential technologies employed in SA. Each layer implements the key technologies that have unique cyber threat risks and impacts on SA (Bui et al. 2024). According to Yazdinejad et al. (2021), IoT and sensor technologies in SA have six critical areas of vulnerability to cyber-attacks, such as malware attacks, attacks on networks and equipment, data and code attacks, and others. Similarly, drone and remote sensing platforms pose confidentiality, integrity and availability-related threats, which eventually reduce the farmer's trust in technology. On the other end, data analytics with AI and Machine learning happens on cloud servers, which is at the risk of a man-in-the-middle attack and has the potential impact of false prediction with AI model manipulation. The other key technologies and related cyber threats are reported in Table 2.

Table 2. Key technologies with cyber threats and their impacts in SA

Key Technologies	Cyber Security Threats	Potential Impact	References
IoT and Sensors	Malware attacks, unauthorised access, data tampering, attacks on the network and related equipment, etc.	Disruption of operations, compromised decision-making, inaccurate data reporting	(Yazdinejad et al., 2021)
Drones	Confidentiality, integrity and availability-related threats such as Data attacks, networking attacks, service composition attacks	Loss of sensitive data, physical disruption, financial loss, possible loss of buyer confidence and revenue	(SP & Balamurugan, 2022)
AI and Big Data	Manipulation of AI models, data breaches, phishing, malicious scripts, DoS/ DDoS	Flawed predictions, accuracy, and operational inefficiencies.	(Zanella et al., 2020)
Cloud Computing	Cloud data leakage	Cloud data centers are distributed across the globe and pose a data leakage risk.	(Gupta et al., 2020)

Figure 2. An illustration of layered architecture for smart agriculture (adapted from (Bapatla et al., 2021))

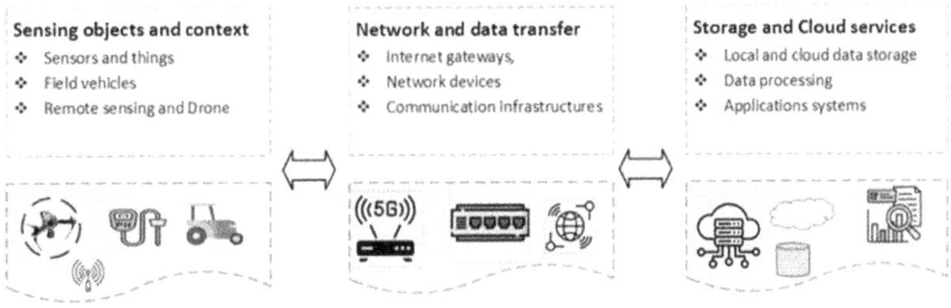

3. IMPACT OF CYBERSECURITY RISKS ON REGIONAL AUSTRALIA

Digital technology is revolutionising agriculture worldwide, including in Australia, by facilitating the adoption of innovative approaches. However, as farming operations are becoming more digitised, concerns about cyber security, including data protection and privacy (KG2, 2024; McCaig et al., 2023) and data security (Rotz et al., 2019), are growing. The cybersecurity risks pose significant challenges to the adoption of new technology and the success of smart agriculture in regional Australia. Below, we discuss the most notable cybersecurity risks and their impact on the agricultural industry.

3.1. Data Leakage and Loss of Confidentiality

Smart Agriculture relies on large datasets like weather patterns and climate change, soil quality and analysis, management of pests, crop production, market trends, tracking of crop development, etc. These large datasets or big data are valuable for modern smart farming, including informed decision-making and driving innovation in agriculture (Karunathilake et al., 2023). For example, agricultural drone spraying relies heavily on large datasets to ensure accuracy and efficient operation, such as minimising waste and avoiding overuse of chemicals in the field (Taseer & Han, 2024). On the other hand, all the smart devices used in SA generate and transmit large volumes of sensitive data such as crop yield, soil conditions, livestock health, operational strategies about the farm, etc. Unauthorised access to this sensitive farming data, i.e. stored in the cloud or on-premises and transmitted through unsecured networks that competitors misuse, sabotages farming operations. For example, GPS Spoofing is another threat in SA where the attacker sends false GPS signals to mislead the agriculture system (Tychola & Rantos, 2025). GPS enables farmers to create precise farm maps, measure field areas, and navigate accurately to specific locations, such as soil sampling and crop monitoring. Additionally, the intruders may maliciously alter critical information such as crop planting schedules and fertilisation periods, reports of pest attacks on crops, and misleading farmers. For instance, Ali et al. (2024) carried out a comprehensive analysis of security and privacy in IoT-based smart farming and concluded that the integration of IoT devices faces significant challenges in SA, including issues related to data protection and security. The consequences of data leakage include financial losses, reputational damage, and reduced trust among stakeholders, suppliers, consumers, etc.

3.2 Compromise of IoT Devices

IoT devices like GPS-enabled equipment, sensors, drones, and automated irrigation systems have revolutionised modern farming. However, they also introduce significant cybersecurity risks, such as many IoT devices lacking robust security protocols, making them vulnerable and easy targets for cyberattacks (Alyahya et al., 2022). For example, hackers can exploit the system's weaknesses by hijacking remotely controlled devices such as sensors and irrigation systems, potentially causing operational disruption. This can lead to crop losses due to flawed or compromised devices. Furthermore, there is a chance of potential environmental damage due to mismanaged irrigation and improper use of pesticides. Tsvetanov and Kuzmanov (2024) demonstrated six cyber security-related issues and communication challenges in smart irrigation systems, including data protection issues, poorly secured data or inappropriate encryption methods that lead to IoT security issues, farmers' underestimation of the importance of cyber security, required high-speed internet access in rural communities and finally physical and hardware issues.

3.3 Economic Losses and Market Instability

Cyber security threats can pose significant risks to national and international stability. Rival countries may launch cyberattacks on agricultural systems to destabilise the economy, targeting critical areas such as production, supply chains and resource management (Hazrati et al., 2022). Such attacks or cyber vulnerabilities disrupt the import and export of agricultural products, which may impact global food security (Gupta et al., 2020). Since around two-thirds of Australia's agricultural products are exported, such disruptions could affect Australia's trade balance and economy.

Ransomware attacks on Australian agriculture and food industry systems have occurred, affecting production and marketing and leading to revenue loss. In 2020, a malware attack disrupted Australian and New Zealand wool sales, and in 2021, a cyber-attack on global meat processing company JBS Foods led to the payment of ransom $ 11 million US dollar (Claughton & Beilharz, 2021; Rees & Rees, 2023). Another example was in 2022 when GrainCrop also faced a cyber-attack. An attacker targeted their grain handling and logistics operation, which caused disruptions to their operation (GrainCrop, 2023).

The growing risk of cyber threats in SA may lead to declining farmers' adoption of emerging technology and hinder the advancement of modern farming practices, research, and innovation. There is a risk of trust in the security of technologies that could decrease the funding for agriculture innovation and development. Small-scale and individual farmers are more vulnerable, facing economic difficulty and thus resourcing, and lacking training and access to skilled resources to invest in cyber security measures and follow the Australian cyber security strategy and compliance.

3.4 Loss of Trust in Technology

Most agribusiness organisations in Australia rely on their own or third-party infrastructure systems for daily business operations, including data storage, procurement and supply chain. Cyber security threats can evoke fear and mistrust among consumers, stakeholders and technology providers (Alqudhaibi et al., 2024). The cyber incident could discourage entire farming communities, especially in regional Australia, which lacks resources and technology expertise. McCaig et al. (2023) interviewed 32 farmers regarding the farmer-centric design thinking principles for smart farming technologies, and technical proficiency and mistrust were impacting factors for farmers of the changing landscape, specifically in IoT adoption. The lack of trust in technology may impact investors' and policymakers' ability to support innovation in agriculture.

4. RISK MITIGATION STRATEGIES AND BEST PRACTICES

Cybersecurity risk has significantly impacted the agriculture and food industry. To safeguard farming operations, protect important data, and ensure continuity in food production, best practices and effective risk mitigation strategies are essential (Hazrati et al., 2022). Below, we discuss these strategies and practices:

4.1 Strengthen IoT Device Security

Strengthening the security of IoT devices is essential for protecting against cyber threats in agriculture. To improve IoT device security, it is crucial to use strong and unique usernames and passwords for each device, updating them regularly to prevent unauthorised access. Additionally, encrypting data sent over insecure channels is crucial to protect sensitive information (Alyahya et al., 2022). Keeping the firmware of devices updated helps fix any security weaknesses that may develop over time. Separating IoT devices from the main network through network segmentation can significantly lower the risk of attacks by limiting access points. Enabling mutual authentication ensures that devices only connect with trusted systems, enhancing overall security (Ma et al., 2023). Other best practices include using secure boot features to prevent unauthorised changes, turning off unused services or ports on IoT devices, and setting up monitoring systems to identify unusual device activity.

4.2 Enhance Network Security

Enhancing network security is a vital strategy for protecting devices from cyberattacks in the agricultural sector. To achieve this, several best practices should be implemented. First, deploying firewalls can help control traffic and monitor the network for suspicious activities through Intrusion Detection Systems (IDS). Implementing network segmentation is also important; this involves separating operational networks, such as IoT devices, from administrative networks like databases to limit potential vulnerabilities. Additionally, encrypted communication protocols such as HTTPS, SSL/TLS, and VPNs are essential to secure data transmission. Access to networks should be restricted through whitelisting and role-based permissions, ensuring that only authorised users can access sensitive information (Gupta et al. 2020).

4.3 Implement Data Security Measures

Implementing data security measures, such as weather patterns, soil quality, crop production, and market trends, is essential for protecting critical datasets in the agriculture industry. According to Oche et al. (2020), their research identified several important measures that should be taken to ensure the security of agricultural data. First, it is crucial to encrypt sensitive data both during transmission over unsecured channels and while it is stored to prevent unauthorised access. To maintain accessibility, organisations should keep up-to-date backups of critical data, securely stored either offline or on separate storage systems. Additionally, implementing role-based access controls is important to limit who can view or modify sensitive information; for example, lower-level employees should have restricted access, while top-level employees may have broader access. Using data anonymisation techniques for shared datasets helps protect personal or sensitive information from exposure. Finally, regularly removing outdated or unnecessary data can further minimise risks associated with data breaches.

4.4 Educate and Raise Awareness Among Farmers

Farmers are important stakeholders in adapting new technologies and promoting SA. Therefore, they need to understand cybersecurity risk and current best practices. Educating farmers on how to use these safely and protect them from unauthorised access is crucial. Russell (2022) states that offering cybersecurity awareness programs specially designed for farmers and agricultural workers can help them learn about possible threats. It is also important to teach them how to recognise and avoid phishing scams and other cyberattacks. Additionally, farmers should be reminded to update their passwords regularly and practice safe browsing habits, like only visiting secure websites (Drape et al., 2021). Providing easy-to-understand guides or visual materials can make these ideas clearer. Finally, training farmers on how to respond to and manage different cyber incidents will help them deal with potential security problems effectively (Hazrati et al., 2022)

4.5 Collaborate with Industry and Government

Fostering collaboration between the agriculture industry and government is necessary to effectively identify and manage cybersecurity risks (Campoverde-Molina et al., 2024). It provides a forum to work together, fighting against cybersecurity threats and sharing threat intelligence and knowledge between industry partners and government agencies. Different groups can learn from each other's experiences and strengthen their defenses against cyber threats (Drape et al., 2021b). Further, collaborating on the development and implementation of sector-specific cybersecurity frameworks and standards can tailor security measures to the unique needs of the agricultural sector. Lastly, participating in regional or national cybersecurity awareness campaigns can raise awareness and educate stakeholders about potential threats and best practices for protection (Shah et al., 2019b).

5. CHALLENGES IN IMPLEMENTING CYBERSECURITY IN REGIONAL AGRICULTURE

Implementing cybersecurity best practices and risk mitigation faced significant challenges in the regional agriculture industry in Australia that hinder the protection of its digital infrastructure, such as IoT devices, cloud computing platforms, Drones and UAVs, GIS and Satellite imagery, smart irrigation systems, Agri digital marketplace, etc. One of the challenges with small farmers is that they do not have a proper infrastructure to protect their farms. Also, they do not have the expertise to operate, maintain or secure digital tools, exacerbating the skills and knowledge gap in the agriculture industry (Drape et al., 2021). The lack of comprehensive cybersecurity policies is another critical challenge for farms vulnerable to cyberattacks, threatening critical operations and data security and causing massive disruption to agribusiness and food production (AgriBusiness, 2025). Mace & Morga's (2022) report highlighted the workshop's findings on the cyber resilience of the business in Agriculture and identified four areas: cybersecurity awareness, backups, password and password management, antivirus protection, and business continuity plan. On the other hand, the Australian Cyber Security Centre (ACSC) has set eight mitigation strategies to protect against all cyber threats for all businesses. The following table highlights the challenges in implementing cybersecurity in regional Agriculture in Australia.

Table 3. Summary of the challenges of implementing cybersecurity

Challenges	Description
High investment in digital technology	Investing in advanced farming technologies such as IoT, drones, AI, and cybersecurity solutions is more expensive for small-scale farmers. However, they can benefit from smart farming tools and technologies, e.g. monitoring cop health via mobile apps.
Lack of cybersecurity policy	Proper cyber security policy and standardised cyber security guidelines make farms more vulnerable to cyber-attacks and operational risks. Also, the lack of proper regulation on security of IoT devices leads to unauthorised access to the system.
Smart agriculture infrastructure deficiencies	Regional Australia also has an inadequate internet connection and limited access to digital infrastructure, which hinders its critical system protection. For example, over the four years, around 50 thousand complaints about mobile and internet services in regional, and remote Australia (McCosker & Burton, 2024).
Resistance to the adoption of smart agriculture technologies	Farmers may resist adopting digital infrastructure because of the complexities of the technology or a lack of expertise in the system. Some of the farmers still used manual methods.
Skills and knowledge gap	Many farmers or workers lack the expertise to operate and maintain, and secure modern systems such as IoT devices, automated machinery, etc. These technologies require operational capabilities and experience to safeguard them against cyber threats. Also, farmers/workers lack formal cybersecurity training on data breaches, ransomware, malware, and phishing attacks.
Management of agricultural data	Managing a large volume of agricultural data is another challenge for farmers. Proper management of a large amount of farm data is vital for making informed decisions, optimising farm operations and maintaining sustainability. The lack of standardised formats and protocols for the different devices and platforms can also be a hurdle for data integration and a challenge in gaining a holistic view of farm operations.
Supply chain risk	Several entities were connected in the supply chain process, such as farmers, suppliers, logistics companies and retailers – these components are always vulnerable to cyberattacks. A security breach in any of these entities can potentially disrupt the entire supply chain and expose vulnerabilities across the system.

6. FUTURE DIRECTIONS, EMERGING TRENDS, AND CONCLUSION

The future of cybersecurity in smart agriculture in regional Australia will be shaped by increasing technological adoption and evolving threats. Federal and regional governments, including agricultural institutions, will likely develop more robust cybersecurity frameworks made-to-order for smart farming technologies. These frameworks will need to focus on securing data transmission, storage, and real-time analytics against cyber threats. Integrated artificial intelligence (AI) and machine learning (ML) in cybersecurity are likely to inform predictive analytics to detect anomalies and potential cyber threats in real-time, thus ensuring proactive security measures. Exploration of blockchain and distributed ledger technology, smart contracts and secure data exchanges will enhance traceability and mitigate cyber risks in agricultural supply chains. Training programs and cybersecurity literacy campaigns will be crucial in reducing vulnerabilities at the user level. Future research and policy development need to prioritise cyber resilience strategies to ensure rapid recovery from cyber incidents and minimise disruptions to agricultural activities.

The cybersecurity threats in smart agriculture in regional Australia are evolving rapidly. Key emerging trends include the increasing use of IoT in agriculture, which is increasing the cyber-attack surface with connected devices for precision farming and livestock monitoring. Cloud-based agriculture management systems are becoming central to data storage but introduce risks such as data breaches and unauthorised access. The integration of 5G and edge computing enhances real-time data analysis while posing new security vulnerabilities. Additionally, the convergence of cyber-physical security through robotics, AI, drones, and autonomous machinery demands comprehensive security solutions. Finally, governments and industry bodies continuously strengthen regulatory and legal frameworks to address sector-specific cybersecurity challenges.

In conclusion, cybersecurity threats in smart agriculture pose significant challenges to regional Australia, given the increasing reliance on digital technologies for efficient and sustainable farming. As cyber risks evolve, the agriculture industry must adopt proactive security measures, including AI-driven threat detection, blockchain security solutions, and robust regulatory frameworks. Farmers and agricultural enterprises must also prioritise cybersecurity education, training and resilience planning to protect critical infrastructure and data. By embracing these future directions and emerging trends, Australia's smart agriculture sector can ensure secure and sustainable activities while mitigating the risks associated with cyber threats.

REFERENCES

ACCC. (2022). *Farm businesses warned as scams targeting the agricultural sector grow*. Retrieved 12 January 2025 from https://www.accc.gov.au/media-release/farm-businesses-warned-as-scams-targeting-agricultural-sector-grow

Adeyemi, O., Grove, I., Peets, S., & Norton, T. (2017). Advanced monitoring and management *systems* for improving sustainability in precision irrigation. *Sustainability (Basel)*, *9*(3), 353. DOI: 10.3390/su9030353

AgriBusiness. (2025). *Boot up to tackle cyber threats*. Retrieved 9 January 2025 from https://agribusinessconnect.com.au/boot-up-to-tackle-cyber-threats/

Alahe, M. A., Wei, L., Chang, Y., Gummi, S. R., Kemeshi, J., Yang, X., Won, K., & Sher, M. (2024). Cyber security in smart agriculture: Threat types, current status, and future trends. *Computers and Electronics in Agriculture*, *226*, 109401. DOI: 10.1016/j.compag.2024.109401

Alfred, R., Obit, J. H., Chin, C. P.-Y., Haviluddin, H., & Lim, Y. (2021). Towards paddy rice smart farming: A review on big data, machine learning, and rice production tasks. *IEEE Access : Practical Innovations, Open Solutions*, *9*, 50358–50380. DOI: 10.1109/ACCESS.2021.3069449

Ali, I. A., Bukhari, W. A., Adnan, M., Kashif, M. I., Danish, A., & Sikander, A. (2024). Security and privacy in IoT-based smart farming: A review. *Multimedia Tools and Applications*, •••, 1–61. DOI: 10.1007/s11042-024-19653-3

Alqudhaibi, A., Krishna, A., Jagtap, S., Williams, N., Afy-Shararah, M., & Salonitis, K. (2024). Cybersecurity 4.0: Safeguarding trust and production in the digital food industry era. *Discover Food*, *4*(1), 2. DOI: 10.1007/s44187-023-00071-7

Alyahya, S., Khan, W. U., Ahmed, S., Marwat, S. N. K., & Habib, S. (2022). Cyber secure framework for smart agriculture: Robust and tamper-resistant authentication scheme for IoT devices. *Electronics (Basel)*, *11*(6), 963. DOI: 10.3390/electronics11060963

Australian Government. (2024). *Australian Government response to the House of Representatives Standing Committee on Agriculture and Industry report: Smart farming - inquiry into agricultural innovation*. Retrieved 24 December 2024 from https://www.agriculture.gov.au/about/reporting/obligations/government-responses/smart-farming-agricultural-innovation-inquiry#daff-page-main

Bapatla, A. K., Mohanty, S. P., & Kougianos, E. (2021). sFarm: A distributed ledger-based remote crop monitoring system for smart farming. IFIP International Internet of Things Conference

Barathkumar, S., Sellamuthu, K., Sathyabama, K., Malathi, P., Kumaraperumal, R., & Devagi, P. (2024). Advancements in soil quality assessment: A comprehensive review of machine learning and AI-driven approaches for nutrient deficiency analysis. *Communications in Soil Science and Plant Analysis*, •••, 1–26.

Basri, R., Islam, F., Shorif, S. B., & Uddin, M. S. (2021). Robots and drones in agriculture—a survey. *Computer vision and machine learning in agriculture*, 9-29.

Bissadu, K. D., Sonko, S., & Hossain, G. (2024). Society 5.0 enabled agriculture: Drivers, enabling technologies, architectures, opportunities, and challenges. *Information Processing in Agriculture*.

Borchi, J., Woodcock, M., Redshaw, M., & Raniga, B. (2021). Cyber security threats-are we prepared? A threat-based assessment of the cyber resilience of the Australian agricultural sector.

Brisco, B., Brown, R., Hirose, T., McNairn, H., & Staenz, K. (1998). Precision agriculture and the role of remote sensing: A review. *Canadian Journal of Remote Sensing*, *24*(3), 315–327. DOI: 10.1080/07038992.1998.10855254

Bui, H. T., Aboutorab, H., Mahboubi, A., Gao, Y., Sultan, N. H., Chauhan, A., & Islam, Z. (2024). Agriculture 4.0 and beyond: Evaluating cyber threat intelligence sources and techniques in smart farming ecosystems. *Computers & Security*, *140*, 103754. DOI: 10.1016/j.cose.2024.103754

Bwambale, E., Abagale, F. K., & Anornu, G. K. (2022). Smart irrigation monitoring and control strategies for improving water use efficiency in precision agriculture: A review. *Agricultural Water Management*, *260*, 107324. DOI: 10.1016/j.agwat.2021.107324

Campoverde-Molina, M., & Luján-Mora, S. (2024). Cybersecurity in smart agriculture: A systematic literature review. *Computers & Security*, •••, 104284.

Claughton, D., & Beilharz, N. (2021). JBS Foods pays $14.2 million ransom to end cyber attack on its global operations. Retrieved 23 December 2024 from https://www.abc.net.au/news/rural/2021-06-10/jbs-foods-pays-14million-ransom-cyber-attack/100204240

Craigen, D., Diakun-Thibault, N., & Purse, R. (2014). Defining cybersecurity. *Technology Innovation Management Review*, *4*(10), 13–21. DOI: 10.22215/timreview/835

Demestichas, K., Peppes, N., & Alexakis, T. (2020). Survey on security threats in agricultural IoT and smart farming. *Sensors (Basel), 20*(22), 6458. DOI: 10.3390/s20226458 PMID: 33198160

Drape, T., Magerkorth, N., Sen, A., Simpson, J., Seibel, M., Murch, R. S., & Duncan, S. E. (2021). Assessing the role of cyberbiosecurity in agriculture: A case study. *Frontiers in Bioengineering and Biotechnology*, *9*, 737927. DOI: 10.3389/fbioe.2021.737927 PMID: 34490231

Gore, S., Patil, D., Mahankale, N., & Gore, S. (2024). Satellite Imaging for Precision Agriculture Enhancing Crop Management, Soil Condition and Yield Prediction. In *Emerging Trends in Smart Societies* (pp. 434–437). Routledge. DOI: 10.4324/9781003489412-104

Goulart, R., Jarvis, D., & Walsh, K. B. (2023). Evaluation of end effectors for robotic harvesting of mango fruit. *Sustainability (Basel), 15*(8), 6769. DOI: 10.3390/su15086769

GrainCrop. (2023). *FY23-Annual Report*. https://www.graincorp.com.au/wp-content/uploads/2023/11/FY23-Annual-Report.pdf

Gupta, M., Abdelsalam, M., Khorsandroo, S., & Mittal, S. (2020). Security and privacy in smart farming: Challenges and opportunities. *IEEE Access : Practical Innovations, Open Solutions*, *8*, 34564–34584. DOI: 10.1109/ACCESS.2020.2975142

Hanif, M. K., Khan, S. Z., & Bibi, M. (2022). Applications of artificial intelligence in pest management. In *Artificial Intelligence and Smart Agriculture Applications* (pp. 277–300). Auerbach Publications. DOI: 10.1201/9781003311782-13

Hazrati, M., Dara, R., & Kaur, J. (2022). On-farm data security: Practical recommendations for securing farm data. *Frontiers in Sustainable Food Systems*, *6*, 884187. DOI: 10.3389/fsufs.2022.884187

Inoue, Y. (2020). Satellite-and drone-based remote sensing of crops and soils for smart farming–a review. *Soil Science and Plant Nutrition*, *66*(6), 798–810. DOI: 10.1080/00380768.2020.1738899

Ivanochko, I., & Melnyk, O. (2024). Smart Farming System Based on Cloud Computing Technologies. *Procedia Computer Science*, *238*, 857–862. DOI: 10.1016/j.procs.2024.06.103

Javaid, M., Haleem, A., Khan, I. H., & Suman, R. (2023). Understanding the potential applications of Artificial Intelligence in Agriculture Sector. *Advanced Agrochem*, *2*(1), 15–30. DOI: 10.1016/j.aac.2022.10.001

KG2. (2024). *Navigating The Digital Divide: Challenges In Adopting Technology In Australian Agriculture*. KG2 Australia Retrieved 20 December 2024 from https://kg2.com.au/navigating-the-digital-divide-challenges-in-adopting-technology-in-australian-agriculture/

Kamilaris, A., Cole, I. R., & Prenafeta-Boldú, F. X. (2021). Blockchain in agriculture. In *Food Technology Disruptions* (pp. 247–284). Elsevier. DOI: 10.1016/B978-0-12-821470-1.00003-3

Karunathilake, E., Le, A. T., Heo, S., Chung, Y. S., & Mansoor, S. (2023). The path to smart farming: Innovations and opportunities in precision agriculture. *Agriculture*, *13*(8), 1593. DOI: 10.3390/agriculture13081593

Khanal, S., Fulton, J., & Shearer, S. (2017). An overview of current and potential applications of thermal remote sensing in precision agriculture. *Computers and Electronics in Agriculture*, *139*, 22–32. DOI: 10.1016/j.compag.2017.05.001

Kumar, S. P., Subeesh, A., Jyoti, B., & Mehta, C. (2023). Applications of drones in smart agriculture. In Pakeerathan, K. (Ed.), *Smart Agriculture for Developing Nations: Status, Perspectives and Challenges* (pp. 33–48). Springer., https://doi.org/https://doi.org/10.1007/978-981-19-8738-0_3 DOI: 10.1007/978-981-19-8738-0_3

Kumari, A., Khan, M. N., & Sinha, A. K. (2024). Harvesting Intelligence: AI and ML Revolutionizing Agriculture. In *Data-Driven Farming* (pp. 126-141). Auerbach Publications.

Lochan, K., Khan, A., Elsayed, I., Suthar, B., Seneviratne, L., & Hussain, I. (2024). Advancements in Precision Spraying of Agricultural Robots: A Comprehensive Review. *IEEE Access : Practical Innovations, Open Solutions*, *12*, 129447–129483. DOI: 10.1109/ACCESS.2024.3450904

Ma, Q., Tan, H., & Zhou, T. (2023). Mutual authentication scheme for smart devices in IoT-enabled smart home systems. *Computer Standards & Interfaces*, *86*, 103743. DOI: 10.1016/j.csi.2023.103743

Mace, G., & Morgan, R. T. (2022). *Cybersecurity in Agriculture – Are the gates locked or left ajar?* Retrieved 9 January 2025 from https://futurebeef.com.au/wp-content/uploads/2022/04/GF-LBR-Narrative_Cybersecurity_FINAL.pdf

Maddikunta, P. K. R., Hakak, S., Alazab, M., Bhattacharya, S., Gadekallu, T. R., Khan, W. Z., & Pham, Q.-V. (2021). Unmanned aerial vehicles in smart agriculture: Applications, requirements, and challenges. *IEEE Sensors Journal*, *21*(16), 17608–17619. DOI: 10.1109/JSEN.2021.3049471

Madeira, M., Porfírio, R. P., Santos, P. A., & Madeira, R. N. (2024). AI-Powered Solution for Plant Disease Detection in Viticulture. *Procedia Computer Science, 238*, 468–475. DOI: 10.1016/j.procs.2024.06.049

McCaig, M., Dara, R., & Rezania, D. (2023). Farmer-centric design thinking principles for smart farming technologies. *Internet of Things : Engineering Cyber Physical Human Systems, 23*, 100898. DOI: 10.1016/j.iot.2023.100898

McCosker, M., & Burton, L. (2024). *Lack of mobile connectivity 'confronting' regional Australians.* ABC News Retrieved 9 January 2025 from https://www.abc.net.au/news/2024-09-03/rural-australians-worse-off-for-mobile-and-internet-connectivity/104287014

Mishra, H., & Mishra, D. (2024). AI for Data-Driven Decision-Making in Smart Agriculture: From Field to Farm Management. In *Artificial Intelligence Techniques in Smart Agriculture* (pp. 173-193). Springer.

Oche, O. E., Nasir, S. M., & Muhammed, A. H. (2020). Internet of Things-based agriculture: A review of security threats and countermeasures. *Int. J. Sci. Res. Publ, 10*(8), 738–748. DOI: 10.29322/IJSRP.10.08.2020.p10494

Pierce, F. J., & Nowak, P. (1999). Aspects of precision agriculture. *Advances in Agronomy, 67*, 1–85. DOI: 10.1016/S0065-2113(08)60513-1

Potts, J. (2019). Blockchain in agriculture. *Available atSSRN* 3397786.

Rathod, M. L., Shivaputra, A., Umadevi, H., Nagamani, K., & Periyasamy, S. (2022). Cloud computing and networking for SmartFarm AgriTech. *Journal of Nanomaterials, 2022*(1), 6491747. DOI: 10.1155/2022/6491747

Rees, J., & Rees, C. J. (2023). Cyber-Security and the Changing Landscape of Critical National Infrastructure: State and Non-state Cyber-Attacks on Organisations, Systems and Services. In *Applications for Artificial Intelligence and Digital Forensics in National Security* (pp. 67-89). Springer.

Rotz, S., Duncan, E., Small, M., Botschner, J., Dara, R., Mosby, I., Reed, M., & Fraser, E. D. (2019). The politics of digital agricultural technologies: A preliminary review. *Sociologia Ruralis, 59*(2), 203–229. DOI: 10.1111/soru.12233

Russell, C. C. (2022). Cyber security in digital agriculture: Investigating farmer perceptions, preferences, & expert knowledge (Doctoral dissertation, University of Guelph).

Shahi, T. B., Sitaula, C., Bhandari, K. P., Poudel, S., Bhandari, R., Mishra, R.,& Mishra, B. (2024). Deep Learning-Based Method for Irrigation Status Detection in Tomato Using Plant Leaves. ESS Open Archive e-prints, 965, 96503836.

Shahi, T. B., Xu, C. Y., Neupane, A., Fresser, D., O'Connor, D., Wright, G., & Guo, W. (2023). A cooperative scheme for late leaf spot estimation in peanut using UAV multispectral images. *PLoS One*, *18*(3), e0282486. DOI: 10.1371/journal.pone.0282486 PMID: 36972266

Singh, P., & Kaur, A. (2022). A systematic review of artificial intelligence in agriculture. In Poonia, R. C., Singh, V., & Nayak, S. R. (Eds.), *Deep learning for sustainable agriculture* (pp. 57–80). Elsevier Inc., https://doi.org/https://doi.org/10.1016/C2020-0-02363-7 DOI: 10.1016/B978-0-323-85214-2.00011-2

Sinha, B. B., & Dhanalakshmi, R. (2022). Recent advancements and challenges of Internet of Things in smart agriculture: A survey. *Future Generation Computer Systems*, *126*, 169–184. DOI: 10.1016/j.future.2021.08.006

Smith, M. J. (2018). Getting value from artificial intelligence in agriculture. *Animal Production Science*, *60*(1), 46–54. DOI: 10.1071/AN18522

SP., P., & Balamurugan, P. (2022). Unmanned aerial vehicle in the smart farming systems: Types, applications and cyber-security threats. *2022 International Conference on Innovative Computing, Intelligent Communication and Smart Electrical Systems (ICSES)*

Taseer, A., & Han, X. (2024). Advancements in variable rate spraying for precise spray requirements in precision agriculture using Unmanned aerial spraying Systems: A review. *Computers and Electronics in Agriculture*, *219*, 108841.

Tsvetanov, F., & Kuzmanov, K. (2024). Communication Challenges in Smart Irrigation Systems. 2024 9th International Conference on Energy Efficiency and Agricultural Engineering (EE&AE), Tychola, K. A., & Rantos, K. (2025). Cyberthreats and Security Measures in Drone-Assisted Agriculture. *Electronics (Basel)*, *14*(1), 149.

Ullo, S. L., & Sinha, G. R. (2021). Advances in IoT and smart sensors for remote sensing and agriculture applications. *Remote Sensing (Basel)*, *13*(13), 2585. DOI: 10.3390/rs13132585

Verma, S., Gala, R., Madhavan, S., Burkule, S., Chauhan, S., & Prakash, C. (2018). An internet of things (IoT) architecture for smart agriculture. 2018 fourth international conference on computing communication control and automation (ICCUBEA)

Virnodkar, S. S., Pachghare, V. K., Patil, V., & Jha, S. K. (2020). Remote sensing and machine learning for crop water stress determination in various crops: A critical review. *Precision Agriculture*, *21*(5), 1121–1155.

Wolfert, S., Ge, L., Verdouw, C., & Bogaardt, M.-J. (2017). Big data in smart farming–a review. *Agricultural Systems*, *153*, 69–80. DOI: 10.1016/j.agsy.2017.01.023

Wu, W., Dawson, D., Fleming-Muñoz, D., Schleiger, E., & Horton, J. (2019). *The future of Australia's Agricultural Workforce*. CSIRO Data61, Canberra, Australia

Yazdinejad, A., Zolfaghari, B., Azmoodeh, A., Dehghantanha, A., Karimipour, H., Fraser, E., Green, A. G., Russell, C., & Duncan, E. (2021). A review on security of smart farming and precision agriculture: Security aspects, attacks, threats and countermeasures. *Applied Sciences (Basel, Switzerland)*, *11*(16), 7518. DOI: 10.3390/app11167518

Zanella, A. R. A., Silva, E., & Albini, L. C. P. (2020). Security challenges to smart agriculture: Current state, key issues, and future directions. *Array (New York, N.Y.)*, *8*, 100048. DOI: 10.1016/j.array.2020.100048

Chapter 11
Comprehensive Research on the Use of Virtual Reality to Find the Potential Buyers of Crops in the Global Market

Aditya Pai H.
https://orcid.org/0000-0002-8186-2467
MIT Art, Design, and Technology University, Pune, India & LIPS Research and DL Card, EIU, Paris, France

J. V. Muruga Lal Jeyan
https://orcid.org/0000-0001-9923-1586
LIPS Research and DL Card, EIU, Paris, India

Ganesh Pathak
https://orcid.org/0009-0001-8698-7648
MIT Art, Design, and Technology University, Pune, India

Sharon Christa
https://orcid.org/0000-0001-6717-2200
MIT Art, Design, and Technology University, Pune, India

Jyoti Agarwal
https://orcid.org/0000-0002-9664-4390
Graphic Era University, Dehradun, India

DOI: 10.4018/979-8-3373-0020-7.ch011

Copyright © 2025, IGI Global Scientific Publishing. Copying or distributing in print or electronic forms without written permission of IGI Global Scientific Publishing is prohibited.

ABSTRACT

Traditional communication methods and dealing with crop buyers have always been very challenging. Thus, there is a need to globalize agricultural trade, which has created significant opportunities for producing good crops. The chapter mainly discusses the innovative virtual reality (VR) method to bridge the gap between potential buyers of crops in the international market and farmers. With virtual reality (VR), farmers can showcase different crop products in real-time, with minute details, creating transparency and building trust with the buyers in the international forum. Additionally, integrating VR with the Internet of Things shows real-time data on crop freshness. Blockchain provides a platform for quality assurance. A VR-based prototype provides the feasibility, scalability, and market impact for the crop trade and the buyer's preferences within a virtual space. The chapter's main objective is to suggest local farmers increase global sales, transforming the agricultural supply chain by providing a sustainable and inclusive solution for crop trading.

INTRODUCTION

The global trade flow of agricultural commodities has accelerated in recent decades due to the increasing demand for food security, sustainability, and quality agricultural production. However, the generalized agricultural trade is riddled with inefficiencies such as lack of transparency, limited market access, logistical problems etc. Such challenges make it difficult to establish relationships between farmers - particularly small and medium size producers- and international buyers. In addition, consumers are demanding transparency and proof of where food is from and how it has been made and raised. The growing consumer awareness demands solutions that will enable increasing progress towards trust, efficiency, and inclusivity across agricultural trade.

Immersive technologies (VR, IoT, Blockchain and Artificial Intelligence(AI)) have the potential to facilitate these steps and help overcome these barriers. You can simulate a near-life experience through a virtual reality (VR) interfaces, where potential purchasers can view crops and interact with farmers without making an actual trip. This deep experience then creates a clear, open and all the verified marketplace. The technology Internet of Things is used to monitor the crops at regular rate, to check the freshness, and also the storage places. Thus, providing the real-time trust while having the VR interactions. Also, the data security and traceability offered by Blockchain helps to develop an assurance and accountability system within the supply chain. The use of AI enhances the ecosystem by bringing in more insights, automating the quality assessments and providing the smart recommendations to the consumers.

This chapter explores how feasible it would be to integrate IoT, VR, Blockchain and AI into agriculture trade. The proposed system adopts the model of "Internet of Everything" to provide farmers with better access to marketplace, to promote sustainable agriculture, and to reduce the carbon footprint of traditional trading. This integration brings alignment with the global trade and the policy frameworks so that international standards are complied with. Through the detailed methodologies, case studies and socio-economic impact analyses, the chapter provides the focus on future technology where it can bridge the gap between farmers and global consumer market, which will foster the economic growth and provide sustainability.

Problem Statement

There are several barriers to efficiency and inclusivity for small and medium-sized producers in the agricultural trade sector. The current system is heavily reliant on the physical interactions and the visual inspections, leading to the delays, and will have lack of transparency between the farmers and the consumers. These inefficiencies will impact the farmers which are in remote and developing cities, that prevent it from participating in the global trade.

With global demand for sustainably and ethically sourced products currently exceeding supply, these gaps represent a timely opportunity for innovative solutions to fill. VR also enables the remote yet realistic inspections of the crops, while IoT ensures that real-time quality tracking, blockchain guarantees the immutable traceability. These technologies, when combine with AI-powered smart analytics, have the potential to bring change in the agricultural commerce, offering an reliable, engaging, data-driven platform for both the farmers and the consumers

Demand for VR Technology in Agriculture and Other Sectors

Application of VR Around the World VR technology has become successful across a range of applications, including gaming, healthcare, educational videos, real estate, and showcasing it own versatility in the simulation world so far. In agriculture, VR was first used to train farmers on crop cultivation, to provide virtual farm tours, and to teach consumers about agricultural practices.

VR is about to reach out more and revolutionize the field of agriculture marketplace. So, in this case, the producers and consumers interact during an abstraction i.e they do not physically spend time together. VR provides after visualisation the buyers the experience of checking the quality of the crop produce, porosity, quality, even the farm conditions and this all is done without travelling. These enhanced

IoT devices create a real-time data experience while using blockchain for product traceability. Therefore, virtual reality is the strongest mean that adds inclusiveness and transparency in the market of brands.

Agriculture has tried to meet all the demands in this complex, connected world. It has also combined traditional practices with futuristic trade. It intends to study how VR can help transform agricultural trade by answering for transparency and offering entry to wider markets through contact with worldwide shoppers.

Objectives

The main aim of the chapter is to explore the solutions provided by virtual reality in the agricultural trade. Virtual reality has offered more transparency, accessibility, and connectivity between farmers and consumers worldwide. VR has played a significant role in removing barriers faced in traditional farming, like logistic issues, geographical barriers, and economic constraints. The chapter objectives are listed below:

1. To discover how VR is vital in providing real-time solutions and bringing immersive communication between the farmers and the consumers.
2. To find the potential benefits, impacts, and technological requirements for a VR-oriented agricultural marketplace.

LITERATURE REVIEW

This section provides related works on using advanced technology like Artificial Intelligence, the Internet of Things, Blockchain and Artificial Intelligence in agriculture. There are some mentions of digital and virtual farming and the barriers faced. Although there are few literary works on using Virtual Reality in agriculture, there is no concrete research on using VR to connect crop producers with global buyers. The following are the literary research works.

An article by A. Gupta et al. (Gupta, A. & Anand, R.,2024) explains how machine learning transforms precision farming. The research demonstrates how ML can be used in IoT-oriented precision farming, thus improving the system's performance and accuracy. The primary use of ML is analyzing the soil components, detecting weather patterns, and providing health information about the crops in real time using advanced ML algorithms and mathematical prediction models. The approach offers support in making accurate decisions, allocating resources, and proactive mediation to improve crop yields. (Sakthipriya S and Naresh R., 2024) The research also discusses how the approach helps tackle changing environmental and economic pressures (Mohyuddin G et.al., 2024) (Reyana A et.al., 2023).

In the article by R. Yadav et al. (Yadav, R. & Kumar Mishra, A., 2024), the author explains that agriculture has played an essential role in the economic development of any country, as agriculture is not only restricted to crop cultivation but also used in raising livestock. There have been many changes in the agriculture sector these days; the farmers face lots of losses due to plant diseases, increased pests, and the absence of moisture in the soil, thus resulting in poor yield of crops. In conventional farming, detecting insects and diseases has been challenging for farmers. Things have been replaced by precision farming to boost productivity. Care has been taken to improve crop yields. The process used in precision farming includes data collection, the use of data to train the system, and the performance of predictions. The technology used in the process comprises IoT, ML, data analytics, mining, etc. (Suchithra, M.S., Pai, M.L., 2020) The author has proposed a model that will predict and detect diseases in the crop on time. (Kasinathan, T et.al., 2021) Also, it determines the soil's moisture level and identifies pests that harm the crops. Thus, increasing the productivity of crop cultivation (Durai, S.K.S..et.al., 2021).

The article by Morepje et al. (Morepje MT et.al., 2024) explains the transformative role played in the Sub-Saharan region of Africa by e-commerce platforms promoting small farmers' sustainable crops. The authors explain how e-commerce has identified the digital market, improved market access, provided economic stability, and adopted sustainable farming practices that present substantial financial and logistical challenges. The research also emphasizes that e-commerce platforms have aided farmers in removing all the traditional barriers to communicating with buyers by introducing direct sales to consumers and removing all the dependencies with intermediaries (Purba, M. .et.al., 2021) (Mangiaracina, R. .et.al., 2019). By providing this direct link, the process enables fairer prices for both. This would expose these smallholder farmers to a larger consumer base, providing them with incentives to adopt sustainable practices to meet the increasing demand for ethically sourced as well as environmentally friendly products. Even so, the study notes that many barriers remain that could impede the more generalized rollout of e-commerce solutions, such as poor connectivity, low digital skills and a lack of delivery mechanisms. Our results imply a high potential for e-commerce platforms to promote sustainable agricultural practices but highlight a crucial need for infrastructural investment and digital training to unlock their potential gains for smallholder farmers in the region (Shuaibu, M. & Nchake, M., 2021).

M.J. Kulkarni et al., in their paper (Kulkarni, M.J. et.al., 2024), describe IoT in agriculture, highlighting mainly its usage in precision farming (Grigorieva, E. et.al., 2023). Precision agriculture facilitated by the IoT includes high-tech sensors, communication technology, and data analysis approaches that maximize total output

(Srinivasarao, C., 2021). The data is collected from the sensors here, and then, using a cloud-based platform, the real-time data analysis is performed. The extracted data is collected and sent to a central system. It mainly gives farmers information regarding pest control, how to fertilize the crop properly, and good irrigation service (Srinivasarao, C. et.al., 2022). It is an IoT-supported innovative farming system integrated with a machine learning-based component. The output indicates that the production increased significantly compared to the traditional method.

Geetha P et al. (Geetha, P. & Karthikeyan, R., 2024) underline that the process-oriented adoption of modern technologies will recognize and drive significant improvements in agriculture through collaborations, will lead towards higher yields, sustainability, and the ability to cope with increasing challenges, stimulating further growth as well as the transformation of the agricultural sector. Their three-part model–survey, development, and warehousing–is supported by automation, precision ag, and IoT-enabled technology integration, giving agriculture a pathway to success (P. Majumdar et.al., 2023) (M. Dhanaraju et.al., 2022). These phases are all part of a more extensive system called the Rule-Based Agriculture System (RBAS), which directs the entire agricultural process from initial soil testing to high-powered technology use. The RBAS Framework incorporates 12 Critical Stages, building blocks to increase productivity and sustainability (V.S. Magomadov, 2019). These phases encompass the full spectrum of aspects of agricultural output, from soil health examination via resource optimization to crop monitoring, with each phase designed to work together seamlessly to create productive high-yield output and resource preservation. Getting farms to operate as a disciplined, rule-abiding unit entices farmers to conform to existing agricultural policies and best practices that lead to better harvest returns within the confines of responsible environmental stewardship. The authors also urge that technology in agriculture, such as IoT, robotics, drones, etc., could radically change the agricultural landscape of the countries. IoT devices enable real-time soil, crop, and environmental monitoring and more efficient planting, harvesting, and pest control with robotics and drones. Combined and used, these technologies create an integrated ecosystem that enhances individual operation precision and efficiency and promotes sustainable practices by reducing waste and using resources more efficiently. Such says the integrated approach showcases how technology can be harnessed to transition agriculture into a more resilient and productive sector.

In Husfarm's article on e-commerce and small-scale farmers, Maria Gonzalez (Maria Gonzalez, 2024) discusses How e-commerce can benefit small-holder farmers for better market access, less intermediary dependency, and more stable income. Secondly, it allows farmers to sell to more remote buyers, which gives them a better price and better access to market information. On the downside, Digital literacy is limited, rural areas have a shortage of internet, and trust and quality assurance in online transactions are required. This makes the digital part very important and highlights the need for digital infrastructure and support for farmers to take full advantage of e-commerce.

The article from A. Pawar et al. (Panwar A et.al., 2023) (Vangala, A et.al., 2021) (Rocha, G.d.S.R. et.al., 2021) (Sharma, R.. et.al., 2021) explains how blockchain improves transparency and traceability in the agricultural supply chain. The blockchain records and verifies every step of farm products—from farm to fork—thus providing assurance of product authenticity, combating fraud, and enhancing food safety (Song, L. et.al., 2021) (Sharma, R. et.al., 2021). This supports more sustainable and transparent agriculture, allowing consumers access to reliable information on the origins of their food (Demestichas, K. et.al., 2021).

As agriculture is getting accustomed to e-commerce, Faojia Fariha in her article (Faojia Fariha, 2023) discusses how this relationship is based on direct communication with consumers instead of transactional communication through wholesalers. The internet minimizes the supply chain for farmers, broadens their customer base, and allows the customers to purchase fresh, prepared foods. This trend was carved out in this context, with the need for ease and security of access to food, brought forth by the COVID-19 outbreak. B2B exchanges can take place seamlessly due to reference checks that e-commerce platforms encourage to flow freely between the two parties, reducing the risk of third-party interference.

This article provides data showing the prolific expansion e-commerce has experienced across sectors. Retail e-commerce worldwide sales were $4.28 trillion by 2020, predicted to hit $5.4 trillion by 2022. The B2B e-commerce market is expected to reach $6.64 trillion by 2020, with a compound annual growth rate (CAGR) of 18.7% from 2021 to 2028. Meanwhile, B2B agriculture e-commerce brought in about 14% of the North American sector in 2020, and the North American food and beverage e-commerce segment earned $18.7 billion in 2020, with projected growth to $25.7 billion by 2025. All these statistics illustrate the potential market expansion for farmers through e-commerce (Faojia Fariha, 2023).

Through online platforms, farmers can showcase diverse products, including fresh produce, organic items, specialty foods like heirloom varieties, dairy, poultry, meat, seafood, and processed foods like jams and sauces. Critical benefits for farmers include expanded market access, increased profit margins from reduced reliance on intermediaries, and greater transparency for consumers, who can trace product origins and learn about farming practices. For consumers, e-commerce platforms offer convenience through simple ordering and delivery options (Faojia Fariha, 2023).

However, the article emphasizes e-commerce in agriculture is not without challenges. The main barrier is still the level of digital literacy; many farmers need some training before they can fully utilize these platforms. Logistics also rank among these factors; it is important to maintain the quality of goods throughout the delivery process. Strong distribution channels need to be built to match customers' expectations. In addition, end users of online shopping should also purchase repeatedly after the initial order as long as product quality and effectiveness are not neglected. This article offers compelling evidence about the prospects of agricultural trade through E-commerce, that such trade can be facilitated more easily and reliably while recognizing the limitations of internationalization in the digital world (Faojia Fariha, 2023).

The article by Modern Farmer (Leah Borts-Kuperman, 2023) discusses the emergence of online farmers' markets because more farmers are trying to sell directly to consumers through the Internet. Click Fork in Ontario and Cow-Op in British Columbia allow many farms to share a single website so that customers can purchase local produce and receive combined shipments. This model enhances market sales and decreases the overproduction of food, as crops are sold before harvesting, giving farmers more flexibility and family time. Farmers feel nostalgic for the Farm to Fork connection, but many farmers' markets are changing without tears thanks to the high accessibility and effectiveness of virtual marketplaces.

Ángel Luis Perales Gómez et al. (Perales Gómez et al., 2023), in their article, have developed a multiple security-oriented frameworks capable of providing solutions for the agri-food supply chain security and traceability issues with cloud-based systems (Aharoni, R. et al., 2019). The proposed system ensures that production, processing, transportation, and even retail activities, which are different supply chain stages, cannot be changed because they are stored in a blockchain (Al-Kofahi et.al., 2020) (Araujo et al., 2019). The proposed method enhances the trust and integrity of data by safeguarding the information from alteration and providing the product's timeline from the farm to the dinner table. As the framework is based on blockchain technology, centralization risks are eliminated, including data tampering and agent abuse of contacts (Alonso R.S. et.al, 2020) (Araujo V. et.al., 2019). The authors note, however, that such openness and traceability prevent not only actors from committing fraud but also encourage consumers who are given trustworthy product information and potential sources of hazards. The model successfully resolves both scalability and security challenges and, therefore, has the potential to modernize and secure agri-food supply chains using cloud-based systems and blockchain.

As stated in the Food Tank article (foodtank.com, 2021), digital farmers' markets are emerging, which include local farmers and consumers over the Internet. Such electronic markets are easy for consumers because they enable local sources of fresh produce and patronage of small-scale agriculture. The transition to electronic platforms enables farmers to cut down on food waste through pre-selling, adjust to consumer trends, and extend their market reach without middlemen. Though facing logistic issues, digital farmers' markets appear to be becoming popular as a viable and sustainable way to solve urban problems.

Digital agricultural marketplaces help support technology integration in practice (Dmytro Lennyi, 2021). These platforms allow farmers to connect with buyers directly, removing intermediaries in the supply chain. Participating in these virtual markets introduces farmers to technology applications that may catalyze the adoption of more sophisticated technologies to increase productivity and efficiency. For instance, Agri Marketplace, based in Portugal, is a B2B business solution that joins farmers to industrial buyers. Maano Virtual Farmers Market in Zambia, created by the UN's World Food Programme, to help market the products of African farmers.)

According to the work of Josalyn Radcliffe et al. (Radcliffe et.al., 2021) in their article, COVID helped the online food spaces to develop more, creating a need for food accessibility and sustainability, where the authors focused their case study on the Yellowknife Farmers Market. Digital solutions are tailored to the specific requirements of the pandemic event with the result of conveying local foods and other sustainability initiatives. Still, the study speaks of "virtual barriers" as well, such as lower levels of internet access, which could frustrate utilization of the benefits in the first place. This case study shows how promising and limiting the devising of digital solutions for local food systems can be (Wills B & Arundel A.,2017) (Ulsperger J & Ulsperger K., 2017) (Klisch S & Soule K, 2020)..

IoT Based Crop Monitoring Using Machine Learning in Precision Agriculture (R. Yadav et.al., 2021), states that Sophisticated devices that reside in all of the Internet of Things (IoT) sensor networks can help monitor the growth of crops and maximize yield with real-time data according to the article. The dataset are used to generate usable information through machine learning techniques to enhance the decision-making processes concerning pest control and water management. While the article notes that IoT by patching up these issues and meeting the agriculture needs can also help achieve sustainable agricultural practices, it mentions data security and infrastructure needs as some of the challenges to be met.

Literature and works indicate the transformability and effectiveness of Artificial Intelligence (AI), IoT, and Blockchain in precision agriculture and supply chain, these are the main areas covered in this study. (Gupta et al., 2024) in the study explains about Data Mining and Machine Learning in Agriculture. The data mining and machine learning need using IoT-based systems for real-time data collection to predict analytic models, which could certainly optimize agriculture activities and resource optimizing, etc. IoT-based real-time detection IoT (Internet of Things) applications in a crop viewpoint. In a similar vein, (Yadav et al., 2024) shows that a combination of IoT and Blockchain technology can be used to monitor the traceability and quality of agricultural products, which addresses logistical challenges while increasing supply chain transparency1.

According to (Kulkarni et al., 2024), precision farming has been improved by integration of IoT and AI, with case studies included showing how by utilizing real-time monitoring, decision making using this technology has resulted in increased productivity as well as sustainability. The use of blockchain in agriculture is exemplified by (Panwar et al., 2023), who discuss how blockchain enables transparency, efficiency, and traceability from farm to fork enabling fair trade and consumer confidence1. Both the studies are extremely relevant and stress on the need to explore IoT and Blockchain as key drivers of the modernization of the agricultural sector but do not take this further with a combined exploration of IoT, Blockchain, and VR for creating immersive and interactive trade platforms.

Advancements of Virtual Reality in Agriculture

A Comparative Review of the Virtual Reality Applications

A comparative analysis of Virtual Reality applications across different industries includes in the area of real estate, health care, academics, which highlights its potential in agriculture. As per the research (Geetha et al., 2024), it highlights what VR can offer during virtual farm tours like immersive training simulations, providing interactive product inspections. Whereas the adoption of the large-scale productions remains largely limited due to infrastructural issues and lack of standardized protocol availability for Virtual Reality. The tabular summary (Table 2) mention the advancements of VR and adoption trends.

Table 1. Summary of findings of virtual reality in agriculture

Research Study	Use of VR in Agriculture	Crucial Findings
Geetha et al. (2024)	VR is used for Virtual farm tours and training purposes	Improvement in the farmer engagement and training efficiency
Morepje et al. (2024)	VR is used for VR-based crop assessment for the trade	Enhanced transparency in trade and less logistical costs
Yadav et al. (2024)	VR is used for the VR-enabled remote quality inspections	Increased buyer trust and market accessibility

METHODOLOGY

Proposed VR based agricultural market place integrates along the features of IoT, Blockchain, AI & VR and delivers transparent, efficient and immersive experience in Trading. Everything begins with using IoT sensors on farms to measure important parameters such as crop freshness, soil moisture, and detailed environmental conditions. Transmitting means of secured data relay systems aggregate and send this real-time data to a central system, passing its accuracy and timeliness to buyer (Figure 1).

Figure 1. IoT data relay process

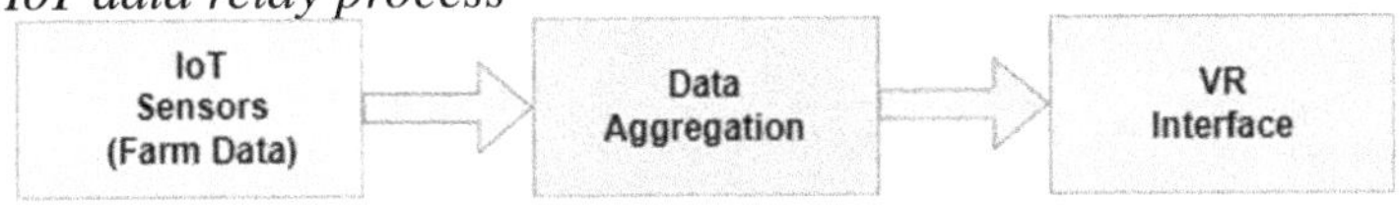

After collecting the data, it is secured with Blockchain because it simply records it in an immutable ledger. This promotes traceability and builds trust among buyers and sellers. Data such as certification or crop quality are thus stored in Blockchain, and they can be accessed directly by the buyer through the VR interface to ensure product authenticity (see Figure 2). Combining these two technologies fosters transparency within the supply chain, preventing fraudulent practices.

Figure 2. Blockchain traceability process

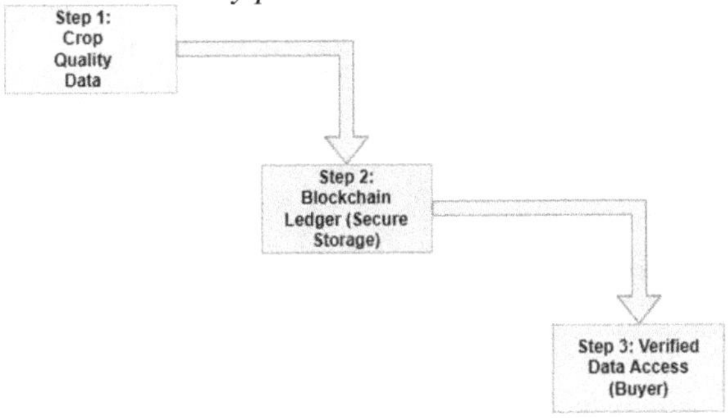

Let us take this an extra mile by automating the recommendation system using AI, like utilizing buyer preferences, past data and market trends to recommend appropriate crops. By doing so, the system processes this information and becomes capable of analyzing the buyer's needs, offering personalized recommendations that ease decision-making and improve user experience (Figure 3).

Figure 3. AI decision-making process

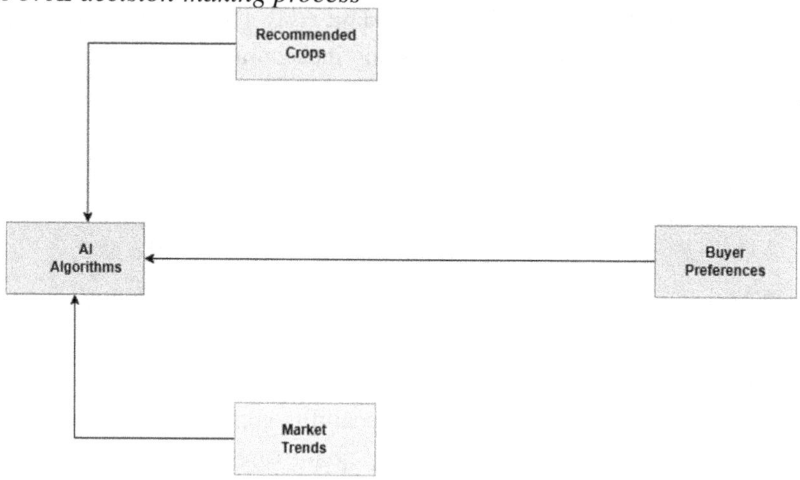

The central component of the methodology is the VR platform, over which buyers can interact with highly detailed 3D models of the crops. Advanced scanning tech creates these optimized models, making them accessible for multiple VR devices. Buyers use the VR interface to walk through virtual fields, assess crops closely, and employ interactive instruments to assess quality. These immersive environments simulate the physical nature of crop inspection while minimizing logistical burdens (Figure 4).

Figure 4. VR marketplace interaction

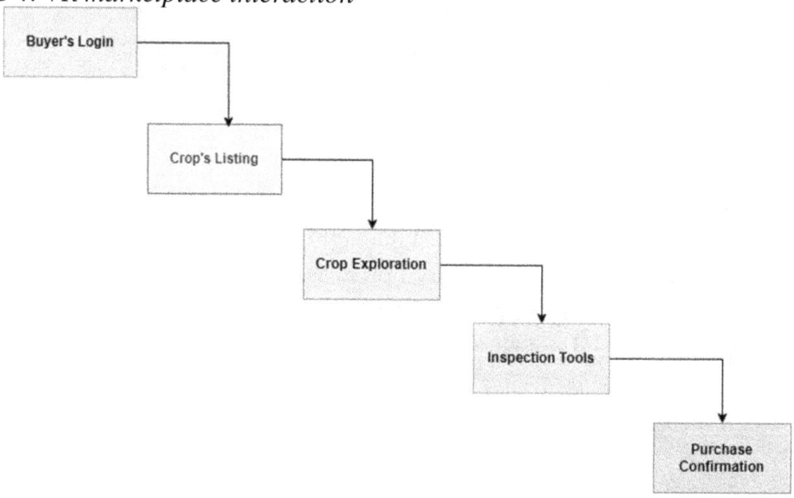

An interruption-free user experience is key to the platform's success (Don, 2020). The user experience is designed to be intuitive and easy to navigate, including real-time negotiation features and feedback collection. Potential buyers participate in virtual farm tours, examine growing conditions, and negotiate terms directly with sellers to create a vibrant and interactive marketplace (Figure 5).

Figure 5. User experience design

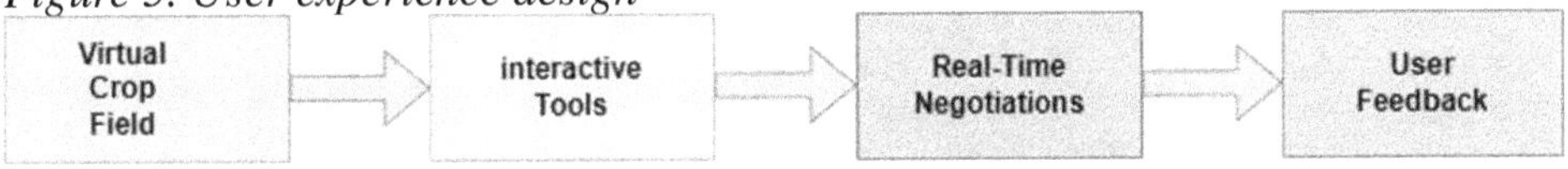

Last, the platform uses substantial data collection and analysis mechanisms to improve its functionality continually. User interactions and preferences data (user engagement data) are analyzed to understand what needs to be improved. User feedback is fed into system updates so that the platform continues to evolve and adapt to the changing needs of its users (Figure 6).

Figure 6. Data collection and refinement

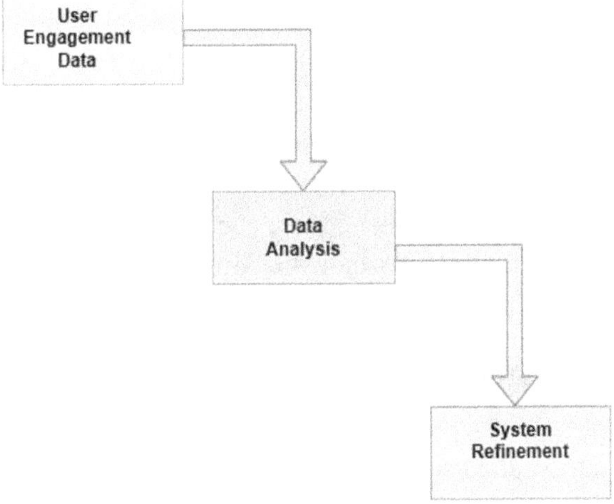

The proposed method utilizes IoT for harvesting data, Blockchain for storing data, AI for customized recommendations, and VR for the marketplace experience. For a more technical audience, each component is described through flow diagrams with illustrations of the processes involved. These technologies, together, provide a breakthrough stage to conquer the challenges that agricultural trade has always faced (Figure 7).

Figure 7. Flow diagram for methodology in VR-based agricultural marketplace

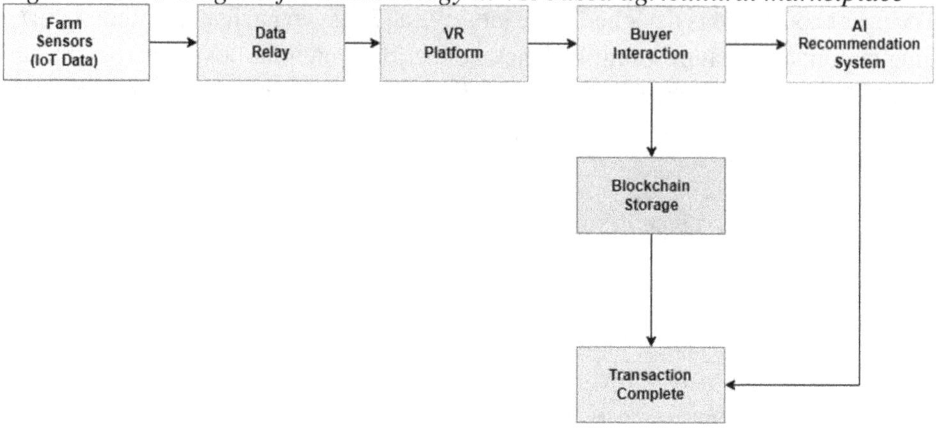

In summary, the data collection is through IoT sensors, that measures the key parameters such as crop freshness, soil moisture level and other environmental conditions. These metrices that are collected in the real time ensures the accurate assessments of the crop and are stored very securely using blockchain. The blockchain stores the batches of crops that are assigned to the digital record. Thus ensuring the immutable traceability from farm to buyer. An Artificial Intelligence algorithms are used to analyze the preferences from the buyers based on the past transactions. It also suggests the trends in the market for customized crop recommendation. The buyers can then use high-resolution 3D model of crops that conducts virtual tour of agricultural farm and also have live interaction with the producers through VR based negotiations. The regulatory compliances are checked using AI that ensures are the transactions made are adhering to the trade standards and sustainability certifications.

DEVELOPMENT AND IMPLEMENTATION

Scalability, security, and user-centric design needs drove the design and implementation of the VR-based agricultural marketplace. A mix of IoT, Blockchain, AI, and VR technologies are combined into one architectural component of the platform's system that enables efficient and transparent agricultural trading overall. (Figure 8) depicts that a cloud-based architecture supports a high data rate and is scalable, where components can easily connect. Sensors for the Internet of Things (IoT) on farms can monitor real-time metrics, including crop freshness and soil moisture. The information gets sent to the cloud, where it is visualized inside the VR platform so that buyers are always up to date.

Figure 8. High-level architecture of the VR-based marketplace

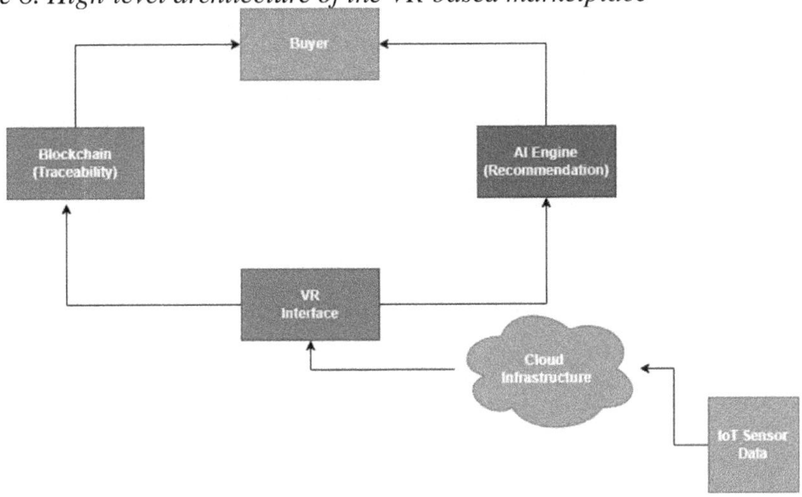

The VR environment is device-agnostic and can be explored via Unity and Unreal Engine engines via easy APIs like WebVR and WebXR. This enables buyers to interact with realistic 3D crop models created with programs such as Blender and Maya. So the models are fine-tuned so that the platform runs well on the different pieces of VR hardware and can get in front of as many people as possible. Various technologies are tailored, customized, and integrated depending on the needs of the individual supply chain — our blockchain technology, for example, where data on crop certifications and traceability records are stored securely, making it transparent and trusted. Through analyzing buyer behavior patterns and prevailing market trends, AI algorithms ensure that personalized suggestions arrive with the guarantee that decision-making has been further set in motion. The core technologies and their roles are summarized in (Table 2).

Table 2. Summary of features and technologies

Feature	Technology Used	Description
Internet of Things Integration	Sensors, Cloud Storage	Gathers real-time information about crop freshness, soil moisture, and environmental conditions.
Virtual Reality Engine	Unity, Unreal Engine	It is an immersive place for buyers to see and use crop models.
Integration with Blockchain	Ethereum, Hyperledger	Data is Secured and Traceability is Set to Immutable Ledgers

continued on following page

Table 2. Continued

Feature	Technology Used	Description
AI Recommendations	Machine Learning Models	Understanding purchaser tastes and marketplace tendencies to recommend suitable plants
Real-Time Interactions	WebVR, WebXR APIs	Let's users interact with the platform through different devices

Figure 9. User journey in the prototype

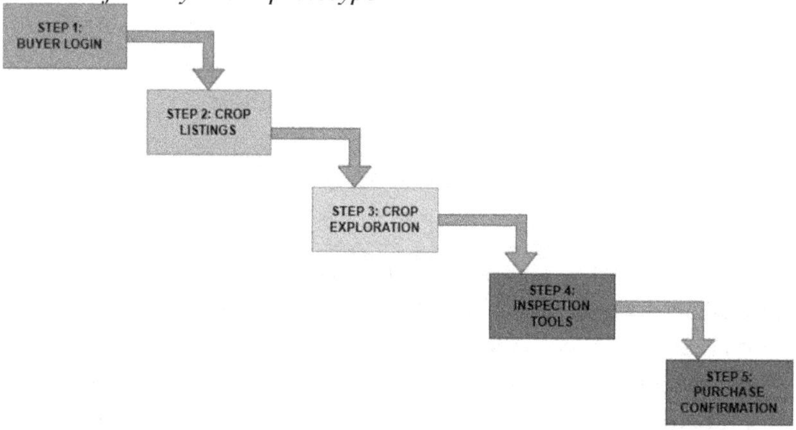

A prototype of the platform was developed to showcase its functionalities. Buyers navigate the VR environment to explore crop listings enriched with real-time IoT data overlays, inspect crops in detail, and verify certifications through Blockchain. (Figure 9) outlines the process, from login and preference setting to crop exploration, inspection, and purchase. The interactive VR environment replicates the physical experience of inspecting crops while eliminating logistical constraints.

The performance and usability of the hybrid were heavily tested using the platform. Performance testing also tested system stability during peak user usage and latency of real-time interaction. Usability studies with varied demographic participants were conducted to ensure the interface was friendly and easy to comprehend. All platform compatibility tests were carried out and confirmed that it works across devices, from high-end VR headsets to smartphone-based systems. Such assessments are performed in order to verify that the system is capable of handling a variety of users while staying both effective and dependable.

This entire development was focused on technology integration to address conventional agriculture trade issues. Based on that paper, IoT sensors collected crop conditions, and data were securely stored and authenticated using blockchain. This

AI engine provided crop recommendations with 10 percent accuracy, and the VR interface transformed the buyer experience with real-time personalization. As holistic finders and combiners that ultimately culminate in playing the new platform, all the figures and tables presented work together to exhibit this process of integration. A platform whose entire development was based on the ideas of transparency, efficiency, and the interests of users formed the basis of a revolutionary agricultural trade system.

CASE STUDIES AND PRACTICAL APPLICATIONS

The concept also presents case studies demonstrating the potentialities of this VR-based agricultural marketplace, thus demonstrating the suggested platform's practical applications and relevance to agriculture overall. The platform enables transparency, accessibility, and market access that have eluded small-scale producers and specialty crop markets for years.

One of the successful examples of the Internet is smallholder farmers, especially in the developing world, who have historically lacked access to the global marketplace, a situation aggravated by geographic isolation and logistic constraints. In Ethiopia, coffee farmers, for example, used the platform to display their beans to buyers in Europe and North America. Buying in the VR space enabled customers to check crop quality and verify freshness and sustainability through farm tours. Farmers were able to have leverage on the fixed price of their own product since they directly connected (without an intermediator) with the buyer and constantly refined their taste in order to meet the buyer's demand by using the data generated by the IoT. Then came blockchain traceability, which enhanced buyer trust through proof of crop origin and quality and ensured that trade was ethical and transparent.

Another use is in organic and specialty crop markets where customers want quality, traceable products. For example, an organic apple producer in New Zealand used the platform to advertise their sustainable practices such as natural fertilizers and renewable energy. So now, premium grocery stores and gourmet cooks could visit the Farm in immersive VR and watch Blockchain-verified certifications in real-time demonstrating organic compliance. This established the uniqueness and authenticity of these products, letting producers sell for more both locally and abroad, and consumers trusted they were purchasing from the right place. Using IoT data overlaid with Blockchain ledgers, the purchasers were able to track the product life cycle from cultivation to delivery, thus strengthening transparency and sustainability.

These use cases showcase the platform's transformative potential for inclusive and efficient trading systems. Since buyers from countries in different parts of the world are posting products on the platform, it also supports global sustainability targets by helping to reduce physical travel and virtual inspections, which reduces the carbon footprint generally connected with traditional agricultural trade. In addition, partnering with agricultural cooperatives, technology vendors, and government agencies could enhance the visibility and adoption of the platform, resulting in additional scaling. These partnerships will drive for both sides by enabling the farmers to utilize the VR hardware, training programs and infrastructure investments, thus further ensuring that the platform serves the right mass of traditional stakeholders with the capacity for systemic changes in the agricultural industry. These case studies cool-shift the VR-based marketplace to establish a wholesome, sustainable, and creative network and are a testament to their own aspiration to be at the forefront of such principles.

BENEFITS AND POTENTIAL IMPACT

Benefits Offered By VR-Based Agricultural Marketplace This approach not only solves important issues in traditional agricultural trade but also creates transparency, efficiency and sustainability. Such issues are found around the platform, which constitute the main advantages offered by the platform in terms of trust, market opportunities, environmental impact as well as economic implications.

The Trust and Transparency in Global Trade

Building trust and transparency into agricultural trade is one of the platform's most powerful contributions. Blockchain Technology Integration allows secure and incontrovertible crop quality, certifications, and transaction data recording. In the virtual reality environment, buyers are able to verify the source of the goods and authenticity so that there is a minimum chance of fraud or false representation. This degree of trust can be augmented with data on crop conditions (real-time freshness of crop and storage parameters) through dedicated IoT devices powered by the Internet of Things. A standard rice buyer in Germany will see live updates and blockchain-verifiable records of the quality and certificates of the batch imported in Thailand without going to its farm directly. Such transparency helps in building trust in the farmer-buyer relationship, facilitating long-term trading.

Expanding Market Opportunities and Inclusion

The platform generates market opportunities for small and medium-sized farmers who have traditionally faced barriers to trade by linking producers with buyers overseas. It helps farmers reach customers in the global market, reducing travel and intermediary expenses. This enables a cocoa producer in Ghana to work with chocolate manufacturers based in Switzerland, negotiating the terms of this exchange directly and, at last, sending the product into a previously unreachable market. Features such as guided tours and translation support integrated into the platform's user-friendly interface are a form of inclusivity for farmers without technical skills, Mariani said. This levels the playing field so that smaller producers can compete with bigger players and enter international markets.

Reducing Carbon Footprint and Promoting Sustainability

Remember that the platform significantly minimizes the environmental impact of conventional agricultural commerce. Virtual inspections and negotiations eliminate travel, a major contributor to the carbon in the atmosphere. A buyer in the United States, for example, can use the technology to take a virtual tour of a tea plantation in India, inspect the quality of the leaves, and close a deal without having to leave their desk. IoT-enabled pre-inspection and Blockchain-backed traceability ensure that only the highest quality goods are introduced into the supply chain, optimizing production and thereby together reducing waste. Manufacturer and seller's eco-friendly policy will support its worldwide sustainability aim, which will tempt eco-friendly buyers and producers.

Economic Implications for Farmers

The econometric returns on the platform are impressive, especially for local farmers. Farmers can negotiate better prices for their crops by cutting out middlemen and connecting directly with customers outside of their country. Using VR to prove the quality, origin, and sustainability of their products adds value that allows them to charge higher prices. As an example, through the platform, an organic avocado grower in Mexico can tap premium Europe markets with soaring revenues. Farmers can also do analysis of the data to keep in check of the production that meet the needs of the market and ensuring max profit.

CHALLENGES AND LIMITATIONS

The VR based cultivation faces some challenges even if it has good scope. Some of them are listed below:

Technical Challenges

One of the major challenge is the technical challenge. The ecosystem is highly dependent on VR, IoT and Blockchain. The usage of these is very challenging especially in the developing countries where there is poor internet connectivity. Hence the real time data collection through VR becomes a challenge. In addition there is a regular need upgradation of VR and IoT devices, which becomes difficult for the small farmers.

Data Security and Privacy Concerns

Although Blockchain technology is used to protect the data, still it faces some privacy concerns. One among them is that since farmers are required to promote their crops hence data profiling is done which will obviously disclose some of the sensitive information about the farmers crops, their farming practices and their actual geographical coordinates. These type of information leakage will make the competitors or unethical buyers to misuse the information. Although use of blockchain makes data manipulation difficult but that doesn't mean that available can be misused. Transparency needs to be combined with meaningful privacy protections that ensure that farmers have control of their data. This requires clear policies regarding data storage, access rights, and whether using Blockchain to record sensitive information is even lawful.

Barriers to Adoption and Training Needs

Moreover, farmers in remote or less developed areas might not be well versed in VR and similar technologies (particularly older farmers), leading to hesitance to use and adopt the platform. VR hardware, IoT devices, and platform subscriptions are steep from the outset, which can act as a disincentive to engaging with these new forms of production and dissemination — a particularly stark challenge for small and medium producers with little cargo space. Its high-speed requirement will make it difficult for non-techies to make their way into the VR, which will further require an understanding of technical skills. But in spite of these challenges, we are dependent upon a good training and support system. Farmers also need hands-on workshops or tutorials to feel more comfortable and proficient using online and in-person platforms. Non-technical users can also use the platform by making the interface voice-friendly, intuitive, and easy to navigate.

Regulatory and Policy Constraints

By adopting the VR-based agricultural trade in align with international policies like trade laws, data protection regulations such as GDPR, CCPA, global e-commerce compliance standards and fair trade certification guidelines. It ensures that these compliance through the blockchain inspections increases the trust and reduces the legal risks by providing an immutable, transparent records for all transactions, certifications and legal documentations.

Moreover, the governments and regulatory bodies must bring in some of the policies that define the legal admissibility of VR-based transactions in agricultural trade. The policy framework includes:

- Establishing the smart contract-based regulatory compliance systems which automatically checks the trade agreements and dispute resolutions.
- To develop the international trade agreements which recognize the blockchain-verified Virtual Reality inspections as the valid quality assessment measures.
- To ensures that cross-border agricultural transactions adhere to the environmental sustainability rules by tracking the carbon footprints through IoT and Blockchain.

Now by integrating all these frameworks, the VR based agricultural markers can gain wider adoption while ensuring the security, data privacy and fair trading practices across the international markets.

CONCLUSION AND FUTURE RESEARCH DIRECTIONS

A VR-based agro marketplace is a breakthrough innovation that can transform the global agriculture trade. By leveraging the power of VR, IoT, Blockchain, and AI technologies, the platform addresses major problems in transparency, market access, and sustainable practices. While this vision may bring about exciting possibilities, it is still far from full realization—and also provides many opportunities for further research and development.

Conclusion

The trust and transparency and practically easy experiences provided by Blockchain on the platform and the use of VR will be the two significant aspects that make this platform a holistic solution for the agriculture sector. The platform fosters equitable trading practices by decreasing reliance on intermediaries and expanding access to global markets for smallholder producers. The sustainable side of this business fits in with environmental goals worldwide, from reducing carbon footprint through trade to improving production through live IoT insights. In such a solution, a market-based integrated approach not only helps farmers gain access to premium markets to negotiate the price of their produce but will also ensure a sustainable business model, thereby maximizing farmers' profitability and initiating a systemic change in the world of agriculture. In terms of limitations: infrastructure, training, and regulatory compliance) However, the platform is a powerful base for use cases to build an inclusive and efficient agricultural ecosystem.

Future Research Directions

Several important parts of future research should be focused on to ensure the scalability and adaptability of the platform:

1. Technology Exploration: Explore the potential addition of advanced VR features like haptic feedback to enrich the buyer's multisensory engagement. Likewise, the investigation of next-gen IoT sensors to capture more detailed data and enhancement in Blockchain for fast and low-energy transactions is equally important.
2. Subsets to Products: Expand the platform's application from crops to livestock, dairy, and processed agricultural products. For example, virtual inspections of dairy farms or livestock health could help ease some trade bottlenecks.
3. AI and Personalization: Work on AI algorithms to personalize crop suggestions and predictive analytics, considering the global market trends and environmental conditions.
4. Standardization in Policy and Regulation: Work with policymakers to design international standards for trade involving VR and related tech for farming. Research of smart contracts and automated compliance mechanisms based on Blockchain could facilitate regulatory compliance.
5. Impact on Agricultural Livelihoods: The economic repercussions of shifting agricultural practices are a natural topic of study; longitudinal studies could determine the platform's influence on farmers' income, the economics of the regions involved, or general consumer trust.
6. Infrastructure Understanding: appealing and affordable VR hardware and faster connection options for country aspects. These challenges could be alleviated through partnerships with technology providers and government initiatives to ensure the availability of talented professionals.

Several issues related to this and more can then be explored in future research to enable the successful global scaling of the platform. Such orientations that advance technological progress constitute one pathway toward a more equitable, sustainable, and productive global agricultural system. Conclusion – The Future of Market Linkage– With the continuous evolution of this system, it will increasingly enable farmers to access markets, and will help drive innovation in agriculture in the future.

REFERENCES

Aharoni, R., Klymiuk, V., Sarusi, B., Young, S., Fahima, T., Fishbain, B., & Kendler, S. (2021). Spectral light-reflection data dimensionality reduction for timely detection of yellow rust. *Precision Agriculture*, *22*(1), 267–286. DOI: 10.1007/s11119-020-09742-2

Al-Kofahi, M. M., & Al-Shorman, M. Y. (2019). Toward ene. Effi. Micro. and inter.-of-thin. systems. *Computers & Electrical Engineering*, *79*, 106457. DOI: 10.1016/j.compeleceng.2019.106457

Al-Qerem, A., Alauthman. (. (2020). IoT trans. Process. Thro. Coop. concur. control on fog-cloud compu. Environ. *Soft Computing*, *24*(8), 5695–5711. DOI: 10.1007/s00500-019-04220-y

Alonso, R. S., & Sittón-Candanedo, I. (2020). An intelli. edge-iot plat. for moni. livestock and crops in a dairy farm. *Scen. Ad Hoc Netw.*, *98*, 102047. DOI: 10.1016/j.adhoc.2019.102047

Araujo, V., Mitra, K., Saguna, S., & Åhlund, C. (2019). Perform. Eval. of fiware: A cloud-based iot plat. for smart cities. *Journal of Parallel and Distributed Computing*, *132*, 250–261. DOI: 10.1016/j.jpdc.2018.12.010

Demestichas, K., Peppes, N., Alexakis, T., & Adamopoulou, E. (2020). E. Blockchain in agriculture traceability systems: A review. *Applied Sciences (Basel, Switzerland)*, *10*(12), 4113. DOI: 10.3390/app10124113

Dhanaraju, M., Chenniappan, P., & Ramalingam, K. (2022). Smart farming: Inter. of things (IoT)-based sust. agri. *Agriculture*, *12*, 1745. DOI: 10.3390/agriculture12101745

Durai, S. K. S., & Shamili, M. D. (2022). Smart farm. using mach. Learn. and deep learn. *Tech. Decis. Anal. J.*, *3*, 100041.

Fariha, F. (2023). E-commerce in Agriculture: Creating Direct Farmer-Buyer Link https://coredevsltd.com/articles/e-commerce-in-agriculture/

Geetha, P., & Karthikeyan, R. (2024). Embra. IoT and Prec. on Agri. for Sust. Crop Yields. In Balasubramanian, S., Natarajan, G., & Chelliah, P. R. (Eds.), *Intell. Robots and Dron. for Prec. Agri. Sign. and Comm. Tech.* Springer., DOI: 10.1007/978-3-031-51195-0_8

Geetha, P., & Karthikeyan, R. (2024). Embr. IoT and Preci. Agri. for Sust. Crop Yields. In *Intell. Robots and Drones for Preci. Agri*. Springer., DOI: 10.1007/978-3-031-51195-0_8

Gonzalez, M. (2024), The Imp. of E-Comm. on Small-Scale Farmers: Lessons from Online Marketplaces, https://husfarm.com/article/the-impact-of-e-commerce-on-small-scale-farmers-lessons-from-online-marketplaces#google_vignette

Grigorieva, E., Livenets, A., & Stelmakh, E. (2023). Adaptation of agriculture to climate change: A scoping review. *Climate (Basel)*, *11*(10), 202. DOI: 10.3390/cli11100202

Gupta, A., Anand, R., & Sindhwani, N. (2024). Perf. and Acc. Enhance. of Mach. Learn. & IoT-based Agri. Precision AI Sys *SN COMP.SCIE*, *5*(930). Advance online publication. DOI: 10.1007/s42979-024-03238-w

Gupta, A., Anand, R., & Sindhwani, N.. (2024). Perform. and Accu. Enhance. of Mach. Learn. & IoT-based Agri. *Precis. AI Syst. SN Computer Science*, *5*, 930. DOI: 10.1007/s42979-024-03238-w

Kasinathan, T., & Singaraju, D. (2021). Insect class. and detect. in fiel. Crop. using modern mach. Learn. Tech. *Information Processing in Agriculture*, *8*(3), 446–457. DOI: 10.1016/j.inpa.2020.09.006

Klisch, S., & Soule, K. (2020). Farmers Markets: Working with comm. Partn. to prov. Essen. services during COVID-19. *Journal of Agriculture, Food Systems, and Community Development*, *9*, 1–5. DOI: 10.5304/jafscd.2020.094.035

Kulkarni, M. J., Ramamoorthy, M., & Ramsankar, G. (2024). Sust. Crop Monit. and Manage.ment for Enha.Agri. Product. Throu. IoT, AI&ML: Case Studies and Innovations. In: Manoharan, S., Tugui, A., Baig, Z. (eds) *Proceedings of 4th Int. Conf. on Arti. Intell. and Smart Ener. ICAIS 2024. Info. Sys. Eng. and Manage., vol 3. Springer, Cham.*https://doi.org/DOI: 10.1007/978-3-031-61471-2_13

Kulkarni, M. J., Ramamoorthy, M., Ramsankar, G., ... (2024). Sust. Crop Moni. and Manage. for Enha. Agri. Prod. Throu. IoT, AI & ML: Case Studie. and Innov. In Proc. of the 4th Intern. Conf. on Arti. Intell. and Smar. Energ., Springer. DOI: 10.1007/978-3-031-61471-2_13

Leah Borts-Kuperman. (2023). The Rise of Virtual Farmers Markets https://modernfarmer.com/2023/01/virtual-farmers-markets/

Lennyi, D. "Online ag marketplaces: An entry point for farmers into the tech ecosystem," https://agfundernews.com/online-ag-marketplaces-an-entry-point-for-farmers-into-the-tech-ecosystem

Magomadov, V. S. (2019). Deep learn. and its role in smart agri. *Journal of Physics: Conference Series, 1399*(4), 044109. DOI: 10.1088/1742-6596/1399/4/044109

Majumdar, P., & Bhattacharya, D. (2023). *Utilities of 5G comm. Techn. for promo. advancement in agri. 4.0: Recent trends, research iss. and rev. of lit., in 5G and Beyond*. Springer Tracts in Elect. and Elect. Engi.

Mangiaracina, R., & Perego, A. (2019). Inno. Solu. to Incr. Last-Mile Delivery Effi. in B2C E-Commerce: A Lit. [Review]. *International Journal of Physical Distribution & Logistics Management, 49*, 901–920. DOI: 10.1108/IJPDLM-02-2019-0048

Mohyuddin, G., Khan, M. A., & Haseeb, A. (2024). Eval. of mach. learn. Appro. for Precision Farm. in Smart Agri. Syst.: A Compr. [Review]. *IEEE Access : Practical Innovations, Open Solutions, 12*, 60155–60184. DOI: 10.1109/ACCESS.2024.3390581

Morepje, M. T., & Sithole, M. Z. (2024). The Infl. of E-Commerce Plat. on Sust. Agri. Practices among Small.holder Farmers in Sub-Saharan Africa. *Sustainability, 16*(15), 6496. DOI: 10.3390/su16156496

Panwar, A., Khari, M., & Misra, S.. (2023). Blockch. in Agri. to Ensu. Trust, Effect., and Traceab. from Farm Fields to Groc. *Future Internet, 15*(12), 404. DOI: 10.3390/fi15120404

Panwar, A., Khari, M., Misra, S., & Sugandh, U. (2023). Blockchain in Agri. to Ensure Trust Effective., and Trace. from Farm Fields to Groceries. *Future Internet, 15*(12), 404. DOI: 10.3390/fi15120404

Perales Gómez, Á. L. López-de-Teruel, (2022) FARMIT: contin. Assess. of crop qual. using mach. learning and deep lear. Tech. for IoT-based smart farming. *Cluster Comput* 25, 2163–2178. https://foodtank.com/news/2021/09/from-farm-to-doorstep-the-rise-of-digital-farmers-markets/DOI: 10.1007/s10586-021-03489-9

Purba, M., Simanjutak, D., Malau, Y., Sholihat, W., & Ahmadi, E. (2021). The Effect of Digital Marketing and E-Commerce on Financial Performance and Business Sustainability of MSMEs During COVID-19 Pandemic in Indonesia. *Int. J. Data Netw. Sci., 5*, 275–282. DOI: 10.5267/j.ijdns.2021.6.006

Radcliffe, J., Skinner, K., & Spring, A. (2021). Virt. barriers: Unpack. the sustain. Implic. of online food spaces and the Yellowknife Farmers Market's response to COVID-19. *Nutrition Journal, 20*, 12. DOI: 10.1186/s12937-021-00664-x PMID: 33514376

Reyana, A., Kautish, S., & Karthik, P. M. S. (2023). Accel. Crop Yield: Multisens. Data Fusi. and Mach. Learn. for *Agric. Text Classi.IEEE Access : Practical Innovations, Open Solutions, 11*, 20795–20805. DOI: 10.1109/ACCESS.2023.3249205

Rocha, G. S. R., de Oliveira, L., & Talamini, E. (2021). E. Blockcha. appl. in agribusi.: A systematic review. *Future Internet*, *13*(4), 95. DOI: 10.3390/fi13040095

Sakthipriya, S., & Naresh, R. (2024). Precision agri. based on convo. neural net. in rice prod. nutrient manage. using mach. learn. genetic algorithm, *Engine.Appl. of Arti. Intell. Elsevier BV*, *130*, 107682. DOI: 10.1016/j.engappai.2023.107682

Sharma, R., Kamble, S. S., Gunasekaran, A., Kumar, V., & Kumar, A. (2020). A systematic litera. Rev. on mach. Learn. Appl. for sust. Agri. supply chain perfo. *Computers & Operations Research*, *119*, 104926. DOI: 10.1016/j.cor.2020.104926

Sharma, R.; Samad, T.A.; Jabbour, C.J.C.; de Queiroz, M.J. (2021) Levera. Blockcha. Techno. for circu. in agri. Supp. Chain.: Evid. from a fast-growing eco. *J. Enterp. Inf. Manag.* ahead-of-print.

Shuaibu, M., & Nchake, M. (2021). Impa. of Credit Market Cond. on Agri. Prod. in Sub-Saharan Africa. Agric. *Financial Review*, *81*, 520–534.

Song, L., & Wang, X.; Merveille. (. (2021). N. Blockchain-Based Flex. Double-Chain Arch. and Perfo. Optimiz. for Better Sust. in Agri. *Computers, Materials & Continua*, *68*, 1429–1446. DOI: 10.32604/cmc.2021.016954

Srinivasarao, C. (2021). Soil degra. Chall. for sust. Agri. in tropical India. *Current Science*, *120*(3), 492–500. DOI: 10.18520/cs/v120/i3/492-500

Srinivasarao, C. (2022). Indian Agri. Tow. 2030. *India Stud.* In *Bus. and Eco.* Springer.

Suchithra, M. S., & Pai, M. L. (2020). Impro. the pred. accu. of soil nutrient class. by optimizing extreme learn. Mach. parameters. *Information Processing in Agriculture*, *7*(1), 72–82. DOI: 10.1016/j.inpa.2019.05.003

Ulsperger, J., & Ulsperger, K. (2017). Community, regional identity, and civic agriculture: A structural ritualization analysis of rural online farmers' market sellers. *Journal of Rural Social Sciences*, *32*(2), 98–124.

Vangala, A., Das, A. K., Kumar, N., & Alazab, M. (2021). Smart Secu. Sens. for IoT-Based Agricu.: Blockcha. Perspective. *IEEE Sensors Journal*, *21*(16), 17591–17607. DOI: 10.1109/JSEN.2020.3012294

Wills, B., & Arundel, A. (2017). Inter.-enabled acc. to alter. food networks: A comp. of on-line and off-line food shop. and their differ. Interpret. of quality. *Agri Hum Vals.*, *34*(3), 701–712. DOI: 10.1007/s10460-017-9771-2

Yadav, R., & Kumar Mishra, A. (2024). IoT and Mach. Learn.-Based Monit. of the Growth of Crops Using Blockchain Techn. In Verma, A., Verma, P., Pattanaik, K. K., Dhurandher, S. K., & Woungang, I. (Eds.), *Advanced Net. Tech. and Intell. Comput. ANTIC 2023. Commun. in Comp. and Info. Science* (Vol. 2090). Springer., DOI: 10.1007/978-3-031-64076-6_11

Yadav, R., & Kumar Mishra, A.. (2024). IoT and Mach. Learn. -Based Monit. of the Grow. of Crops Using Block.chain Tech. In *Adv. Net. Tech. and Intell. Comp.* Springer., DOI: 10.1007/978-3-031-64076-6_11

Yadav, R., Saroj, R., & Kumar Verma, A. (2023). *A Survey of IoT and Mach. Learn. Based Mon. of the Growth of Crops Using Blockchain Tech. 2023 Intern. Conf. on IoT, Comm. and Auto. Tech.* ICICAT., DOI: 10.1109/ICICAT57735.2023.10263755

Compilation of References

Aaronson, S. (2008). The limits of quantum. *Scientific American*, *298*(3), 62–69. DOI: 10.1038/scientificamerican0308-62 PMID: 18357822

Aaronson, S. (2015). Read the fine print. *Nature Physics*, *11*(4), 291–293. DOI: 10.1038/nphys3272

Abbasi, R., Maritinez, P., & Ahmad, R. (2022). The digitization of agricultural industry – a systematic literature review on agriculture 4.0. *Smart Agricultural Technology*, *2*(1), 1–10. DOI: 10.1016/j.atech.2022.100042

Abdolmaleky, M., Naseri, M., Batle, J., Farouk, A., & Gong, L.-H. (2017). Red-Green-Blue multi-channel quantum representation of digital images. *Optik (Stuttgart)*, *128*, 121–132. DOI: 10.1016/j.ijleo.2016.09.123

Abiri, R., Rizan, N., Balasundram, A. K., Shahbazi, A. B., & Abdul-Hamid, H. (2023). Application of digital technologies for ensuring agricultural productivity. *Heliyon*, *9*(12), 1–10. DOI: 10.1016/j.heliyon.2023.e22601 PMID: 38125472

ACCC. (2022). *Farm businesses warned as scams targeting the agricultural sector grow*. Retrieved 12 January 2025 from https://www.accc.gov.au/media-release/farm-businesses-warned-as-scams-targeting-agricultural-sector-grow

Addison, M., Bonuedi, I., Arhin, A. A., Wadei, B., Owusu-Addo, E., Antoh, E. F., & Mensah-Odum, N. (2024). Exploring the impact of agricultural digitalization on smallholder farmers' livelihoods in Ghana. *Heliyon*, *10*(6), 1–12. DOI: 10.1016/j.heliyon.2024.e27541 PMID: 38533044

Adefila, A. O., Ajayi, O. O., Toromade, A. S., & Sam-Bulya, N. J. (2024). Integrating traditional knowledge with modern agricultural practices: A sociocultural framework for sustainable development. *World Journal of Pharmaceutical and Life Sciences*, *10*(4), 1–10.

Adeyemi, O., Grove, I., Peets, S., & Norton, T. (2017). Advanced monitoring and management *systems* for improving sustainability in precision irrigation. *Sustainability (Basel)*, *9*(3), 353. DOI: 10.3390/su9030353

Afonso, A. M., Antunes, M. D., Cruz, S., Cavaco, A. M., & Guerra, R. (2022). Non-destructive follow-up of 'Jintao'kiwifruit ripening through VIS-NIR spectroscopy–individual vs. average calibration model's predictions. *Postharvest Biology and Technology*, *188*, 111895. DOI: 10.1016/j.postharvbio.2022.111895

AgriBusiness. (2025). *Boot up to tackle cyber threats*. Retrieved 9 January 2025 from https://agribusinessconnect.com.au/boot-up-to-tackle-cyber-threats/

Aharoni, R., Klymiuk, V., Sarusi, B., Young, S., Fahima, T., Fishbain, B., & Kendler, S. (2021). Spectral light-reflection data dimensionality reduction for timely detection of yellow rust. *Precision Agriculture*, *22*(1), 267–286. DOI: 10.1007/s11119-020-09742-2

Akbar, J. U. M., Kamarulzaman, S. F., Muzahid, A. J. M., Rahman, Md. A., & Uddin, M. (2024). A Comprehensive Review on Deep Learning Assisted Computer Vision Techniques for Smart Greenhouse Agriculture. *IEEE Access, 12*, 4485–4522. IEEE Access. DOI: 10.1109/ACCESS.2024.3349418

Åkerfeldt, M. P., Gunnarsson, S., Bernes, G., & Blanco-Penedo, I. (2021). Health and welfare in organic livestock production systems—A systematic mapping of current knowledge. *Organic Agriculture*, *11*(1), 105–132. DOI: 10.1007/s13165-020-00334-y

Akintuyi, O. B. (2024). *AI in agriculture: A comparative review of developments in the USA and Africa*. Open Access Research Journal of Science and Technology., DOI: 10.53022/oarjst.2024.10.2.0051

Al-Adhami, A., Hasoun, R. K., Gbashi, E. K., & Hassan, S. (2022). A secure communication protocol for civil drones. *Indonesian Journal of Electrical Engineering and Computer Science*, *27*(3). Advance online publication. https://ijeecs.iaescore.com/index.php/IJEECS/article/view/26098. DOI: 10.11591/ijeecs.v27.i3.pp1490-1501

Alahe, M. A., Wei, L., Chang, Y., Gummi, S. R., Kemeshi, J., Yang, X., Won, K., & Sher, M. (2024). Cyber security in smart agriculture: Threat types, current status, and future trends. *Computers and Electronics in Agriculture*, *226*, 109401. DOI: 10.1016/j.compag.2024.109401

Alfaro, C.. (2020). Solar-powered autonomous robots in agriculture: Sustainability and challenges. *Renewable Energy*, *145*, 2685–2696. DOI: 10.1016/j.renene.2020.06.024

Alfred, R., Obit, J. H., Chin, C. P.-Y., Haviluddin, H., & Lim, Y. (2021). Towards paddy rice smart farming: A review on big data, machine learning, and rice production tasks. *IEEE Access : Practical Innovations, Open Solutions*, *9*, 50358–50380. DOI: 10.1109/ACCESS.2021.3069449

Al-Hraishawi, H., Rehman, J. U., Razavi, M., & Chatzinotas, S. (2024). Characterizing and utilizing the interplay between quantum technologies and non-terrestrial networks. *IEEE Open Journal of the Communications Society*, *5*, 1937–1957. DOI: 10.1109/OJCOMS.2024.3380508

Ali, E., Gninigue̊, M., & Awade, N. E. (2023). Sectoral value chains and environmental pollution in Africa: Can development policies target digitalization and structural transformation to enhance environmental governance? *Journal of Environmental Economics and Policy*, *12*(2), 229–247. DOI: 10.1080/21606544.2022.2110163

Ali, I. A., Bukhari, W. A., Adnan, M., Kashif, M. I., Danish, A., & Sikander, A. (2024). Security and privacy in IoT-based smart farming: A review. *Multimedia Tools and Applications*, •••, 1–61. DOI: 10.1007/s11042-024-19653-3

Ali, M. Z., Abohmra, A., Usman, M., Zahid, A., Heidari, H., Imran, M. A., & Abbasi, Q. H. (2023). Quantum for 6G communication: A perspective. *IET Quantum Communication*, *4*(3), 112–124. DOI: 10.1049/qtc2.12060

Ali, S. M., Dash, N., & Pradhan, A. (2012). Role of renewable energy on agriculture. *International Journal of Engineering Sciences & Emerging Technologies*, *4*(1), 51–57.

Alkhammash, E. H. (2025). Multi-Classification Using YOLOv11 and Hybrid YOLO11n-MobileNet Models: A Fire Classes Case Study. *Fire (Basel, Switzerland)*, *8*(1), 1. Advance online publication. DOI: 10.3390/fire8010017

Al-Kofahi, M. M., & Al-Shorman, M. Y. (2019). Toward ene. Effi. Micro. and inter.-of-thin. systems. *Computers & Electrical Engineering*, *79*, 106457. DOI: 10.1016/j.compeleceng.2019.106457

Almaraz, M., Simmonds, M., Boudinot, F. G., Di, V. A. V., Bingham, N., Khalsa, S. D. S., Ostoja, S., Scow, K., Jones, A., Holzer, I., Manaigo, E., Geoghegan, E., Goertzen, H., & Silver, W. L. (2023). Soil carbon sequestration in global working lands as a gateway for negative emission technologies. *Global Change Biology*, *29*(21), 5988–5998. DOI: 10.1111/gcb.16884 PMID: 37476859

Alonso, R. S., & Sittón-Candanedo, I. (2020). An intelli. edge-iot plat. for moni. livestock and crops in a dairy farm. *Scen. Ad Hoc Netw.*, *98*, 102047. DOI: 10.1016/j.adhoc.2019.102047

Al-Qerem, A., Alauthman. (. (2020). IoT trans. Process. Thro. Coop. concur. control on fog-cloud compu. Environ. *Soft Computing*, *24*(8), 5695–5711. DOI: 10.1007/s00500-019-04220-y

Alqudhaibi, A., Krishna, A., Jagtap, S., Williams, N., Afy-Shararah, M., & Salonitis, K. (2024). Cybersecurity 4.0: Safeguarding trust and production in the digital food industry era. *Discover Food*, *4*(1), 2. DOI: 10.1007/s44187-023-00071-7

Alsamhi, S. H., Ma, O., Ansari, M. S., & Gupta, S. K. (2019). Collaboration of drone and Internet of Public Safety Things in smart cities: An overview of QoS and network performance optimization. *Drones (Basel)*, *3*(1), 13. Advance online publication. DOI: 10.3390/drones3010013

Alyahya, S., Khan, W. U., Ahmed, S., Marwat, S. N. K., & Habib, S. (2022). Cyber secure framework for smart agriculture: Robust and tamper-resistant authentication scheme for IoT devices. *Electronics (Basel)*, *11*(6), 963. DOI: 10.3390/electronics11060963

Alzubaidi, L., Zhang, J., Humaidi, A. J., Al-Dujaili, A., Duan, Y., Al-Shamma, O., Santamaría, J., Fadhel, M. A., Al-Amidie, M., & Farhan, L. (2021). Review of deep learning: Concepts, CNN architectures, challenges, applications, future directions. *Journal of Big Data*, *8*(1), 53. DOI: 10.1186/s40537-021-00444-8 PMID: 33816053

Amaral, M. H., McConchie, C., Dickinson, G., & Walsh, K. B. (2023). Growing degree day targets for fruit development of Australian mango cultivars. *Horticulturae*, *9*(4), 489. DOI: 10.3390/horticulturae9040489

Anderson, N. T., Subedi, P. P., & Walsh, K. B. (2017). Manipulation of mango fruit dry matter content to improve eating quality. *Scientia Horticulturae*, *226*, 316–321. DOI: 10.1016/j.scienta.2017.09.001

Anderson, N. T., Underwood, J. P., Rahman, M. M., Robson, A., & Walsh, K. B. (2018). Estimation of fruit load in mango orchards: Tree sampling considerations and use of machine vision and satellite imagery [journal article]. *Precision Agriculture*. Advance online publication. DOI: 10.1007/s11119-018-9614-1

Anderson, N. T., & Walsh, K. B. (2022). The evolution of chemometrics coupled with near infrared spectroscopy for fruit quality evaluation. *Journal of Near Infrared Spectroscopy*, *30*(1), 3–17. DOI: 10.1177/09670335211057235

Anderson, N. T., Walsh, K. B., Koirala, A., Wang, Z., Amaral, M. H., Dickinson, G. R., Sinha, P., & Robson, A. J. (2021). Estimation of Fruit Load in Australian Mango Orchards Using Machine Vision. *Agronomy (Basel)*, *11*(9), 1711. https://www.mdpi.com/2073-4395/11/9/1711. DOI: 10.3390/agronomy11091711

Anderson, N. T., Walsh, K. B., & Wulfsohn, D. (2021). Technologies for Forecasting Tree Fruit Load and Harvest Timing—From Ground, Sky and Time. *Agronomy (Basel), 11*(7), 1409. https://www.mdpi.com/2073-4395/11/7/1409. DOI: 10.3390/agronomy11071409

Anderson, N., Walsh, K., Flynn, J., & Walsh, J. P. (2021). Achieving robustness across season, location and cultivar for a NIRS model for intact mango fruit dry matter content. II. Local PLS and nonlinear models. *Postharvest Biology and Technology, 171*, 111358. DOI: 10.1016/j.postharvbio.2020.111358

Anderson, N., Walsh, K., Subedi, P., & Hayes, C. (2020). Achieving robustness across season, location and cultivar for a NIRS model for intact mango fruit dry matter content. *Postharvest Biology and Technology, 168*, 111202. DOI: 10.1016/j.postharvbio.2020.111202

Annamalai, P., & Suk Lee, W. (2003). *Citrus Yield Mapping System Using Machine Vision* St. Joseph, MI. https://elibrary.asabe.org/abstract.asp?aid=13701&t=5

Annosi, M. C., Brunetta, F., Capo, F., & Heideveld, L. (2020). Digitalization in the agri-food industry: The relationship between technology and sustainable development. *Management Decision, 58*(8), 1737–1757. DOI: 10.1108/MD-09-2019-1328

Aoki, H., Tanabe, T., & Akinaga, T. (2004). Pre and postharvest data collecting system using the compact NIR analyzer. III International Symposium on Tropical and Subtropical Fruits 864, Assimakopoulos, F., Vassilakis, C., Margaris, D., Kotis, K., & Spiliotopoulos, D. (2024). Artificial Intelligence Tools for the Agriculture Value Chain: Status and Prospects. *Electronics (Basel), 13*(22), 4362. https://www.mdpi.com/2079-9292/13/22/4362

Araujo, V., Mitra, K., Saguna, S., & Åhlund, C. (2019). Perform. Eval. of fiware: A cloud-based iot plat. for smart cities. *Journal of Parallel and Distributed Computing, 132*, 250–261. DOI: 10.1016/j.jpdc.2018.12.010

Archana, R., & Jeevaraj, P. S. E. (2024). Deep learning models for digital image processing: A review. *Artificial Intelligence Review, 57*(1), 11. DOI: 10.1007/s10462-023-10631-z

Arora, S., & Wadhwa, S. (2014). *FDI in Agricultural sector in India: status and challenges.* In book: Foreign Direct Investment Chapter: FDI in the agricultural sector in India: status and challenges. *Avon Publications*, New Delhi.

Arya, S., Yang, J., & Wang, Y. (2023). Towards the designing of low-latency SAGIN: Ground-to-UAV communications over interference channel. *Drones (Basel), 7*(7), 479. Advance online publication. DOI: 10.3390/drones7070479

Ashraf, S. N., & Singh, A. K. (2021). Impact of technological change on growth and agricultural sector in Gujarat state of India: A time-series data study. *Asian Development Policy Review*, *9*(3), 144–160. DOI: 10.18488/journal.107.2021.93.144.160

Ashraf, S. N., & Singh, A. K. (2022). Implications of appropriate technology and farm inputs in the agricultural sector of Gujarat: Empirical analysis based on primary data. *Agricultural Economics Research Review*, *35*(2), 59–77. DOI: 10.5958/0974-0279.2022.00031.3

Ashurov, A., Al-Gaashani, M. S. A. M., Samee, N. A., Alkanhel, R., Atteia, G., Abdallah, H. A., & Muthanna, M. S. A. (2025). Enhancing plant disease detection through deep learning: A Depthwise CNN with squeeze and excitation integration and residual skip connections. *Frontiers in Plant Science*, *15*, 1505857. Advance online publication. DOI: 10.3389/fpls.2024.1505857 PMID: 39925367

Australian Government. (2024). *Australian Government response to the House of Representatives Standing Committee on Agriculture and Industry report: Smart farming - inquiry into agricultural innovation.* Retrieved 24 December 2024 from https://www.agriculture.gov.au/about/reporting/obligations/government-responses/smart-farming-agricultural-innovation-inquiry#daff-page-main

Avaliani, A., “Quantum computers,” arXiv preprint cs/0405004, 2004.

Banzi, M., & Shiloh, M. (2014). *Getting started with Arduino: The open source electronics prototyping platform* (3rd ed.). Maker Media, Inc.

Bao, Z. (2024). The UAV Target Detection Algorithm Based on Improved YOLO V8. *Proceedings of the International Conference on Image Processing, Machine Learning and Pattern Recognition*, 264–269. DOI: 10.1145/3700906.3700949

Bapatla, A. K., Mohanty, S. P., & Kougianos, E. (2021). sFarm: A distributed ledger-based remote crop monitoring system for smart farming. IFIP International Internet of Things Conference

Barathkumar, S., Sellamuthu, K., Sathyabama, K., Malathi, P., Kumaraperumal, R., & Devagi, P. (2024). Advancements in soil quality assessment: A comprehensive review of machine learning and AI-driven approaches for nutrient deficiency analysis. *Communications in Soil Science and Plant Analysis*, •••, 1–26.

Barnas, A. F., Chabot, D., Hodgson, A. J., Johnston, D. W., Bird, D. M., & Ellis-Felege, S. N. (2020). A standardized protocol for reporting methods when using drones for wildlife research. *Journal of Unmanned Vehicle Systems*, *8*(2), 89–98. DOI: 10.1139/juvs-2019-0011

Basri, R., Islam, F., Shorif, S. B., & Uddin, M. S. (2021). Robots and drones in agriculture—a survey. *Computer vision and machine learning in agriculture*, 9-29.

Bay, H., Ess, A., Tuytelaars, T., & Van Gool, L. (2008). Speeded-up robust features (SURF). *Computer Vision and Image Understanding*, *110*(3), 346–359. DOI: 10.1016/j.cviu.2007.09.014

Behjati, M., Mohd Noh, A. B., Alobaidy, H. A. H., Zulkifley, M. A., Nordin, R., & Abdullah, N. F. (2021). LoRa communications as an enabler for internet of drones towards large-scale livestock monitoring in rural farms. *Sensors (Basel)*, *21*(15), 5044. Advance online publication. DOI: 10.3390/s21155044 PMID: 34372281

Beig, B., Niazi, M. B. K., Jahan, Z., Zia, M., Shah, G. A., & Iqbal, Z. (2023). Biodegradable polymer encapsulated nickel nanoparticles for slow release urea promotes rhode grass yield and nitrogen recovery. *Journal of Polymers and the Environment*, *31*(5), 1866–1883. DOI: 10.1007/s10924-022-02729-2

Bertone, E., Venturello, A., Leardi, R., & Geobaldo, F. (2012). Prediction of the optimum harvest time of 'Scarlet' apples using DR-UV–Vis and NIR spectroscopy. *Postharvest Biology and Technology*, *69*, 15–23. DOI: 10.1016/j.postharvbio.2012.02.009

Bhandari, M., Shahi, T. B., Neupane, A. S., & Walsh, K. B. (2023). BotanicX-AI: Identification of Tomato Leaf Diseases Using an Explanation-Driven Deep-Learning Model. *Journal of Imaging*, *9*(2), 53. DOI: 10.3390/jimaging9020053 PMID: 36826972

Bhat, R., Gupta, M., Choudhary, T. F., & Singh, A. K. (2024). Modelling the effects of climatic variables and political instability on the agricultural sector in South Asian countries. In Ordóñez de Pablos, P. (Ed.), *Building Climate Neutral Economies Through Digital Business and Green Skills* (pp. 1–36). IGI Global Scientific Publishing., DOI: 10.4018/979-8-3693-5673-9.ch001

Bhuiyan, M. R., & Wree, P. (2023). Animal Behavior for Chicken Identification and Monitoring the Health Condition Using Computer Vision: A Systematic Review. *IEEE Access, 11*, 126601–126610. IEEE Access. DOI: 10.1109/ACCESS.2023.3331092

Bian, L., Zhang, H., Ge, Y., Čepl, J., Stejskal, J., & EL-Kassaby, Y. A. (2022). Closing the gap between phenotyping and genotyping: Review of advanced, image-based phenotyping technologies in forestry. *Annals of Forest Science*, *79*(1), 22. DOI: 10.1186/s13595-022-01143-x

Biró, K., & Csete, M. S. (2023). How can development strategies foster agri-digitalisation in the era of climate change? a common agriculture policy based consistency analysis in Hungary. *Periodica Polytechnica Social and Management Sciences*, *31*(1), 9–18. DOI: 10.3311/PPso.20411

Bissadu, K. D., Sonko, S., & Hossain, G. (2024). Society 5.0 enabled agriculture: Drivers, enabling technologies, architectures, opportunities, and challenges. *Information Processing in Agriculture.*

Bist, R. B., Bist, K., Poudel, S., Subedi, D., Yang, X., Paneru, B., Mani, S., Wang, D., & Chai, L. (2024). Sustainable poultry farming practices: A critical review of current strategies and future prospects. *Poultry Science*, *103*(12), 104295. DOI: 10.1016/j.psj.2024.104295 PMID: 39312848

Bist, R. B., Subedi, S., Yang, X., & Chai, L. (2023). A Novel YOLOv6 Object Detector for Monitoring Piling Behavior of Cage-Free Laying Hens. *AgriEngineering*, *5*(2), 2. Advance online publication. DOI: 10.3390/agriengineering5020056

Boadi, S. A., Olwig, M. F., Asare, R., Bosselmann, A. S., & Owusu, K. (2022). The role of innovation in sustainable cocoa cultivation: Moving beyond mitigation and adaptation. *Climate-Induced Innovation: Mitigation and Adaptation to Climate Change*, pp. 47-80. DOI: DOI: 10.1007/978-3-031-01330-0_3

Bobier, C. A. (2021). What Would the Virtuous Person Eat? The Case for Virtuous Omnivorism. *Journal of Agricultural & Environmental Ethics*, *34*(3), 3. Advance online publication. DOI: 10.1007/s10806-021-09860-5

Boev, A. S. (2022). D. N. Trubnikov D. Bokhan, A. S. Mastiukova and A. K. Fedorov, "Multiclass classification using quantum convolutional neural networks with hybrid quantum-classical learning,". *Frontiers in Physics*, *10*, 1173.

Bonnefous, C., Collin, A., Guilloteau, L. A., Guesdon, V., Filliat, C., Réhault-Godbert, S., Rodenburg, T. B., Tuyttens, F. A. M., Warin, L., Steenfeldt, S., Baldinger, L., Re, M., Ponzio, R., Zuliani, A., Venezia, P., Väre, M., Parrott, P., Walley, K., Niemi, J. K., & Leterrier, C. (2022). Welfare issues and potential solutions for laying hens in free range and organic production systems: A review based on literature and interviews. *Frontiers in Veterinary Science*, *9*, 952922. Advance online publication. DOI: 10.3389/fvets.2022.952922 PMID: 35990274

Borchi, J., Woodcock, M., Redshaw, M., & Raniga, B. (2021). Cyber security threats-are we prepared? A threat-based assessment of the cyber resilience of the Australian agricultural sector.

Bova, F., Goldfarb, A., & Melko, R. G. (2021). Commercial applications of quantum computing. *EPJ Quantum Technology*, *8*(1), 2. DOI: 10.1140/epjqt/s40507-021-00091-1 PMID: 33569545

Brisco, B., Brown, R., Hirose, T., McNairn, H., & Staenz, K. (1998). Precision agriculture and the role of remote sensing: A review. *Canadian Journal of Remote Sensing*, *24*(3), 315–327. DOI: 10.1080/07038992.1998.10855254

Brownlee, J. (2018). *Machine learning mastery with Python: Machine learning for big data: From conventional ML to incremental learning*. Machine Learning Press.

Bshouty, H. N., & Jackson, J. C. "Learning DNF over the uniform distribution using a quantum example oracle", Proceedings of the eighth annual conference on Computational learning theory,vol.,no.,pp.118-127,1995. DOI: 10.1145/225298.225312

Budi Rahmadya. (2021). *Drone-based wireless sensor systems using Zigbee and RFID technology* (Ph.D. dissertation). Ibaraki University.

Bui, H. T., Aboutorab, H., Mahboubi, A., Gao, Y., Sultan, N. H., Chauhan, A., & Islam, Z. (2024). Agriculture 4.0 and beyond: Evaluating cyber threat intelligence sources and techniques in smart farming ecosystems. *Computers & Security*, *140*, 103754. DOI: 10.1016/j.cose.2024.103754

Bwambale, E., Abagale, F. K., & Anornu, G. K. (2022). Smart irrigation monitoring and control strategies for improving water use efficiency in precision agriculture: A review. *Agricultural Water Management*, *260*, 107324. DOI: 10.1016/j.agwat.2021.107324

Campbell, M., Miller, P., Díaz-Chito, K., Hong, X., McLaughlin, N., Parvinzamir, F., Martínez Del Rincón, J., & O'Connell, N. (2024). A computer vision approach to monitor activity in commercial broiler chickens using trajectory-based clustering analysis. *Computers and Electronics in Agriculture*, *217*, 108591. DOI: 10.1016/j.compag.2023.108591

Campoverde-Molina, M., & Luján-Mora, S. (2024). Cybersecurity in smart agriculture: A systematic literature review. *Computers & Security*, •••, 104284.

Carion, N., Massa, F., Synnaeve, G., Usunier, N., Kirillov, A., & Zagoruyko, S. (2020). End-to-end object detection with transformers. In *Proceedings of the European Conference on Computer Vision* (pp. 213–229). https://doi.org/DOI: 10.1007/978-3-030-58452-8_13

Caruso, A., Chessa, S., Escolar, S., Barba, J., & López, J. C. (2021). Collection of data with drones in precision agriculture: Analytical model and LoRa case study. *IEEE Internet of Things Journal*, *8*(22), 16692–16704. DOI: 10.1109/JIOT.2021.3075561

Cavaco, A., Antunes, M., Guerra, R., Rosendo, M., Pires, R., Brázio, A., Silva, L., Afonso, A., & Panagopoulos, T. (2016). Fruit quality survey techniques to optimize crop management in citrus orchards. VIII International Postharvest Symposium: Enhancing Supply Chain and Consumer Benefits-Ethical and Technological Issues 1194

Cavaco, A. M., Pires, R., Antunes, M. D., Panagopoulos, T., Brázio, A., Afonso, A. M., Silva, L., Lucas, M. R., Cadeiras, B., & Cruz, S. P. (2018). Validation of short wave near infrared calibration models for the quality and ripening of 'Newhall'orange on tree across years and orchards. *Postharvest Biology and Technology*, *141*, 86–97. DOI: 10.1016/j.postharvbio.2018.03.013

Chalidabhongse, T., Yimyam, P., & Sirisomboon, P. (2006). 2D/3D Vision-Based Mango's Feature Extraction and Sorting. *9th International Conference on Control, Automation, Robotics and Vision*

Chandio, A. A., Amin, A., Khan, I., Rehman, A., & Memon, A. (2024). Can digitalization improve agriculture? Exploring the impact of ICT on grain food production in SAARC countries. *Information Development*, *10*(10), 1–14. DOI: 10.1177/02666669231225945

Chandrappa, V. Y., Ray, B., Ashwatha, N., & Shrestha, P. (2023). Spatiotemporal modeling to predict soil moisture for sustainable smart irrigation. *Internet of Things : Engineering Cyber Physical Human Systems*, *21*, 100671. DOI: 10.1016/j.iot.2022.100671

Chaudhary, A. (2016). Role of FDI in the growth of Indian agricultural sector: A post-reform study. *Global Journal of Finance and Management*, *8*(2), 119–129.

Chel, A., & Kaushik, G. (2011). Renewable energy for sustainable agriculture. *Agronomy for Sustainable Development*, *31*(1), 91–118. DOI: 10.1051/agro/2010029

Cheng, S., Qi, H., Sun, Y., Liu, L., & Xie, S. (2024). Toward quantum unmanned systems. *Science China. Technological Sciences*, •••, 67.

Chen, J. (2021). Review on quantum communication and quantum computation. *Journal of Physics: Conference Series*, *1865*(2), 022008. DOI: 10.1088/1742-6596/1865/2/022008

Chen, Y., & Li, M. (2024). How does the digital transformation of agriculture affect carbon emissions? Evidence from China's provincial panel data. *Humanities & Social Sciences Communications*, *11*(1), 1–18. DOI: 10.1057/s41599-024-03223-x

Chen, Z.-P., Morris, J., & Martin, E. (2005). Correction of temperature-induced spectral variations by loading space standardization. *Analytical Chemistry*, *77*(5), 1376–1384.

Chiti, F., Picchi, R., & Pierucci, L. (2022). Metropolitan quantum-drone networking and computing: A software-defined perspective. *IEEE Access : Practical Innovations, Open Solutions*, *10*, 126062–126073. Advance online publication. DOI: 10.1109/ACCESS.2022.3226127

Choi, J.-H., Chen, P.-A., Lee, B., Yim, S.-H., Kim, M.-S., Bae, Y.-S., Lim, D.-C., & Seo, H.-J. (2017). Portable, non-destructive tester integrating VIS/NIR reflectance spectroscopy for the detection of sugar content in Asian pears. *Scientia Horticulturae*, *220*, 147–153. DOI: 10.1016/j.scienta.2017.03.050

Choudhary, T. F., & Gupta, M. (2023). Analyzing the long-run and short-run impacts of climate change on wheat and maize yield in the western Himalayan region of India. *Climate Change Economics (Singapore)*, 2350019. DOI: 10.1142/S2010007823500197

Cioanca, O., Lungu, I.-I., Mita-Baciu, I., Robu, S., Burlec, A. F., Hancianu, M., & Crivoi, F. (2024). Extraction and Purification of Catechins from Tea Leaves: An Overview of Methods, Advantages, and Disadvantages. *Separations*, *11*(6), 6. Advance online publication. DOI: 10.3390/separations11060171

Claughton, D., & Beilharz, N. (2021). JBS Foods pays $14.2 million ransom to end cyber attack on its global operations. Retrieved 23 December 2024 from https://www.abc.net.au/news/rural/2021-06-10/jbs-foods-pays-14million-ransom-cyber-attack/100204240

Comes, J. D. G., Poniman, M., van Oosten, L., Doets, K., de Cloe, S., Geertsema, C., & Pijlman, G. P. (2024). Infectious clone of a contemporary Tembusu virus and replicons expressing reporter genes or heterologous antigens from poultry viruses. *Biotechnology Journal*, *19*(1), 2300254. DOI: 10.1002/biot.202300254 PMID: 37750498

Craigen, D., Diakun-Thibault, N., & Purse, R. (2014). Defining cybersecurity. *Technology Innovation Management Review*, *4*(10), 13–21. DOI: 10.22215/timreview/835

Cravero, A., Pardo, S., Sepúlveda, S., & Muñoz, L. (2022). Challenges to Use Machine Learning in Agricultural Big Data: A Systematic Literature Review. *Agronomy (Basel)*, *12*(3), 748. DOI: 10.3390/agronomy12030748

Crawford, S. E., Shugayev, R. A., Paudel, H. P., Lu, P., Syamlal, M., Ohodnicki, P. R., Chorpening, B., Gentry, R., & Duan, Y. (2021). Quantum sensing for energy applications: Review and perspective. *Advanced Quantum Technologies*, *4*(8), 2100049. DOI: 10.1002/qute.202100049

Dambal, V. A., Mohadikar, S., Kumbhar, A., & Guvenc, I. (2019). Improving LoRa signal coverage in urban and sub-urban environments with UAVs. *2019 International Workshop on Antenna Technology (iWAT)*, 210–213. https://api.semanticscholar.org/CorpusID:86764193

Dargan, S., Kumar, M., Ayyagari, M. R., & Kumar, G. (2020). A Survey of Deep Learning and Its Applications: A New Paradigm to Machine Learning. *Archives of Computational Methods in Engineering*, *27*(4), 1071–1092. DOI: 10.1007/s11831-019-09344-w

David, . (2016). Milestones toward Majorana-based quantum computing. *Physical Review X*, *6*(3), 031016. DOI: 10.1103/PhysRevX.6.031016

Dawn, N., Ghosh, T., Ghosh, S., Saha, A., Mukherjee, P., Sarkar, S., & Sanyal, T. (2023). Implementation of Artificial Intelligence, Machine Learning, and Internet of Things (IoT) in revolutionizing Agriculture: A review on recent trends and challenges. *International Journal of Experimental Research and Review*, *30*, 190–218.

Demestichas, K., Peppes, N., & Alexakis, T. (2020). Survey on security threats in agricultural IoT and smart farming. *Sensors (Basel)*, *20*(22), 6458. DOI: 10.3390/s20226458 PMID: 33198160

Demestichas, K., Peppes, N., Alexakis, T., & Adamopoulou, E. (2020). E. Blockchain in agriculture traceability systems: A review. *Applied Sciences (Basel, Switzerland)*, *10*(12), 4113. DOI: 10.3390/app10124113

Dendukuri, A., & Luu, K. Image processing in quantum computers. arXiv preprint arXiv:1812.11042, 2018.

Dey, S. K., Thakur, U., Upadhyay, A., & Kabir, Md. S., Sintheia, T. S., & Raju, V. (2024). Improving Guava Disease Detection using Inception Mobilenet. *2024 International Conference on Advances in Computing, Communication and Applied Informatics (ACCAI)*, 1–7. DOI: 10.1109/ACCAI61061.2024.10601750

Dhanaraju, M., Chenniappan, P., & Ramalingam, K. (2022). Smart farming: Inter. of things (IoT)-based sust. agri. *Agriculture*, *12*, 1745. DOI: 10.3390/agriculture12101745

Dhanya, V. G., Subeesh, A., Kushwaha, N. L., Vishwakarma, D. K., Nagesh Kumar, T., Ritika, G., & Singh, A. N. (2022). Deep learning based computer vision approaches for smart agricultural applications. *Artificial Intelligence in Agriculture*, *6*, 211–229. DOI: 10.1016/j.aiia.2022.09.007

Dhonju, H. K., Walsh, K. B., & Bhattarai, T. (2024). Management information systems for tree fruit—1: A review. *Horticulturae*, *10*(1), 108. DOI: 10.3390/horticulturae10010108

Dhungana, B. R., & Ghimire, R. (2013). The role of FDI on agriculture sector in India. In *Conference Paper: Foreign Direct Investment in India's Retail Sector: Issues and Options, National Seminar organized by Department of Business Administration,* Deen Dayal Upadhyaya University, Gorakhpur, India.

Dibbern, T., Ramani, L. A. S., & Massruhá, S. M. F. S. (2024). Main drivers and barriers to the adoption of digital agriculture technologies. *Smart Agricultural Technology*, *8*(1), 1–12. DOI: 10.1016/j.atech.2024.100459

Diwan, T., Anirudh, G., & Tembhurne, J. V. (2023). Object detection using YOLO: Challenges, architectural successors, datasets and applications. *Multimedia Tools and Applications*, *82*(6), 9243–9275. DOI: 10.1007/s11042-022-13644-y PMID: 35968414

Dong, P. Q. (2011). Le and Hirota, K., "A flexible representation of quantum images for polynomial preparation, image compression, and processing operations,". *Quantum Information Processing*, *10*(1), 63–84. DOI: 10.1007/s11128-010-0177-y

Dorj, U.-O., Lee, M., & Yun, S. (2017). An yield estimation in citrus orchards via fruit detection and counting using image processing. *Computers and Electronics in Agriculture*, *140*, 103–112. https://doi.org/https://doi.org/10.1016/j.compag.2017.05.019. DOI: 10.1016/j.compag.2017.05.019

Drape, T., Magerkorth, N., Sen, A., Simpson, J., Seibel, M., Murch, R. S., & Duncan, S. E. (2021). Assessing the role of cyberbiosecurity in agriculture: A case study. *Frontiers in Bioengineering and Biotechnology*, *9*, 737927. DOI: 10.3389/fbioe.2021.737927 PMID: 34490231

Du, S., Zhang, B., Zhang, P., & Xiang, P. (2021). An Improved Bounding Box Regression Loss Function Based on CIOU Loss for Multi-scale Object Detection. *2021 IEEE 2nd International Conference on Pattern Recognition and Machine Learning (PRML)*, 92–98. DOI: 10.1109/PRML52754.2021.9520717

Duarah, P., Joardar, S., Debnath, B., & Purkait, M. K. (2024). Optimized extraction of polyphenols from tea factory waste and cost-effective drying methods for sustainable utilization. *Bioresource Technology Reports*, *26*, 101833. DOI: 10.1016/j.biteb.2024.101833

Du, K., Li, P., & Yan, Z. (2019). Do green technology innovations contribute to carbon dioxide emission reduction? Empirical evidence from patent data. *Technological Forecasting and Social Change, 146*(1), 297–303. DOI: 10.1016/j.techfore.2019.06.010

Durai, S. K. S., & Shamili, M. D. (2022). Smart farm. using mach. Learn. and deep learn. *Tech. Decis. Anal. J., 3*, 100041.

Đuric', I. (2020). *Digital technology and agricultural markets – Background paper for The State of Agricultural Commodity Markets (SOCO) 2020*. FAO., DOI: 10.4060/cb0701en

Dutta, G., & Goswami, P. (2020). Application of drone in agriculture: A review. *International Journal of Chemical Studies, 8*(8), 181–187. DOI: 10.22271/chemi.2020.v8.i5d.10529

Du, X., Wang, X., & Hatzenbuehler, P. (2022). Digital technology in agriculture: A review of issues, applications and methodologies. *China Agricultural Economic Review, 15*(1), 95–108. DOI: 10.1108/CAER-01-2022-0009

Eckert, K., Hyllus, P., Brüß, D., Poulsen, U. V., Lewenstein, M., Jentsch, C., Müller, T., Rasel, E. M., & Ertmer, W. (2006). Differential atom interferometry beyond the standard quantum limit. *Physical Review A, 73*(1), 013814. DOI: 10.1103/PhysRevA.73.013814

Ehrlich, P. R.. (2015). Global environmental change and sustainability. *Global Change Biology, 21*(12), 4417–4428. DOI: 10.1111/gcb.13030

Elfring, J., Van De Molengraft, M. J. G., & Steinbuch, M. (2013). Learning motion and sensor models for mobile robot localization. *Robotics and Autonomous Systems, 61*(3), 223–236. DOI: 10.1016/j.robot.2012.11.008

Eyuboglu, K., & Uzar, U. (2020). Examining the roles of renewable energy consumption and agriculture on CO2 emission in lucky-seven countries. *Environmental Science and Pollution Research International, 27*(36), 45031–45040. DOI: 10.1007/s11356-020-10374-2 PMID: 32772294

Farid, A., Hussain, F., Khan, K., Shahzad, M., Khan, U., & Mahmood, Z. (2023). A Fast and Accurate Real-Time Vehicle Detection Method Using Deep Learning for Unconstrained Environments. *Applied Sciences (Basel, Switzerland), 13*(5), 5. Advance online publication. DOI: 10.3390/app13053059

Fariha, F. (2023). E-commerce in Agriculture: Creating Direct Farmer-Buyer Link https://coredevsltd.com/articles/e-commerce-in-agriculture/

Feng, J., McGlone, A. V., Currie, M., Clark, C. J., & Jordan, B. R. (2011). Assessment of yellow-fleshed kiwifruit (actinidia chinensis 'hort16a') quality in pre-and post-harvest conditions using a portable near-infrared spectrometer. *HortScience*, *46*(1), 57–63. DOI: 10.21273/HORTSCI.46.1.57

Feynman, R. P. (2018). *"Simulating physics with computers," in Feynman and computation*. CRC Press.

Fouda, A., Ibrahim, A. S., Güvenç, Í., & Ghosh, M. (2019). Interference management in UAV-assisted integrated access and backhaul cellular networks. *IEEE Access : Practical Innovations, Open Solutions*, *7*, 104553–104566. DOI: 10.1109/ACCESS.2019.2927176

Fountas, S.. (2006). Precision agriculture: A methodological framework for assessing implementation and impact. *Agricultural Systems*, *91*(1–2), 159–182. DOI: 10.1016/j.agsy.2006.03.005

Francis, R. J., Brandis, K. J., & McCann, J. A. (2022). Offline imagery checks for remote drone usage. *Drones (Basel)*, *6*(12), 395. Advance online publication. DOI: 10.3390/drones6120395

Gabriel, A., & Gandorfer, M. (2023). Adoption of digital technologies in agriculture—An inventory in a european small-scale farming region. *Precision Agriculture*, *24*(1), 68–91. DOI: 10.1007/s11119-022-09931-1

Gadade, H. D., & Kirange, D. K. (2021). Machine Learning Based Identification of Tomato Leaf Diseases at Various Stages of Development. *International Conference Computing Methodologies and Communication*, 814–819. https://doi.org/DOI: 10.1109/ICCMC51019.2021.9418263

Ganesh, N. B. (2025). Plant Disease Detection by Image Processing. *Indian Scientific Journal Of Research In Engineering And Management*, *09*(01), 1–9. DOI: 10.55041/IJSREM40956

Gao, Y., Zhang, Y., Lu, K., & Wang, M. (2013). NEQR: A novel enhanced quantum representation of digital images. *Quantum Information Processing*, *12*(8), 2833–2860. DOI: 10.1007/s11128-013-0567-z

Gebbers, R., & Adamchuk, V. I. (2010). Precision agriculture and food security. *Science*, *327*(5967), 828–831. DOI: 10.1126/science.1183899 PMID: 20150492

Gebeyehu, S. T. (2021, May 17). *Genetic mapping on local Swedish chicken breeds* [Second cycle, A2E]. SLU, Dept. of Animal Breeding and Genetics (until 231231). https://stud.epsilon.slu.se/16684/

Gebregiorgis, G., Zeweld, W., Hdgot, A., Gebregziabher, D., Azadi, H., Van Huylenbroeck, G., Tesfay, G., & Speelman, S. (2024). *Lessons Learned from Adopting Sustainable Agricultural Practices to Address Food Insecurity and Climate Change Risks: Opportunities, Constraints, and Limitations.* Research on World Agricultural Economy., DOI: 10.36956/rwae.v5i3.1131

Geetha, P., & Karthikeyan, R. (2024). Embra. IoT and Prec. on Agri. for Sust. Crop Yields. In Balasubramanian, S., Natarajan, G., & Chelliah, P. R. (Eds.), *Intell. Robots and Dron. for Prec. Agri. Sign. and Comm. Tech.* Springer., DOI: 10.1007/978-3-031-51195-0_8

Geng, W., Liu, L., Zhao, J., Kang, X., & Wang, W. (2024). Digital technologies adoption and economic benefits in agriculture: A mixed methods approach. *Sustainability (Basel), 16*(11), 1–15. DOI: 10.3390/su16114431

George, D. A. S., & George, A. S. H. (2023). Optimizing Poultry Production Through Advanced Monitoring and Control Systems. *Partners Universal International Innovation Journal, 1*(5), 5. Advance online publication. DOI: 10.5281/zenodo.10050352

Ghamari, M., Rangel, P., Mehrubeoglu, M., Tewolde, G. S., & Sherratt, R. S. (2022). Unmanned aerial vehicle communications for civil applications: A review. *IEEE Access : Practical Innovations, Open Solutions, 10*, 102492–102531. DOI: 10.1109/ACCESS.2022.3208571

Ghazali, M. H. M., Teoh, K., & Rahiman, W. (2021). A systematic review of real-time deployments of UAV-based LoRa communication network. *IEEE Access : Practical Innovations, Open Solutions, 9*, 124817–124830. DOI: 10.1109/ACCESS.2021.3110872

Giménez-Gallego, J., Martinez-del-Rincon, J., González-Teruel, J. D., Navarro-Hellín, H., Navarro, P. J., & Torres-Sánchez, R. (2024). On-tree fruit image segmentation comparing Mask R-CNN and Vision Transformer models. Application in a novel algorithm for pixel-based fruit size estimation. *Computers and Electronics in Agriculture, 222*, 109077. https://doi.org/https://doi.org/10.1016/j.compag.2024.109077. DOI: 10.1016/j.compag.2024.109077

Giovannetti, V., Lloyd, S., & Maccone, L. (2008). Architectures for a quantum random access memory. *Physical Review A, 78*(5), 052310. DOI: 10.1103/PhysRevA.78.052310

Giovannetti, V., Lloyd, S., & Maccone, L. (2008). Quantum random access memory. *Physical Review Letters, 100*(16), 160501. DOI: 10.1103/PhysRevLett.100.160501 PMID: 18518173

Girshick, R., Donahue, J., Darrell, T., & Malik, J. (2016). Region-Based Convolutional Networks for Accurate Object Detection and Segmentation. *IEEE Transactions on Pattern Analysis and Machine Intelligence*, *38*(1), 142–158. DOI: 10.1109/TPAMI.2015.2437384 PMID: 26656583

Givens, D. I. (2024). Animal board invited review: Dietary transition from animal to plant-derived foods: Are there risks to health? *Animal*, *18*(9), 101263. DOI: 10.1016/j.animal.2024.101263 PMID: 39121724

Glorot, X., & Bengio, Y. (2010). Understanding the difficulty of training deep feedforward neural networks. Proceedings of the thirteenth international conference on artificial intelligence and statistics, Gongal, A., Amatya, S., Karkee, M., Zhang, Q., & Lewis, K. (2015). Sensors and systems for fruit detection and localization: A review. *Computers and Electronics in Agriculture*, *116*, 8–19. https://doi.org/http://dx.doi.org/10.1016/j.compag.2015.05.021

Gong, D., Yang, J., Liu, L., Zhang, Y., Reid, I., Shen, C., van den Hengel, A., & Shi, Q. (2017). *From Motion Blur to Motion Flow: A Deep Learning Solution for Removing Heterogeneous Motion Blur*. 2319–2328. https://openaccess.thecvf.com/content_cvpr_2017/html/Gong_From_Motion_Blur_CVPR_2017_paper.html

Gonzalez, M. (2024), The Imp. of E-Comm. on Small-Scale Farmers: Lessons from Online Marketplaces, https://husfarm.com/article/the-impact-of-e-commerce-on-small-scale-farmers-lessons-from-online-marketplaces#google_vignette

Gonzalez, J., Fernandez, P., & Lopez, D. (2022). Hybrid AI-IoT smart irrigation: A cloud-edge approach for real-time water management. *IEEE Internet of Things Journal*, *9*(2), 3124–3138.

Gore, S., Patil, D., Mahankale, N., & Gore, S. (2024). Satellite Imaging for Precision Agriculture Enhancing Crop Management, Soil Condition and Yield Prediction. In *Emerging Trends in Smart Societies* (pp. 434–437). Routledge. DOI: 10.4324/9781003489412-104

Goulart, R., Jarvis, D., & Walsh, K. B. (2023). Evaluation of end effectors for robotic harvesting of mango fruit. *Sustainability (Basel)*, *15*(8), 6769. DOI: 10.3390/su15086769

GrainCrop. (2023). *FY23-Annual Report*. https://www.graincorp.com.au/wp-content/uploads/2023/11/FY23-Annual-Report.pdf

Grandin, T., & Deesing, M. J. (2022). Genetics and animal welfare. In Grandin, T. (Ed.), *Genetics and the Behavior of Domestic Animals* (3rd ed., pp. 507–548). Academic Press., DOI: 10.1016/B978-0-323-85752-9.00013-5

Grigorieva, E., Livenets, A., & Stelmakh, E. (2023). Adaptation of agriculture to climate change: A scoping review. *Climate (Basel), 11*(10), 202. DOI: 10.3390/cli11100202

Grigoryan, A. M., & Agaian, S. S. (2020). New look on quantum representation of images: Fourier transform representation. *Quantum Information Processing, 19*(5), 148. DOI: 10.1007/s11128-020-02643-3

Grilo, A. B., Kerenidis, I., & Zijlstra, T. (2019). Learning-with- errors problem is easy with quantum samples. *Physical Review. A, 99*(3), 032314. DOI: 10.1103/PhysRevA.99.032314

Grover, L. K. "A fast quantum mechanical algorithm for database search," in *Proceedings of the twenty-eighth annual ACM symposium on Theory of computing*, 1996, pp. 212–219. DOI: 10.1145/237814.237866

Guan, L. (2023). Weight Prediction Boosts the Convergence of AdamW. In Kashima, H., Ide, T., & Peng, W.-C. (Eds.), *Advances in Knowledge Discovery and Data Mining* (pp. 329–340). Springer Nature Switzerland., DOI: 10.1007/978-3-031-33374-3_26

Guerra, J. P., Cuevas, F., Guerra, J. P., & Cuevas, F. (2024). Application of Digital Image Processing Techniques for Agriculture: A Review. In *Digital Image Processing—Latest Advances and Applications*. IntechOpen., DOI: 10.5772/intechopen.1004767

Guerreschi, G. G., Mandra, S., & Aspuru-Guzik, A. (2016). Faster than classical quantum algorithm for dense formulas of exact satisfiability and occupation problems. *New Journal of Physics, 18*(7), 073003. DOI: 10.1088/1367-2630/18/7/073003

Guo, X., Ho, C.-T., Schwab, W., & Wan, X. (2021). Effect of the roasting degree on flavor quality of large-leaf yellow tea. *Food Chemistry, 347*, 129016. DOI: 10.1016/j.foodchem.2021.129016 PMID: 33486364

Gupta, A., Anand, R., & Sindhwani, N. (2024). Perf. and Acc. Enhance. of Mach. Learn. & IoT-based Agri. Precision AI Sys *SN COMP.SCIE, 5*(930). Advance online publication. DOI: 10.1007/s42979-024-03238-w

Gupta, A., Chug, A., & Singh, A. P. (2024). Processing and optimized learning for improved classification of categorical plant disease datasets. *Intelligent Data Analysis, 28*(6), 1–25. DOI: 10.3233/IDA-230651

Gupta, M., Abdelsalam, M., Khorsandroo, S., & Mittal, S. (2020). Security and privacy in smart farming: Challenges and opportunities. *IEEE Access : Practical Innovations, Open Solutions, 8*, 34564–34584. DOI: 10.1109/ACCESS.2020.2975142

Gupta, M., Sharma, B., & Singh, A. K. (2025). Impact of digital technologies on tourist arrival in selected economies. In Lytras, M., Alkhaldi, A., & Ordóñez de Pablos, P. (Eds.), *Harnessing AI, Blockchain, and Cloud Computing for Enhanced e-Government Services* (pp. 441–474). IGI Global Scientific Publishing., DOI: 10.4018/979-8-3693-7678-2.ch014

Guthrie, J., Walsh, K., Reid, D., & Liebenberg, C. (2005). Assessment of internal quality attributes of mandarin fruit. 1. NIR calibration model development. *Australian Journal of Agricultural Research, 56*(4), 405–416. DOI: 10.1071/AR04257

Gutiérrez, S., Wendel, A., & Underwood, J. (2019). Ground based hyperspectral imaging for extensive mango yield estimation. *Computers and Electronics in Agriculture, 157*, 126–135. https://doi.org/https://doi.org/10.1016/j.compag.2018.12.041. DOI: 10.1016/j.compag.2018.12.041

Haider, S. K., Nauman, A., Jamshed, M. A., Jiang, A., Batool, S., & Kim, S. W. (2022). Internet of drones: Routing algorithms, techniques and challenges. *Mathematics, 10*(9), 1488. Advance online publication. DOI: 10.3390/math10091488

Hanif, M. K., Khan, S. Z., & Bibi, M. (2022). Applications of artificial intelligence in pest management. In *Artificial Intelligence and Smart Agriculture Applications* (pp. 277–300). Auerbach Publications. DOI: 10.1201/9781003311782-13

Hannan, M., Burks, T., & Bulanon, D. M. (2009). A machine vision algorithm combining adaptive segmentation and shape analysis for orange fruit detection. *Agricultural Engineering International: CIGR Journal*.

Haque, E., Paul, M., Ulhaq, A., & Debnath, T. "Advance quantum image representation and compression using DCT-EFRQI approach," Scientific Report, vol. 13, 2023.

Hasan, S. R., Chowdhury, M. Z., Saiam, M., & Jang, Y. M. (2023). Quantum communication systems: Vision, protocols, applications, and challenges. *IEEE Access : Practical Innovations, Open Solutions, 11*, 15855–15877. DOI: 10.1109/ACCESS.2023.3244395

Hayati, M., Muchtar, K., Roslidar, , Maulina, N., Syamsuddin, I., Elwirehardja, G. N., & Pardamean, B. (2023). Impact of CLAHE-based image enhancement for diabetic retinopathy classification through deep learning. *Procedia Computer Science, 216*, 57–66. DOI: 10.1016/j.procs.2022.12.111

Hazrati, M., Dara, R., & Kaur, J. (2022). On-farm data security: Practical recommendations for securing farm data. *Frontiers in Sustainable Food Systems, 6*, 884187. DOI: 10.3389/fsufs.2022.884187

He, K., Zhang, X., Ren, S., & Sun, J. (2016). Deep residual learning for image recognition. *Proceedings of the IEEE conference on computer vision and pattern recognition*

He, L., Fang, W., Zhao, G., Wu, Z., Fu, L., Li, R., Majeed, Y., & Dhupia, J. (2022). Fruit yield prediction and estimation in orchards: A state-of-the-art comprehensive review for both direct and indirect methods. *Computers and Electronics in Agriculture, 195*, 106812. https://doi.org/https://doi.org/10.1016/j.compag.2022.106812 DOI: 10.1016/j.compag.2022.106812

Hemmati, A., Zarei, M., & Souri, A. (2023). UAV-based Internet of Vehicles: A systematic literature review. *Intelligent Systems with Applications, 18*, 200226. Advance online publication. DOI: 10.1016/j.iswa.2023.200226

Hill, A. D., Chapman, J., Herndon, K., Chopp, C., Gauthier, D. J., & Kwiat, P. (2017). Drone-based quantum key distribution. *Urbana (Caracas, Venezuela), 51*, 61801–63003.

Holt, J. S.. (2017). Impact of herbicide use in modern agriculture. *Journal of Environmental Management, 201*, 113–119. DOI: 10.1016/j.jenvman.2017.06.048

Hosen, M. D., & Islam, M. H. (2025). Aggrotech: Leveraging Deep Learning for Sustainable Tomato Disease Management. arXiv preprint arXiv:2501.12052.

Hou, F., Lei, W., Li, S., Xi, J., Xu, M., & Luo, J. (2021). Improved Mask R-CNN with distance guided intersection over union for GPR signature detection and segmentation. *Automation in Construction, 121*, 103414. DOI: 10.1016/j.autcon.2020.103414

Hu, W. W., Zhou, R. G., El-Rafei, A., & Jiang, S. X. (2019). Quantum image watermarking algorithm based on Haar wavelet transform. IEEE Access, 7, 121303 121320.

Huangfu, Y., Huang, Z., Yang, X., Zhang, Y., Li, W., Shi, J., & Yang, L. (2024). HHS-RT-DETR: A Method for the Detection of Citrus Greening Disease. *Agronomy (Basel), 14*(12), 2900. DOI: 10.3390/agronomy14122900

Hughes, D. P., & Salathé, M. (2016). An open access repository of images on plant health to enable the development of mobile disease diagnostics. arXiv. https://arxiv.org/abs/1511.08060

Hu, H. Y. (2018). Dang, N. Jiang, X. Lu and Y. Cai, "A novel quantum image compression method based on jpeg,". *International Journal of Theoretical Physics, 57*(3), 611–636. DOI: 10.1007/s10773-017-3593-2

Hurtik, P., Tomasiello, S., Hula, J., & Hynar, D. (2022). Binary cross-entropy with dynamical clipping. *Neural Computing & Applications*, *34*(14), 12029–12041. DOI: 10.1007/s00521-022-07091-x

Hussain, M. (2023). YOLO-v1 to YOLO-v8, the rise of YOLO and its complementary nature toward digital manufacturing and industrial defect detection. *Machines (Basel)*, *11*(7), 677. DOI: 10.3390/machines11070677

Hussain, M., & Khanam, R. (2024). In-Depth Review of YOLOv1 to YOLOv10 Variants for Enhanced Photovoltaic Defect Detection. *Solar*, *4*(3), 3. Advance online publication. DOI: 10.3390/solar4030016

Ibrahim, I. A., & Truby, J. M. (2023). Farm tech: Regulating the use of digital technologies in the agricultural sector. *Food and Energy Security*, *12*(4), 1–12. DOI: 10.1002/fes3.483

Igne, B., Allen, R., Gahkani, A., Kojic, D., Kovacs, Z., McCormick, R., & Mohan, S. (2019). Summary of the 2018 International Diffuse Reflectance Conference (IDRC) software shoot-out. In: SAGE Publications Sage UK: London, England.

Iliyasu, F. (2016). Yan and Venegas-Andraca. S. E., "A survey of quantum image representations,". *Quantum Information Processing*, *15*(1), 1–35. DOI: 10.1007/s11128-015-1195-6

Information Processing, 2014. 13: p. 1353-1379.

Ingram, J., & Maye, D. (2020). What are the implications of digitalisation for agricultural knowledge? *Frontiers in Sustainable Food Systems*, *4*(1), 1–13. DOI: 10.3389/fsufs.2020.00066

Inoue, Y. (2020). Satellite-and drone-based remote sensing of crops and soils for smart farming–a review. *Soil Science and Plant Nutrition*, *66*(6), 798–810. DOI: 10.1080/00380768.2020.1738899

Ishaq, M. N., & Waqas, M. (2023). Early Detection of Late Blight Tomato Disease using Histogram Oriented Gradient based Support Vector Machine.*arXiv.Org*. https://doi.org//arXiv.2306.08326DOI: 10.48550

Islam, M. T., & Ali, A. (2024). Sustainable green energy transition in Saudi Arabia: Characterizing policy framework, interrelations and future research directions. *Next Energy*, *5*, 100161. DOI: 10.1016/j.nxener.2024.100161

Ivanochko, I., & Melnyk, O. (2024). Smart Farming System Based on Cloud Computing Technologies. *Procedia Computer Science*, *238*, 857–862. DOI: 10.1016/j.procs.2024.06.103

Jabbar, W. A., Tan, M. T., Hamidun, M. F. I., Che Kamarudin, A. H., Wu, W., Sultan, J., Alsewari, A. A., & Ali, M. A. H. (2024). Development of LoRaWAN-based IoT system for water quality monitoring in rural areas. *Expert Systems with Applications, 242*, 122862. Advance online publication. DOI: 10.1016/j.eswa.2023.122862

Jacobs, I. S. (1963). *Fine particles, thin films and exchange anisotropy*. Magnetism.

Jang, E.-K., Lim, E. M., Kim, J., Kang, M.-J., Choi, G., & Moon, J. (2023). Risk management of methane reduction clean development mechanism projects in rice paddy fields. *Agronomy (Basel), 13*(6), 1–11. DOI: 10.3390/agronomy13061639

Jatav, S. S., & Kalu, N. (2023). Measuring the agricultural sustainability of India: an application of pressure-state-response (PSR) model. *Regional Sustainability*, 4(3), 218-234. https://www.sciencedirect.com/science/article/pii/S2666660X2300035X

Jatav, S. S. (2024). Development of multidimensional food security index for Rajasthan, India: A district-level analysis. *Local Development & Society*, *5*(2), 221–243. https://www.tandfonline.com/doi/abs/10.1080/26883597.2022.2151373. DOI: 10.1080/26883597.2022.2151373

Javaid, M., Haleem, A., Khan, I. H., & Suman, R. (2023). Understanding the potential applications of Artificial Intelligence in Agriculture Sector. *Advanced Agrochem*, *2*(1), 15–30. DOI: 10.1016/j.aac.2022.10.001

Jenni, S., Stewart, K. A., Bourgeois, G., & Cloutier, D. C. (1996). Nondestructive volume estimation for growth analysis of eastern-type muskmelon ovaries. *HortScience*, *31*(7), 1236–1236. DOI: 10.21273/HORTSCI.31.7.1236

Jenni, S., Stewart, K. A., Bourgeois, G., & Cloutier, D. C. (1998). Predicting yield and time to maturity of muskmelons from weather and crop observations. *Journal of the American Society for Horticultural Science*, *123*(2), 195–201. DOI: 10.21273/JASHS.123.2.195

Jeremy., ... "Advances in quantum machine learning",arXiv preprint arXiv:1512.02900,vol.,no.4,pp., 2015.

Jiang, K., Itoh, H., Oda, M., Okumura, T., Mori, Y., Misawa, M., Hayashi, T., Kudo, S.-E., & Mori, K. (2023). Gaussian affinity and GIoU-based loss for perforation detection and localization from colonoscopy videos. *International Journal of Computer Assisted Radiology and Surgery*, *18*(5), 795–805. DOI: 10.1007/s11548-022-02821-x PMID: 36913126

Jiang, N., Hu, H., Dang, Y., & Zhang, W. (2017). Quantum point cloud and its compression. *International Journal of Theoretical Physics*, *56*(10), 3147–3163. DOI: 10.1007/s10773-017-3483-7

Jiang, N., & Wang, L. (2015). Quantum image scaling using nearest neighbor interpolation. *Quantum Information Processing*, *14*(5), 1559–1571. DOI: 10.1007/s11128-014-0841-8

Jiang, P., Ergu, D., Liu, F., Cai, Y., & Ma, B. (2022). A Review of Yolo Algorithm Developments. *Procedia Computer Science*, *199*, 1066–1073. DOI: 10.1016/j.procs.2022.01.135

Jiang, Y., Li, C., Paterson, A. H., Sun, S., Xu, R., & Robertson, J. (2018). Quantitative Analysis of Cotton Canopy Size in Field Conditions Using a Consumer-Grade RGB-D Camera [Original Research]. *Frontiers in Plant Science*, *8*(2233), 2233. Advance online publication. DOI: 10.3389/fpls.2017.02233 PMID: 29441074

Jocher, G., Chaurasia, A., Qiu, J., & Stoken, A. (2023). YOLO by Ultralytics. GitHub Repository. Retrieved from https://github.com/ultralytics/yolov8

Jocher, G., Chaurasia, A., Stoken, A., Borovec, J., Kwon, Y., Fang, J., Michael, K., Montes, D., Nadar, J., & Skalski, P. (2022). ultralytics/yolov5: v6. 1-tensorrt, tensorflow edge tpu and openvino export and inference. Zenodo.

Jones, R., & Taylor, M. (2020). Rooting characteristics of turf grasses in tropical climates. *Journal of Tropical Ecology*, *45*(2), 89–105. DOI: 10.1080/02664670.2020.12.345

Jrondi, Z., Moussaid, A., & Hadi, M. Y. (2024). Exploring End-to-End object detection with transformers versus YOLOv8 for enhanced citrus fruit detection within trees. *Systems and Soft Computing, 6*, 200103. https://doi.org/https://doi.org/10.1016/j.sasc.2024.200103

Jyoti, B., Singh, A. K., & Ashraf, S. N. (2023). Appropriate technologies and their implications in the agricultural sector. In Alex Khang, P. H. (Ed.), *Advanced Technologies and AI-Equipped IoT Applications in High-Tech Agriculture* (pp. 65–87). IGI Global., DOI: 10.4018/978-1-6684-9231-4.ch004

Kabilesh, S. K., Mohanapriya, D., Suseendhar, P., Indra, J., Gunasekar, T., & Senthilvel, N. (2023, April 03). MohanaPriya, D., Suseendhar, P., Indra, J., Gunasekar, T., & Senthilvel, N. (2023). Research on Artificial Intelligence based Fruit Disease Identification System (AI-FDIS) with the Internet of Things (IoT). *Journal of Intelligent & Fuzzy Systems*, *44*(4), 6593–6608. Advance online publication. DOI: 10.3233/JIFS-222017

Kabir, M. M., Hafiz, M. S., Bandyopadhyaa, S., Jim, J. R., & Mridha, F. (2024). *TeaLeafAgeQuality: Age-Stratified Tea Leaf Quality Classification Dataset. 1*. DOI: 10.17632/7t964jmmy3.1

Kalonia, M., & Gupta, T. "Review on applications of quantum image processing," in Thirteenth International Conference on Digital Image Processing (ICDIP 2021). SPIE, 2021, vol. 11878, pp. 519–528. DOI: 10.1117/12.2601025

Kamilaris, A., Cole, I. R., & Prenafeta-Boldú, F. X. (2021). Blockchain in agriculture. In *Food Technology Disruptions* (pp. 247–284). Elsevier. DOI: 10.1016/B978-0-12-821470-1.00003-3

Kantsepolsky, B., Aviv, I., Weitzfeld, R., & Bordo, E. (2023). Exploring quantum sensing potential for systems applications. *IEEE Access : Practical Innovations, Open Solutions*, *11*, 31569–31582. DOI: 10.1109/ACCESS.2023.3262506

Karunathilake, E., Le, A. T., Heo, S., Chung, Y. S., & Mansoor, S. (2023). The path to smart farming: Innovations and opportunities in precision agriculture. *Agriculture*, *13*(8), 1593. DOI: 10.3390/agriculture13081593

Kasinathan, T., & Singaraju, D. (2021). Insect class. and detect. in fiel. Crop. using modern mach. Learn. Tech. *Information Processing in Agriculture*, *8*(3), 446–457. DOI: 10.1016/j.inpa.2020.09.006

Kataria, S. (2023). *Robust Speaker Recognition using Perceptual and Adversarial Speech Enhancement* [Johns Hopkins University]. http://jhir.library.jhu.edu/handle/1774.2/68958

Kazuhiro, G. (2015). "Quantum computing: Limits, options and applications," 2020, Forbes Technology Council. Jiang, N., J. Wang, and Y. Mu, Quantum image scaling up based on nearest-neighbor interpolation with integer scaling ratio. *Quantum Information Processing*, *14*(11), 4001–4026.

Kernot, I., Meurant, N., Holmes, R., MacLeod, N., Fullelove, G., & Bally, I. (1999). *Mango information kit.* Queensland Department of Primary Industries.

Kestur, R., Meduri, A., & Narasipura, O. (2019). MangoNet: A deep semantic segmentation architecture for a method to detect and count mangoes in an open orchard. *Engineering Applications of Artificial Intelligence*, *77*, 59–69. https://doi.org/https://doi.org/10.1016/j.engappai.2018.09.011. DOI: 10.1016/j.engappai.2018.09.011

KG2. (2024). *Navigating The Digital Divide: Challenges In Adopting Technology In Australian Agriculture.* KG2 Australia Retrieved 20 December 2024 from https://kg2.com.au/navigating-the-digital-divide-challenges-in-adopting-technology-in-australian-agriculture/

Khanal, S., Fulton, J., & Shearer, S. (2017). An overview of current and potential applications of thermal remote sensing in precision agriculture. *Computers and Electronics in Agriculture*, *139*, 22–32. DOI: 10.1016/j.compag.2017.05.001

Khan, N. A., Jhanjhi, N. Z., Brohi, S. N., & Nayyar, A. (2020). Chapter three - Emerging use of UAVs: Secure communication protocol issues and challenges. In Al-Turjman, F. (Ed.), *Drones in smart-cities* (pp. 37–55). Elsevier., DOI: 10.1016/B978-0-12-819972-5.00003-3

Khan, R. A. (2019). An improved flexible representation of quantum images. *Quantum Information Processing*, *18*(7), 1–19. DOI: 10.1007/s11128-019-2306-6

Khoshnoud, F., Esat, I. I., de Silva, C. W., & Quadrelli, M. B. (2019). Quantum network of cooperative unmanned autonomous systems. *Unmanned Systems*, *7*(02), 137–145. DOI: 10.1142/S2301385019500055

Khun, A. T. P., Shan, L., Lim, Y., & Tan, Y. (2023). MCST scheme for UAV systems over LoRa networks. *Drones (Basel)*, *7*(6), 371. DOI: 10.3390/drones7060371

Klerkx, L., Jakku, E., & Labarthe, P. (2019). A review of social science on digital agriculture, smart farming and agriculture 4.0: New contributions and a future research agenda. *NJAS Wageningen Journal of Life Sciences*, *90-91*(1), 1–16. DOI: 10.1016/j.njas.2019.100315

Klisch, S., & Soule, K. (2020). Farmers Markets: Working with comm. Partn. to prov. Essen. services during COVID-19. *Journal of Agriculture, Food Systems, and Community Development*, *9*, 1–5. DOI: 10.5304/jafscd.2020.094.035

Knill, E., Laflamme, R., & Viola, L. (2000). Theory of quantum error correction for general noise. *Physical Review Letters*, *84*(11), 2525–2528. DOI: 10.1103/PhysRevLett.84.2525 PMID: 11018926

Kodaira, V., Siriani, A. L. R., Medeiros, H. P., De Moura, D. J., & Pereira, D. F. (2023). Assessment of Preference Behavior of Layer Hens under Different Light Colors and Temperature Environments in Long-Time Footage Using a Computer Vision System. *Animals (Basel)*, *13*(15), 15. Advance online publication. DOI: 10.3390/ani13152426 PMID: 37570237

Koirala, A., Walsh, K. B., Wang, Z., & Anderson, N. (2020). Deep Learning for Mango (Mangifera indica) Panicle Stage Classification. *Agronomy (Basel)*, *10*(1), 143. https://www.mdpi.com/2073-4395/10/1/143. DOI: 10.3390/agronomy10010143

Koirala, A., Walsh, K. B., Wang, Z., & McCarthy, C. (2019). Deep learning for real-time fruit detection and orchard fruit load estimation: Benchmarking of 'MangoYOLO' [journal article]. *Precision Agriculture*, *20*(6), 1–29. DOI: 10.1007/s11119-019-09642-0

Ko, Y., Kim, J., Duguma, D. G., Astillo, P. V., You, I., & Pau, G. (2021). Drone secure communication protocol for future sensitive applications in military zone. *Sensors (Basel)*, *21*(6), 2057. Advance online publication. DOI: 10.3390/s21062057 PMID: 33804023

Krishana, N. G. (2024). A dataset of field plant images for plant disease detection and classification with deep learning. *Indian Scientific Journal Of Research In Engineering And Management*, *08*(04), 1–5. DOI: 10.55041/IJSREM32034

Krizhevsky, A., Sutskever, I., & Hinton, G. E. (2012). Imagenet classification with deep convolutional neural networks. *Advances in Neural Information Processing Systems*.

Kuang, Y. C., Streeter, L., Cree, M. J., & Ooi, M. P. L. (May 2019). Evaluation of Deep Neural Network and Alternating Decision Tree for Kiwifruit Detection. 2019 IEEE International Instrumentation and Measurement Technology Conference (I2MTC)

Kulkarni, M. J., Ramamoorthy, M., & Ramsankar, G. (2024). Sust. Crop Monit. and Manage.ment for Enha.Agri. Product. Throu. IoT, AI&ML: Case Studies and Innovations. In: Manoharan, S., Tugui, A., Baig, Z. (eds) *Proceedings of 4th Int. Conf. on Arti. Intell. and Smart Ener. ICAIS 2024. Info. Sys. Eng. and Manage., vol 3. Springer, Cham.*https://doi.org/DOI: 10.1007/978-3-031-61471-2_13

Kumar, R., Gupta, P., & Singh, P. (2021). A Device for Effective Weed Removal for Smart Agriculture Using Convolutional Neural Network. Springer. https://link.springer.com/article/10.1007/s13198-021-01441-z

Kumar, A., Ahmad, M. M., & Sharma, P. (2017). Influence of climatic and non-climatic factors on sustainable food security in India: A statistical investigation. *International Journal of Sustainable Agricultural Management and Informatics*, *3*(1), 1–30. DOI: 10.1504/IJSAMI.2017.082917

Kumar, A., & Ambika, R. (2020). Drone integrated weather sensors for agriculture purpose. *International Journal of Electrical Engineering & Technology*, *11*(5), 83–90.

Kumar, A., Bhatia, S., Kaushik, K., Gandhi, S. M., Devi, S. G., De, J., Pacheco, D. A., & Mashat, A. (2021). Survey of promising technologies for quantum drones and networks. *IEEE Access : Practical Innovations, Open Solutions*, *9*, 125868–125911. DOI: 10.1109/ACCESS.2021.3109816

Kumar, A., de Jesus Pacheco, D. A., Kaushik, K., & Rodrigues, J. J. P. C. (2022). Futuristic view of the Internet of quantum drones: Review, challenges and research agenda. *Vehicular Communications*, *36*, 100487. DOI: 10.1016/j.vehcom.2022.100487

Kumari, A., Khan, M. N., & Sinha, A. K. (2024). Harvesting Intelligence: AI and ML Revolutionizing Agriculture. In *Data-Driven Farming* (pp. 126-141). Auerbach Publications.

Kumar, K. A., & Verma, S. (2024). Harnessing Computer Vision for Agricultural Transformation: Insights, Techniques, and Applications. In *Applications of Computer Vision and Drone Technology in Agriculture 4.0* (pp. 111–131). Springer Nature Singapore.

Kumar, P. P., Kumar, T. A., Muthu, P., Rajendirane, R., & Samuel, R. D. J. (2022). Security analysis of UAV communication protocols: Solutions, prospects, and encounters. In *The Internet of Drones* (pp. 167–190). Apple Academic Press. DOI: 10.1201/9781003277491-9

Kumar, P., & Kumar, M. (2018). *Leaf Curl Disease: A Significant Constraint in the Production of Tomato in India*. IntechOpen., DOI: 10.5772/intechopen.76049

Kumar, P., Ojha, N. K., Altaf, F., & Maity, S. (2023). Quantum secured 6G technology-based applications in Internet of Everything. *Telecommunication Systems*, *82*(2), 315–344. DOI: 10.1007/s11235-022-00979-y

Kumar, P., Pruthvi, C. P., Preethi, H., & Punyashree, T. (2024). Comprehensive Review on Plant Disease Detection Techniques. *Indian Scientific Journal Of Research In Engineering And Management*, *08*(12), 1–9. DOI: 10.55041/IJSREM40039

Kumar, S. P., Subeesh, A., Jyoti, B., & Mehta, C. (2023). Applications of drones in smart agriculture. In Pakeerathan, K. (Ed.), *Smart Agriculture for Developing Nations: Status, Perspectives and Challenges* (pp. 33–48). Springer., https://doi.org/https://doi.org/10.1007/978-981-19-8738-0_3 DOI: 10.1007/978-981-19-8738-0_3

Kumar, S., & Singh, A. K. (2023). Modeling the effects of climate change on agricultural productivity: Evidence from Himachal Pradesh, India. *Asia-Pacific Journal of Regional Science*, *7*(2), 1–28. DOI: 10.1007/s41685-023-00291-w

Kumar, V., Sharma, K. V., Kedam, N., Patel, A., Kate, T. R., & Rathnayake, U. (2024). A comprehensive review on smart and sustainable agriculture using IoT technologies. *Smart Agricultural Technology*, *8*, 100487. DOI: 10.1016/j.atech.2024.100487

Kunduracıoğlu, İ. (2024). Utilizing ResNet Architectures for Identification of Tomato Diseases. *Utilizing ResNet Architectures for Identification of Tomato Diseases.*, *1*, 104–119. DOI: 10.59543/jidmis.v1i.11949

Latorre, J. I., "Image compression and entanglement," arXiv preprint quant ph/0510031, 2005.

Laurel, C. O., Dong, S., & Cruz-Irisson, M. (2015). Equivalence of a bit pixel image to a quantum pixel image. *Communications in Theoretical Physics*, *64*(5), 501–506. DOI: 10.1088/0253-6102/64/5/501

Lawal, O. M. (2021). YOLOMuskmelon: Quest for Fruit Detection Speed and Accuracy Using Deep Learning. *IEEE Access, 9*, 15221–15227. IEEE Access. DOI: 10.1109/ACCESS.2021.3053167

Lawal, M. O. (2021). Tomato detection based on modified YOLOv3 framework. *Scientific Reports*, *11*(1), 1447. DOI: 10.1038/s41598-021-81216-5 PMID: 33446897

Lawal, R. A., Martin, S. H., Vanmechelen, K., Vereijken, A., Silva, P., Al-Atiyat, R. M., Aljumaah, R. S., Mwacharo, J. M., Wu, D.-D., Zhang, Y.-P., Hocking, P. M., Smith, J., Wragg, D., & Hanotte, O. (2020). The wild species genome ancestry of domestic chickens. *BMC Biology*, *18*(1), 1. Advance online publication. DOI: 10.1186/s12915-020-0738-1 PMID: 32050971

Leah Borts-Kuperman. (2023). The Rise of Virtual Farmers Markets https://modernfarmer.com/2023/01/virtual-farmers-markets/

Léchaudel, M., & Joas, J. (2007). An overview of preharvest factors influencing mango fruit growth, quality and postharvest behaviour. *Brazilian Journal of Plant Physiology*, *19*(4), 287–298.

LeCun, Y., Bengio, Y., & Hinton, G. (2015). Deep learning. *Nature*, *521*(7553), 436–444. DOI: 10.1038/nature14539 PMID: 26017442

LeCun, Y., Bottou, L., Bengio, Y., & Haffner, P. (1998). Gradient-based learning applied to document recognition. *Proceedings of the IEEE*, *86*(11), 2278–2324. DOI: 10.1109/5.726791

Lennyi, D. "Online ag marketplaces: An entry point for farmers into the tech ecosystem," https://agfundernews.com/online-ag-marketplaces-an-entry-point-for-farmers-into-the-tech-ecosystem

Lesnoff, M. (2024). Averaging a local PLSR pipeline to predict chemical compositions and nutritive values of forages and feed from spectral near infrared data. *Chemometrics and Intelligent Laboratory Systems*, *244*, 105031. DOI: 10.1016/j.chemolab.2023.105031

Lesnoff, M., Metz, M., & Roger, J. M. (2020). Comparison of locally weighted PLS strategies for regression and discrimination on agronomic NIR data. *Journal of Chemometrics*, *34*(5), e3209. DOI: 10.1002/cem.3209

Li, L., Wang, Z., Hou, W., Zhou, Z., & Xue, H. (2024). A Recognition Method for Aggressive Chicken Behavior Based on Machine Learning. *IEEE Access, 12*, 24762–24775. IEEE Access. DOI: 10.1109/ACCESS.2024.3365552

Li, X., Wang, W., Wu, L., Chen, S., Hu, X., Li, J., Tang, J., & Yang, J. (2020). Generalized Focal Loss: Learning Qualified and Distributed Bounding Boxes for Dense Object Detection. *Advances in Neural Information Processing Systems, 33*, 21002–21012. https://proceedings.neurips.cc/paper/2020/hash/f0bda020d2470f2e74990a07a607ebd9-Abstract.html

Li, F., Yan, H., & Shi, L. (2024). Multi-scale coupled attention for visual object detection. *Scientific Reports, 14*(1), 11191. DOI: 10.1038/s41598-024-60897-8 PMID: 38755252

Li, G., Hui, X., Lin, F., & Zhao, Y. (2020). Developing and Evaluating Poultry Preening Behavior Detectors via Mask Region-Based Convolutional Neural Network. *Animals (Basel), 10*(10), 10. Advance online publication. DOI: 10.3390/ani10101762 PMID: 32998372

Li, H.-S., Chen, X., Xia, H., Liang, Y., & Zhou, Z. (2018). A quantum image representation based on bitplanes. *IEEE Access : Practical Innovations, Open Solutions, 6*, 62396–62404. DOI: 10.1109/ACCESS.2018.2871691

Li, H.-S., Qingxin, Z., Lan, S., Shen, C.-Y., Zhou, R., & Mo, J. (2013). Image storage, retrieval, compression and segmentation in a quantum system. *Quantum Information Processing, 12*(6), 2269–2290. DOI: 10.1007/s11128-012-0521-5

Li, H.-S., Zhu, Q., Zhou, R.-G., Song, L., & Yang, X. (2014). Multi-dimensional color image storage and retrieval for a normal arbitrary quantum superposition state. *Quantum Information Processing, 13*(4), 991–1011. DOI: 10.1007/s11128-013-0705-7

Li, M. (2014). *H. Ian H. Li, Q. Zhu et al., "Multidimensional color image storage, retrieval, and compression based on quantum amplitudes and phases* (Vol. 273). Information Sciences.

Li, M., Chen, M., Zhang, Y., Fu, C., Xing, B., Li, W., Qian, J., Li, S., Wang, H., Fan, X., Yan, Y., Wang, Y., & Yang, X. (2015). Apple Fruit Diameter and Length Estimation by Using the Thermal and Sunshine Hours Approach and Its Application to the Digital Orchard Management Information System. *PLoS One, 10*(4), e0120124. DOI: 10.1371/journal.pone.0120124 PMID: 25831065

Lin, C., & Chen, Z. (2019). Implementing adaptive irrigation control through IoT and machine learning. *Journal of Agricultural Science and Technology, 21*(1), 213–227.

Lin, T.-Y., Maire, M., Belongie, S., Hays, J., Perona, P., Ramanan, D., Dollár, P., & Zitnick, C. L. (2014). Microsoft coco: Common objects in context. Computer Vision–ECCV 2014: *13th European Conference*, Zurich, Switzerland, September 6-12, 2014, Proceedings, Part V 13

Li, P., Xiao, H., & Li, B. (2016). Quantum representation and watermark strategy for color images based on the controlled rotation of qubits. *Quantum Information Processing, 15*(11), 4415–4440. DOI: 10.1007/s11128-016-1413-x

Liu, C., S. Lu J. Su, X. Guo and L. Li, "An improved novel quantum image representation and its experimental test on ibm quantum experience," Scientific

Liu, H., & Ying, Y. (2015). Evaluation of sugar content of Huanghua Pear on trees by visible/near infrared spectroscopy. *Guang pu xue yu Guang pu fen xi= Guang pu, 35*(11), 3078-3081.

Liu, W., Anguelov, D., Erhan, D., Szegedy, C., Reed, S., Fu, C.-Y., & Berg, A. C. (2016). SSD: Single Shot MultiBox Detector. In B. Leibe, J. Matas, N. Sebe, & M. Welling, *Computer Vision – ECCV 2016* Cham.

Liu, H.-Y., Tian, X.-H., Gu, C., Fan, P., Ni, X., Yang, R., Zhang, J.-N., Hu, M., Guo, J., Cao, X., Hu, X., Zhao, G., Lu, Y.-Q., Gong, Y.-X., Xie, Z., & Zhu, S.-N. (2020). Drone-based entanglement distribution towards mobile quantum networks. *National Science Review, 7*(5), 921–928. DOI: 10.1093/nsr/nwz227 PMID: 34691535

Liu, K., Zhang, Y., Lu, K., Wang, X., & Wang, X. (2018). An optimized quantum representation for color digital images. *International Journal of Theoretical Physics, 57*(10), 2938–2948. DOI: 10.1007/s10773-018-3813-4

Li, X., & Gallagher, K. P. (2022). Assessing the climate change exposure of foreign direct investment. *Nature Communications, 13*(1), 1451. DOI: 10.1038/s41467-022-28975-5 PMID: 35304443

Li, Y., & Qian, Y. (2019). Noise reduction techniques in environmental sensor data. *Sensors (Basel), 19*(10), 2270. PMID: 31100901

Lochan, K., Khan, A., Elsayed, I., Suthar, B., Seneviratne, L., & Hussain, I. (2024). Advancements in Precision Spraying of Agricultural Robots: A Comprehensive Review. *IEEE Access : Practical Innovations, Open Solutions, 12*, 129447–129483. DOI: 10.1109/ACCESS.2024.3450904

Lowe, D. G. (1999). Object recognition from local scale-invariant features. iccv, Luo, L., Tang, Y., Lu, Q., Chen, X., Zhang, P., & Zou, X. (2018). A vision methodology for harvesting robot to detect cutting points on peduncles of double overlapping grape clusters in a vineyard. *Computers in Industry*, *99*, 130–139. https://doi.org/ https://doi.org/10.1016/j.compind.2018.03.017

Mace, G., & Morgan, R. T. (2022). *Cybersecurity in Agriculture – Are the gates locked or left ajar?* Retrieved 9 January 2025 from https://futurebeef.com.au/wp -content/uploads/2022/04/GF-LBR-Narrative_Cybersecurity_FINAL.pdf

Mac, T. T., Nguyen, T. D., Dang, H. K., Nguyen, D.-T., & Nguyen, X.-T. (2024). Intelligent agricultural robotic detection system for greenhouse tomato leaf diseases using soft computing techniques and deep learning. *Scientific Reports*, *14*(1), 23887. DOI: 10.1038/s41598-024-75285-5 PMID: 39396063

Maddikunta, P. K. R., Hakak, S., Alazab, M., Bhattacharya, S., Gadekallu, T. R., Khan, W. Z., & Pham, Q.-V. (2021). Unmanned aerial vehicles in smart agriculture: Applications, requirements, and challenges. *IEEE Sensors Journal*, *21*(16), 17608–17619. DOI: 10.1109/JSEN.2021.3049471

Madeira, M., Porfírio, R. P., Santos, P. A., & Madeira, R. N. (2024). AI-Powered Solution for Plant Disease Detection in Viticulture. *Procedia Computer Science*, *238*, 468–475. DOI: 10.1016/j.procs.2024.06.049

Maeda-Gutiérrez, V., Galvan-Tejada, C. E., Zanella-Calzada, L. A., Celaya-Padilla, J. M., Galván-Tejada, J. I., Gamboa-Rosales, H., Luna-Garcia, H., Magallanes-Quintanar, R., Guerrero Mendez, C. A., & Olvera-Olvera, C. A. (2020). Comparison of convolutional neural network architectures for classification of tomato plant diseases. *Applied Sciences (Basel, Switzerland)*, *10*(4), 1245. DOI: 10.3390/app10041245

Magomadov, V. S. (2019). Deep learn. and its role in smart agri. *Journal of Physics: Conference Series*, *1399*(4), 044109. DOI: 10.1088/1742-6596/1399/4/044109

Mahabub, S., Jahan, I., Hasan, M. N., Islam, M. S., Akter, L., Musfiqur, M., & Onik, M. K. R. (2024). *Efficient detection of tomato leaf diseases using optimized Compact Convolutional Transformers (CCT)*. Model.

Maharjan, K. L., Singh, M., & Gonzalvo, C. M. (2018). Drivers of environmental conservation agriculture and women farmer empowerment in Namobuddha municipality, Nepal. *Journal of Agriculture and Food Research*, *13*(1), 1–16. DOI: 10.1016/j.jafr.2023.100631

Mahaur, B., Mishra, K. K., & Kumar, A. (2023). An improved lightweight small object detection framework applied to real-time autonomous driving. *Expert Systems with Applications*, *234*, 121036. DOI: 10.1016/j.eswa.2023.121036

Maher, A., Inoue, M., Watanbe, A., Yokemura, T., Ogyu, F., & Iida, H. (2020). UAV communication system integrated into network traversal with mobility. *SN Applied Sciences*, *2*(6), 1057. Advance online publication. DOI: 10.1007/s42452-020-2749-5

Majumdar, P., & Bhattacharya, D. (2023). *Utilities of 5G comm. Techn. for promo. advancement in agri. 4.0: Recent trends, research iss. and rev. of lit., in 5G and Beyond*. Springer Tracts in Elect. and Elect. Engi.

Ma, N., Wu, Y., Bo, Y., & Yan, H. (2024). Chili Pepper Object Detection Method Based on Improved YOLOv8n. *Plants*, *13*(17), 17. Advance online publication. DOI: 10.3390/plants13172402 PMID: 39273886

Mangiaracina, R., & Perego, A. (2019). Inno. Solu. to Incr. Last-Mile Delivery Effi. in B2C E-Commerce: A Lit. [Review]. *International Journal of Physical Distribution & Logistics Management*, *49*, 901–920. DOI: 10.1108/IJPDLM-02-2019-0048

Ma, Q., Tan, H., & Zhou, T. (2023). Mutual authentication scheme for smart devices in IoT-enabled smart home systems. *Computer Standards & Interfaces*, *86*, 103743. DOI: 10.1016/j.csi.2023.103743

Maraveas, C., Konar, D., Michopoulos, D. K., Arvanitis, K. G., & Peppas, K. P. (2024). Harnessing quantum computing for smart agriculture: Empowering sustainable crop management and yield optimization. *Computers and Electronics in Agriculture*, *218*, 108680. DOI: 10.1016/j.compag.2024.108680

Marini, R. P. (2001). Estimating mean fruit weight and mean fruit value for apple trees: Comparison of two sampling methods with the true mean. *Journal of the American Society for Horticultural Science*, *126*(4), 503–510. DOI: 10.21273/JASHS.126.4.503

Marinoudi, V., Sørensen, C. G., Pearson, S., & Bochtis, D. (2019). Robotics and labour in agriculture. A review. *Biosystems Engineering*, *184*, 111–121. DOI: 10.1016/j.biosystemseng.2019.06.013

McBratney, A. B., Whelan, B., Ancev, T., & Bouma, J. (2005). Future directions of precision agriculture. *Precision Agriculture*, *6*(1), 7–23. DOI: 10.1007/s11119-005-0681-8

McCaig, M., Dara, R., & Rezania, D. (2023). Farmer-centric design thinking principles for smart farming technologies. *Internet of Things : Engineering Cyber Physical Human Systems*, *23*, 100898. DOI: 10.1016/j.iot.2023.100898

McCosker, M., & Burton, L. (2024). *Lack of mobile connectivity 'confronting' regional Australians.* ABC News Retrieved 9 January 2025 from https://www.abc.net.au/news/2024-09-03/rural-australians-worse-off-for-mobile-and-internet-connectivity/104287014

Mekyassi, H., & Kızıldeniz, T. (2024). The role of nutrition-sensitive climate-smart agriculture in ensuring global food security. *BIO Web of Conferences*, 85. DOI: DOI: 10.1051/bioconf/20248501055

Mhaned, A., Salma, M., Haji, M., & Benhra, J. (2025). Plant disease detection using vision transformers. *International Journal of Power Electronics and Drive Systems, 15*(2), 2334. DOI: 10.11591/ijece.v15i2.pp2334-2344

Mhlanga, D., & Ndhlovu, E. (2023). Digital technology adoption in the agriculture sector: Challenges and complexities in Africa. *Human Behavior and Emerging Technologies*, *1*(1), 1–10. DOI: 10.1155/2023/6951879

Millner, N., Cunliffe, A. M., Mulero-Pázmány, M., Newport, B., Sandbrook, C., & Wich, S. A. (2023). Exploring the opportunities and risks of aerial monitoring for biodiversity conservation. *Global Social Challenges Journal.*https://api.semanticscholar.org/CorpusID:259430259

Mimma, N.-E.-A., Ahmed, S., Rahman, T., & Khan, R. (2022). Fruits classification and detection application using deep learning. *Scientific Programming*, *2022*(1), 4194874. DOI: 10.1155/2022/4194874

Mirhaji, H., Soleymani, M., Asakereh, A., & Abdanan Mehdizadeh, S. (2021). Fruit detection and load estimation of an orange orchard using the YOLO models through simple approaches in different imaging and illumination conditions. *Computers and Electronics in Agriculture*, *191*, 106533. https://doi.org/https://doi.org/10.1016/j.compag.2021.106533. DOI: 10.1016/j.compag.2021.106533

Mishra, H., & Mishra, D. (2024). AI for Data-Driven Decision-Making in Smart Agriculture: From Field to Farm Management. In *Artificial Intelligence Techniques in Smart Agriculture* (pp. 173-193). Springer.

Misle, E., Kahlaoui, B., Hachicha, M., & Alvarado, P. (2013). Leaf area estimation in muskmelon by allometry. *Photosynthetica*, *51*(4), 613–620. DOI: 10.1007/s11099-013-0062-x

Mitchell, P. (1986). Pear fruit growth and the use of diameter to estimate fruit volume and weight. *HortScience*, *21*(4), 1003–1005. DOI: 10.21273/HORTSCI.21.4.1003

MIYAMOTO, K. (2007). Near infrared spectroscopy for bio-monitoring and bio-diagnosis of a fruit tree. *JOURNAL of the JAPANESE SOCIETY of AGRICULTURAL MACHINERY, 69*(3), 11-14.

Mohyuddin, G., Khan, M. A., & Haseeb, A. (2024). Eval. of mach. learn. Appro. for Precision Farm. in Smart Agri. Syst.: A Compr. [Review]. *IEEE Access : Practical Innovations, Open Solutions, 12*, 60155–60184. DOI: 10.1109/ACCESS.2024.3390581

Mokrane, A., Braham, A. C., & Cherki, B. (2019). UAV coverage path planning for supporting autonomous fruit counting systems. *2019 International Conference on Applied Automation and Industrial Diagnostics (ICAAID)*

Moll, J.. (2019). The impact of mechanized weed control in large-scale farming. *Agricultural Systems*, *167*, 112–121. DOI: 10.1016/j.agsy.2018.09.006

Moreda, G. P., Ortiz-Cañavate, J., García-Ramos, F. J., & Ruiz-Altisent, M. (2009). Non-destructive technologies for fruit and vegetable size determination – A review. *Journal of Food Engineering*, *92*(2), 119–136. https://doi.org/http://dx.doi.org/10.1016/j.jfoodeng.2008.11.004

Morepje, M. T., & Sithole, M. Z. (2024). The Infl. of E-Commerce Plat. on Sust. Agri. Practices among Small.holder Farmers in Sub-Saharan Africa. *Sustainability*, *16*(15), 6496. DOI: 10.3390/su16156496

Morimoto, S. (2002). Non-destructive near infrared spectrometers: Development of portable fruit quality meters. Near Infrared Spectroscopy, *Proceedings of the 10th International Conference*, NIR Publications, West Sussex, UK.

Moustaine, M. (2023). *Symptomology of the Main Fungal Diseases of the Tomato (Lycopersicon Esculentum) and Its Management*. Austin Journal of Plant Biology., DOI: 10.26420/austinjplantbiol.2023.1035

Mozaffari, M. (2018). *Wireless communications and networking with unmanned aerial vehicles: Fundamentals, deployment, and optimization*. Virginia Tech.

Murshed, M. (2021). LPG consumption and environmental Kuznets curve hypothesis in South Asia: A time-series ARDL analysis with multiple structural breaks. *Environmental Science and Pollution Research International*, *28*(7), 8337–8372. DOI: 10.1007/s11356-020-10701-7 PMID: 33058083

Mustafa, N. E., & Alizadeh, F. (2024). Unmanned aerial vehicle (UAV) images of road vehicles dataset. *Data in Brief*, *54*, 110264. Advance online publication. DOI: 10.1016/j.dib.2024.110264 PMID: 38516279

Naidu, G., Zuva, T., & Sibanda, E. M. (2023). A Review of Evaluation Metrics in Machine Learning Algorithms. In Silhavy, R., & Silhavy, P. (Eds.), *Artificial Intelligence Application in Networks and Systems* (pp. 15–25). Springer International Publishing., DOI: 10.1007/978-3-031-35314-7_2

NASA. (2020). Quantum technologies are changing the face of unmanned aircraft communications. https://www.nasa.gov/aeronautics/quantum-technologies-are-changing-the-face-of-unmanned-aircraft-communications/ (Accessed: 2024-08-21).

Natij, Y., El Karch, H., Maafiri, A., & Mezouari, A. (2024). Evaluating the Performance of YOLO Object Detectors for Plant Disease Detection. *2024 11th International Conference on Wireless Networks and Mobile Communications (WINCOM)*, 1–6. DOI: 10.1109/WINCOM62286.2024.10656918

Nawaz, S. J., Sharma, S. K., Wyne, S., Patwary, M. N., & Asaduzzaman, M. (2019). Quantum machine learning for 6G communication networks: State-of-the-art and vision for the future. *IEEE Access : Practical Innovations, Open Solutions*, *7*, 46317–46350. DOI: 10.1109/ACCESS.2019.2909490

Naylor, R.. (2018). The economics of manual weeding in small-scale farms. *Economic Botany*, *72*(4), 425–432. DOI: 10.1007/s12231-018-9435-6

Ncama, K., Tesfay, S. Z., Fawole, O. A., Opara, U. L., & Magwaza, L. S. (2018). Non-destructive prediction of 'Marsh'grapefruit susceptibility to postharvest rind pitting disorder using reflectance Vis/NIR spectroscopy. *Scientia Horticulturae*, *231*, 265–271.

Neethirajan, S. (2024). *Rethinking Poultry Welfare—Multifactorial and Digital Approaches to Enhance Animal Well-Being* (SSRN Scholarly Paper 5034840). Social Science Research Network. DOI: 10.2139/ssrn.5034840

Neupane, C., Walsh, K. B., Goulart, R., & Koirala, A. (2024). Developing Machine Vision in Tree-Fruit Applications—Fruit Count, Fruit Size and Branch Avoidance in Automated Harvesting. *Sensors (Basel)*, *24*(17), 5593. https://www.mdpi.com/1424-8220/24/17/5593. DOI: 10.3390/s24175593 PMID: 39275504

Newman, L., Newel, R., Dring, C., Glaros, A., Fraser, E., Mendly-Zambo, Z., Green, A. G., & Kc, K. B. (2023). Agriculture for the Anthropocene: Novel applications of technology and the future of food. *Food Security*, *15*(3), 613–627. DOI: 10.1007/s12571-023-01356-6

Ngouajio, M., Kirk, W., & Goldy, R. (2003). A simple model for rapid and nondestructive estimation of bell pepper fruit volume. *HortScience*, *38*(4), 509–511. DOI: 10.21273/HORTSCI.38.4.509

Ngugi, L. C., Abelwahab, M., & Abo-Zahhad, M. (2021). Recent advances in image processing techniques for automated leaf pest and disease recognition – A review. *Information Processing in Agriculture*, *8*(1), 27–51. DOI: 10.1016/j.inpa.2020.04.004

Nielson, M. Quirk: quantum circuit simulator, https://algassert.com/quirk, accessed on 14.03.2025.

Niklas, K. J. (1994). *Plant allometry: the scaling of form and process*. University of Chicago Press.

Nisha, S. (2018). Image classification of agricultural data using supervised learning techniques: A survey. *International Journal of Advanced Research in Computer Science*, *9*(2), 188–192. DOI: 10.26483/ijarcs.v9i2.5681

Oche, O. E., Nasir, S. M., & Muhammed, A. H. (2020). Internet of Things-based agriculture: A review of security threats and countermeasures. *Int. J. Sci. Res. Publ*, *10*(8), 738–748. DOI: 10.29322/IJSRP.10.08.2020.p10494

Octavia, D., Suharti, S., Murniati, , Dharmawan, I. W. S., Nugroho, H. Y. S. H., Supriyanto, B., Rohadi, D., Njurumana, G. N., Yeny, I., Hani, A., Mindawati, N., Suratman, , Adalina, Y., Prameswari, D., Hadi, E. E. W., & Ekawati, S. (2022). Mainstreaming smart agroforestry for social forestry implementation to support sustainable development goals in Indonesia: A review. *Sustainability (Basel)*, *14*(15), 93–103. DOI: 10.3390/su14159313

Ojala, T., Pietikainen, M., & Maenpaa, T. (2002). Multiresolution gray-scale and rotation invariant texture classification with local binary patterns. *IEEE Transactions on Pattern Analysis and Machine Intelligence*, *24*(7), 971–987. DOI: 10.1109/TPAMI.2002.1017623

Okubo, S., Tsunewaki, K., Kimura, T., Shimizu, H., Hiraguri, T., Shibasaki, A., Shimada, T., & Takemura, Y. (2024). Pear Blossom Counting System with Drones Using YOLO and Deep SORT. *2024 International Conference on Image Processing and Robotics (ICIPRoB)*

Olarewaju, O. O., Bertling, I., & Magwaza, L. S. (2016). Non-destructive evaluation of avocado fruit maturity using near infrared spectroscopy and PLS regression models. *Scientia Horticulturae*, *199*, 229–236.

Olson, J. P., Romero, J., & Aspuru-Guzik, A. (2017). Quantum autoencoders for efficient compression of quantum data. *Quantum Science and Technology*, *2*(4), 045001. DOI: 10.1088/2058-9565/aa8072

Pandi, S. S., Senthilselvi, A., & Gitanjali, J., ArivuSelvan, K., Gopal, J., & Vellingiri, J. (2024). Tomato leaf disease detection technique using VGG-19. *2024 Second International Conference on Advances in Information Technology (ICAIT)*, 1–6. https://doi.org/DOI: 10.1109/ICAIT54416.2024.00012

Panwar, A., Khari, M., & Misra, S.. (2023). Blockch. in Agri. to Ensu. Trust, Effect., and Traceab. from Farm Fields to Groc. *Future Internet, 15*(12), 404. DOI: 10.3390/fi15120404

Papageorgiou, M., Goliomytis, M., Tzamaloukas, O., Miltiadou, D., & Simitzis, P. (2023). Positive Welfare Indicators and Their Association with Sustainable Management Systems in Poultry. *Sustainability (Basel), 15*(14), 14. Advance online publication. DOI: 10.3390/su151410890

Pasquini, C. (2018). Near infrared spectroscopy: A mature analytical technique with new perspectives–A review. *Analytica Chimica Acta, 1026*, 8–36. DOI: 10.1016/j.aca.2018.04.004 PMID: 29852997

Patel, A., & Kumar, V. (2023). Optimizing irrigation efficiency using deep reinforcement learning in the field. ArXiv Preprint. Retrieved from https://arxiv.org/abs/2302.67890

Patel, R., & Mehta, S. (2021). Data quality improvement for smart irrigation using noise detection algorithms. *Computers and Electronics in Agriculture, 180*, 105937.

Patwal, P., Chauhan, R., Bhatt, C., & Devliyal, S. (2024). Automated tomato disease detection and classification using image processing and machine learning for precision agriculture. In *Challenges in Information* (pp. 481–486). Communication and Computing Technology., DOI: 10.1201/9781003559085-84

Paul, A., & Machavaram, R. (2024). *Advancements in Capsicum Detection in Thermal Greenhouse Imaging: Transitioning from Yolov10 to Sam-Dino* (SSRN Scholarly Paper 4917658). Social Science Research Network. https://papers.ssrn.com/abstract=4917658 DOI: 10.2139/ssrn.4917658

Pawlak, K., & Kołodziejczak, M. (2020). The Role of Agriculture in Ensuring Food Security in Developing Countries: Considerations in the Context of the Problem of Sustainable Food Production. *Sustainability (Basel), 12*(13), 5488. Advance online publication. DOI: 10.3390/su12135488

Payne, A., & Walsh, K. (2014). Machine vision in estimation of fruit crop yield. In S. Dutta Gupta & Y. Ibaraki (Eds.), *Plant image analysis: fundamentals and applications* (pp. 329-374). CRC Press. DOI: 10.1201/b17441-17

Payne, A. B., Walsh, K. B., Subedi, P. P., & Jarvis, D. (2013). Estimation of mango crop yield using image analysis – Segmentation method. *Computers and Electronics in Agriculture, 91*, 57–64. https://doi.org/http://dx.doi.org/10.1016/j.compag.2012.11.009. DOI: 10.1016/j.compag.2012.11.009

Payne, A., Walsh, K., Subedi, P., & Jarvis, D. (2014). Estimating mango crop yield using image analysis using fruit at 'stone hardening' stage and night time imaging. *Computers and Electronics in Agriculture, 100*, 160–167. https://doi.org/http://dx.doi.org/10.1016/j.compag.2013.11.011. DOI: 10.1016/j.compag.2013.11.011

Pazienza, P. (2015). The relationship between CO2 and foreign direct investment in the agriculture and fishing sector of OECD countries: Evidence and policy considerations. *Intelektinė ekonomika, 9*(1), 55-66.

Peirs, A., Lammertyn, J., Ooms, K., & Nicolaï, B. M. (2001). Prediction of the optimal picking date of different apple cultivars by means of VIS/NIR-spectroscopy. *Postharvest Biology and Technology, 21*(2), 189–199. DOI: 10.1016/S0925-5214(00)00145-9

Peirs, A., Schenk, A., & Nicolaï, B. M. (2005). Effect of natural variability among apples on the accuracy of VIS-NIR calibration models for optimal harvest date predictions. *Postharvest Biology and Technology, 35*(1), 1–13. DOI: 10.1016/j.postharvbio.2004.05.010

Perales Gómez, Á. L. López-de-Teruel, (2022) FARMIT: contin. Assess. of crop qual. using mach. learning and deep lear. Tech. for IoT-based smart farming. *Cluster Comput* 25, 2163–2178. https://foodtank.com/news/2021/09/from-farm-to-doorstep-the-rise-of-digital-farmers-markets/DOI: 10.1007/s10586-021-03489-9

Pereira, D. S., De Morais, M. R., Nascimento, L. B. P., Alsina, P. J., Santos, V. G., Fernandes, D. H. S., & Silva, M. R. (2020). Zigbee protocol-based communication network for multi-unmanned aerial vehicle networks. *IEEE Access : Practical Innovations, Open Solutions, 8*, 57762–57771. DOI: 10.1109/ACCESS.2020.2982402

Pierce, F. J., & Nowak, P. (1999). Aspects of precision agriculture. *Advances in Agronomy, 67*, 1–85. DOI: 10.1016/S0065-2113(08)60513-1

Pierce, S., Maffi, D., Faoro, F., Cerabolini, B. E. L., & Spada, A. (2022). The leaf anatomical trade-offs associated with plant ecological strategy variation. *Plant Ecology, 223*(10), 1233–1246. DOI: 10.1007/s11258-022-01270-5

Pires, R., Guerra, R., Cruz, S. P., Antunes, M. D., Brázio, A., Afonso, A. M., Daniel, M., Panagopoulos, T., Gonçalves, I., & Cavaco, A. M. (2022). Ripening assessment of 'Ortanique'(Citrus reticulata Blanco x Citrus sinensis (L) Osbeck) on tree by SW-NIR reflectance spectroscopy-based calibration models. *Postharvest Biology and Technology, 183,* 111750. DOI: 10.1016/j.postharvbio.2021.111750

Pius, L. O., Strausz, P., & Kusza, S. (2021). Overview of Poultry Management as a Key Factor for Solving Food and Nutritional Security with a Special Focus on Chicken Breeding in East African Countries. *Biology (Basel), 10*(8), 8. Advance online publication. DOI: 10.3390/biology10080810 PMID: 34440042

Pore, P., Teli, S., Ghuge, S., & Patil, N. (2023). Leaf Disease Detection. *International Journal for Research in Applied Science and Engineering Technology, 11*(5), 1767–1770. DOI: 10.22214/ijraset.2023.51405

Potter, B., Valentino, G., Yates, L., Benzing, T., & Salman, A. (2019). Environmental monitoring using a drone-enabled wireless sensor network. In *2019 Systems and Information Engineering Design Symposium (SIEDS)* (pp. 1-6). IEEE. DOI: 10.1109/SIEDS.2019.8735615

Potts, J. (2019). Blockchain in agriculture. *Available atSSRN* 3397786.

Purba, M., Simanjutak, D., Malau, Y., Sholihat, W., & Ahmadi, E. (2021). The Effect of Digital Marketing and E-Commerce on Financial Performance and Business Sustainability of MSMEs During COVID-19 Pandemic in Indonesia. *Int. J. Data Netw. Sci., 5,* 275–282. DOI: 10.5267/j.ijdns.2021.6.006

Purcell, W., Neubauer, T., & Mallinger, K. (2023). Digital twins in agriculture: Challenges and opportunities for environmental sustainability. *Current Opinion in Environmental Sustainability, 61*(1), 1–12. DOI: 10.1016/j.cosust.2022.101252

Qing-Bo, Y., Pan-Pan, N., Xiang-Yang, W., & Li-Ran, S. (2016). A new robust color image watermarking method for multimedia technology enhanced learning protection. *Journal of Intelligent & Fuzzy Systems, 31*(5), 2553–2564. DOI: 10.3233/JIFS-169096

Qureshi, W. S., Payne, A., Walsh, K. B., Linker, R., Cohen, O., & Dailey, M. N. (2017). Machine vision for counting fruit on mango tree canopies [journal article]. *Precision Agriculture, 18*(2), 224–244. DOI: 10.1007/s11119-016-9458-5

Rabhi, B., Dhahri, H., Jdey, I., & Alhajlah, O. (2025). IoT-Enhanced Tomato Leaf Disease Identification Using MLP-Mixer in Agricultural Environments. In: Czarnowski, I., Howlett, R.J., C. Jain, L. (eds) Intelligent Decision Technologies. KESIDT 2024. Smart Innovation, Systems and Technologies, vol 411. Springer, Singapore. https://doi.org/DOI: 10.1007/978-981-97-7419-7_37

Radcliffe, J., Skinner, K., & Spring, A. (2021). Virt. barriers: Unpack. the sustain. Implic. of online food spaces and the Yellowknife Farmers Market's response to COVID-19. *Nutrition Journal, 20*, 12. DOI: 10.1186/s12937-021-00664-x PMID: 33514376

Rai, A. (2023). Digitalization in agriculture for innovative transformation in India. *Indian Farming*, *73*(6), 1–13. https://epubs.icar.org.in/index.php/IndFarm/article/view/138320

Raihan, A. (2023). An econometric evaluation of the effects of economic growth, energy use, and agricultural value added on carbon dioxide emissions in Vietnam. *Asia-Pacific Journal of Regional Science*, *7*(3), 665–696. DOI: 10.1007/s41685-023-00278-7

Raihan, A., Begum, R. A., Nizam, M., Said, M., & Pereira, J. J. (2022). Dynamic impacts of energy use, agricultural land expansion, and deforestation on CO2 emissions in Malaysia. *Environmental and Ecological Statistics*, *29*(3), 477–507. DOI: 10.1007/s10651-022-00532-9

Rajitha, A., Swathi, M., Ananya, K., Amruthavarshini, R., & Akhila, D. (2025). Plant Disease Detection Using Deep Learning. *International Journal for Research in Applied Science and Engineering Technology*, *13*(1), 1320–1323. DOI: 10.22214/ijraset.2025.66199

Ralegankar, V. K., Bagul, J., Thakkar, B., Gupta, R., Tanwar, S., Sharma, G., & Davidson, I. E. (2022). Quantum cryptography-as-a-service for secure UAV communication: Applications, challenges, and case study. *IEEE Access : Practical Innovations, Open Solutions*, *10*, 1475–1492. DOI: 10.1109/ACCESS.2021.3138753

Ramya, J. (2023). Crop recommendation using supervised learning techniques. In *Proceedings of the ICOSEC Conference* (pp. 1076-1084). https://doi.org/DOI: 10.1109/ICOSEC58147.2023.10276199

Rathod, M. L., Shivaputra, A., Umadevi, H., Nagamani, K., & Periyasamy, S. (2022). Cloud computing and networking for SmartFarm AgriTech. *Journal of Nanomaterials*, *2022*(1), 6491747. DOI: 10.1155/2022/6491747

Redmon, J., Divvala, S., Girshick, R., & Farhadi, A. (2016). You only look once: Unified, real-time object detection. *Proceedings of the IEEE conference on computer vision and pattern recognition*

Rees, J., & Rees, C. J. (2023). Cyber-Security and the Changing Landscape of Critical National Infrastructure: State and Non-state Cyber-Attacks on Organisations, Systems and Services. In *Applications for Artificial Intelligence and Digital Forensics in National Security* (pp. 67-89). Springer.

Rehman, S.. (2019). Drone-based weed detection and management in precision agriculture. *Computers and Electronics in Agriculture*, *156*, 482–492. DOI: 10.1016/j.compag.2018.11.020

Relese, I. T. Advantages and disadvantages of quantum computers. Accessed on 01, 2023. Rehman, J., "Advantages and disadvantages of quantum computers," Learn about IT beyond distance, https://www.itrelease.com/2020/10/advantages-and-disadvantages-of-quantum-computers/

Ren, S., He, K., Girshick, R., & Sun, J. (2015). Faster r-cnn: Towards real-time object detection with region proposal networks. *Advances in Neural Information Processing Systems.*

Reyana, A., Kautish, S., & Karthik, P. M. S. (2023). Accel. Crop Yield: Multisens. Data Fusi. and Mach. Learn. for *Agric. Text Classi.IEEE Access : Practical Innovations, Open Solutions*, *11*, 20795–20805. DOI: 10.1109/ACCESS.2023.3249205

Reyes, J. F., Contreras, E., Correa, C., & Melin, P. (2021). Image analysis of real-time classification of cherry fruit from colour features. *Journal of Agricultural Engineering*, *52*(4). Advance online publication. https://doi.org/ DOI: 10.4081/jae.2021.1160

Reza, S. K., Chattaraj, S., Mukhopadhyay, S., Daripa, A., Saha, S., & Ray, S. K. (2024). Delineation of high-resolution soil carbon management zones using digital soil mapping: A step towards mitigating climate change in the Northeastern Himalayas, India. *Soil Use and Management*, *40*(1), 1–13. DOI: 10.1111/sum.12995

Rocha, G. S. R., de Oliveira, L., & Talamini, E. (2021). E. Blockcha. appl. in agribusi.: A systematic review. *Future Internet*, *13*(4), 95. DOI: 10.3390/fi13040095

Roger, J.-M., Chauchard, F., & Bellon-Maurel, V. (2003). EPO–PLS external parameter orthogonalisation of PLS application to temperature-independent measurement of sugar content of intact fruits. *Chemometrics and Intelligent Laboratory Systems*, *66*(2), 191–204. DOI: 10.1016/S0169-7439(03)00051-0

Romero, J., & Hernandez, C. (2021). Optimization of irrigation scheduling using genetic algorithms for water conservation. *Environmental Modelling & Software, 148*, 105226.

Rong, W., Li, Z., Zhang, W., & Sun, L. (2014). An improved Canny edge detection algorithm. *2014 IEEE International Conference on Mechatronics and Automation*, 577–582. DOI: 10.1109/ICMA.2014.6885761

Rossi, V., Caffi, T., Salotti, I., & Fedele, G. (2023). Sharing decision-making tools for pest management may foster implementation of Integrated Pest Management. *Food Security, 15*(1), 1459–1474. DOI: 10.1007/s12571-023-01402-3

Rotz, S., Duncan, E., Small, M., Botschner, J., Dara, R., Mosby, I., Reed, M., & Fraser, E. D. (2019). The politics of digital agricultural technologies: A preliminary review. *Sociologia Ruralis, 59*(2), 203–229. DOI: 10.1111/soru.12233

Ruan, Y., Xue, X., & Shen, Y. (2021). Quantum image processing: Opportunities and challenges. *Mathematical Problems in Engineering, 2021*(1), 6671613.

Russell, C. C. (2022). Cyber security in digital agriculture: Investigating farmer perceptions, preferences, & expert knowledge (Doctoral dissertation, University of Guelph).

Şahin, E. and I. Yilmaz, QRMW: quantum representation of multi wavelength images.

Şahin, E., & Yilmaz, İ. (2018). QRMW: Quantum representation of multi wavelength images. *Turkish Journal of Electrical Engineering and Computer Sciences, 26*(2), 768–779. DOI: 10.3906/elk-1705-396

Sa, I., Ge, Z., Dayoub, F., Upcroft, B., Perez, T., & McCool, C. (2016). DeepFruits: A Fruit Detection System Using Deep Neural Networks. *Sensors (Basel), 16*(8), 1222. DOI: 10.3390/s16081222 PMID: 27527168

Saikai, Y., Peake, A., & Chenu, K. (2023). Deep reinforcement learning for irrigation scheduling using high-dimensional sensor feedback. Retrieved from https://journals.plos.org/water/article?id=10.1371/journal.pwat.0000169

Sakib, S. M. (2022). LiDAR technology—An overview. *IUP Journal of Electrical & Electronics Engineering, 15*(1), 10–20.

Sakthipriya, S., & Naresh, R. (2024). Precision agri. based on convo. neural net. in rice prod. nutrient manage. using mach. learn. genetic algorithm, *Engine.Appl. of Arti. Intell. Elsevier BV, 130*, 107682. DOI: 10.1016/j.engappai.2023.107682

Salihu, S., Jauro, S. S., & Salihu, M. N. (n.d.-a). *Enhancing Poultry Welfare and Production through Advanced Chicken Detection, Tracking and Counting: A Deep Learning Approach.*

Salihu, S., Jauro, S. S., & Salihu, M. N. (n.d.-b). *Enhancing Poultry Welfare and Production through Advanced Chicken Detection, Tracking and Counting: A Deep Learning Approach.*

Sang, J., & Shen, W. (2017). A novel quantum representation of color digital images. *Quantum Information Processing, 16*(2), 1–14. DOI: 10.1007/s11128-016-1463-0

Santosh, R., & Rao, M. (2021). Fuzzy logic-based irrigation control for sustainable water management in smart farming. *Computers and Electronics in Agriculture, 180*, 105937.

Saranwong, I., Sornsrivichai, J., & Kawano, S. (2003). On-tree evaluation of harvesting quality of mango fruit using a hand-held NIR instrument. *Journal of Near Infrared Spectroscopy, 11*(4), 283–293. DOI: 10.1255/jnirs.374

Saravanan, C. (2010). Color Image to Grayscale Image Conversion. *2010 Second International Conference on Computer Engineering and Applications, 2*, 196–199. DOI: 10.1109/ICCEA.2010.192

Sarkar, S., Basak, J. K., Moon, B. E., & Kim, H. T. (2020). A comparative study of PLSR and SVM-R with various preprocessing techniques for the quantitative determination of soluble solids content of hardy kiwi fruit by a portable Vis/NIR spectrometer. *Foods, 9*(8), 1078. DOI: 10.3390/foods9081078 PMID: 32784804

Schirber, M. (2021). Quantum drones take flight. *Physics*. https://physics.aps.org/articles/v14/7

Schlechtendahl, E.. (2020). Autonomous weed control: Opportunities and challenges in precision agriculture. *Robotics and Autonomous Systems, 127*, 103–118. DOI: 10.1016/j.robot.2019.11.011

Schulze, K., Nagle, M., Spreer, W., Mahayothee, B., & Müller, J. (2015). Development and assessment of different modeling approaches for size-mass estimation of mango fruits (Mangifera indica L., cv. 'Nam Dokmai'). *Computers and Electronics in Agriculture, 114*, 269–276. https://doi.org/https://doi.org/10.1016/j.compag.2015.04.013. DOI: 10.1016/j.compag.2015.04.013

Seeger, R. M., Hagerman, A. D., Johnson, K. K., Pendell, D. L., & Marsh, T. L. (2021). When poultry take a sick leave: Response costs for the 2014–2015 highly pathogenic avian influenza epidemic in the USA. *Food Policy, 102*, 102068. DOI: 10.1016/j.foodpol.2021.102068

Sekhar, M., Rastogi, M., Rajesh, M., Saikanth, D. R. K., Rout, S., Kumar, S., & Patel, A. K. (2024). *Exploring Traditional Agricultural Techniques Integrated with Modern Farming for a Sustainable Future: A Review*. Journal of Scientific Research and Reports. https://api.semanticscholar.org/CorpusID:267714435

Šermukšnytė-Alešiūnienė, K., & Melnikienė, R. (2024). The effects of digitalization on the sustainability of small farms. *Sustainability (Basel)*, *16*(10), 1–12. DOI: 10.3390/su16104076

Serna-Escolano, V., Giménez, M. J., Zapata, P. J., Cubero, S., Blasco, J., & Munera, S. (2024). Non-destructive assessment of'Fino'lemon quality through ripening using NIRS and chemometric analysis. *Postharvest Biology and Technology*, *212*, 112870. DOI: 10.1016/j.postharvbio.2024.112870

Shafiee, M., Zhou, Z., Mei, L., Dinmohammadi, F., Karama, J., & Flynn, D. (2021). Unmanned aerial drones for inspection of offshore wind turbines: A mission-critical failure analysis. *Robotics (Basel, Switzerland)*, *10*(1), 26. Advance online publication. DOI: 10.3390/robotics10010026

Shafi, U., Mumtaz, R., García-Nieto, J., Hassan, S. A., Zaidi, S. A. R., & Iqbal, N. (2019). Precision Agriculture Techniques and Practices: From Considerations to Applications. *Sensors (Basel)*, *19*(17), 3796. DOI: 10.3390/s19173796 PMID: 31480709

Shahi, T. B., Sitaula, C., Bhandari, K. P., Poudel, S., Bhandari, R., Mishra, R.,& Mishra, B. (2024). Deep Learning-Based Method for Irrigation Status Detection in Tomato Using Plant Leaves. ESS Open Archive e-prints, 965, 96503836.

Shahi, T. B., Xu, C. Y., Neupane, A., Fresser, D., O'Connor, D., Wright, G., & Guo, W. (2023). A cooperative scheme for late leaf spot estimation in peanut using UAV multispectral images. *PLoS One*, *18*(3), e0282486. DOI: 10.1371/journal.pone.0282486 PMID: 36972266

Shakoor, M. T., Rahman, K., Rayta, S. N., & Chakrabarty, A. (2017). Agricultural production output prediction using Supervised Machine Learning techniques. *2017 1st International Conference on Next Generation Computing Applications (NextComp)*, 182-187.

Sharma, R.; Samad, T.A.; Jabbour, C.J.C.; de Queiroz, M.J. (2021) Levera. Blockcha. Techno. for circu. in agri. Supp. Chain.: Evid. from a fast-growing eco. *J. Enterp. Inf. Manag.* ahead-of-print.

Sharma, A., Vanjani, P., Paliwal, N., Basnayaka, C. M. W., Jayakody, D. N. K., Wang, H.-C., & Muthuchidambaranathan, P. (2020). Communication and networking technologies for UAVs: A survey. *Journal of Network and Computer Applications*, *168*, 102739. DOI: 10.1016/j.jnca.2020.102739

Sharma, N., Kumar, A., Pervaiz, H., Magarini, M., Musavian, L., Alam, M., Jindal, A., & Imran, M. (2021). Aerial base station assisted cellular communication: Performance and trade-off. *IEEE Transactions on Network Science and Engineering*, *8*(4), 1. DOI: 10.1109/TNSE.2021.3052984

Sharma, P., Kapoor, A., & Verma, S. (2020). Automatic Weed Removing Robot Using IoT Technology. [IJCRT]. *International Journal of Creative Research Thoughts*, *8*(9), 350–358. https://ijcrt.org/papers/IJCRT2009133.pdf

Sharma, R., Kamble, S. S., Gunasekaran, A., Kumar, V., & Kumar, A. (2020). A systematic litera. Rev. on mach. Learn. Appl. for sust. Agri. supply chain perfo. *Computers & Operations Research*, *119*, 104926. DOI: 10.1016/j.cor.2020.104926

Shetty, P., Kumar, A., Rajesh, B. S., Balipa, M., Kanchan, G., & Kamath, C. S. (2022, December). Tomato Leaf Disease Detection Using Multiple Classifier System. In *2022 International Conference on Artificial Intelligence and Data Engineering (AIDE)* (pp. 316-321). IEEE.

Shor, P. W. "Algorithms for quantum computation: discrete logarithms and factoring," in Proceedings 35th annual symposium on foundations of computer science. Ieee, 1994, pp. 124 134. DOI: 10.1109/SFCS.1994.365700

Shrestha, R. K., Jacinthe, P.-A., Lal, R., Lorenz, K., Singh, M. P., Demyan, S. M., Ren, W., & Lindsey, L. E. (2023). Biochar as a negative emission technology: A synthesis of field research on greenhouse gas emissions. *Journal of Environmental Quality*, *52*(4), 769–798. DOI: 10.1002/jeq2.20475 PMID: 36905388

Shuaibu, M., & Nchake, M. (2021). Impa. of Credit Market Cond. on Agri. Prod. in Sub-Saharan Africa. Agric. *Financial Review*, *81*, 520–534.

Siddiqui, M. H., & Khan, A. R. (2020). Automated Computer Vision Based Weed Removal Bot. IEEE Xplore. https://ieeexplore.ieee.org/document/9198515

Sidhu, J. S., Joshi, S. K., Gündoğan, M., Brougham, T., Lowndes, D., Mazzarella, L., Krutzik, M., Mohapatra, S., Dequal, D., Vallone, G., Villoresi, P., Ling, A., Jennewein, T., Mohageg, M., Rarity, J. G., Fuentes, I., Pirandola, S., & Oi, D. K. L. (2021). Advances in space quantum communications. *IET Quantum Communication*, *2*(4), 182–217. DOI: 10.1049/qtc2.12015

Simonyan, K., & Zisserman, A. (2014). Very deep convolutional networks for large-scale image recognition. *arXiv preprint arXiv:1409.1556.*

Singh, I. (2022). US is developing drones with 'unhackable' quantum communication technology. https://dronedj.com/2022/05/26/quantum-drones/

Singh, A. (2021). Real-time adaptation in IoT-based irrigation systems. *IEEE Internet of Things Journal, 8*(5), 3124–3132.

Singh, A. K. (2025). Importance of digital technologies in economic sector of selected countries. In Ahmadi, A., Mna, A., & Slimani, A. (Eds.), *Renewable Energy and the Economic Welfare of Society* (pp. 335–370). IGI Global Scientific Publishing., DOI: 10.4018/979-8-3693-7580-8.ch015

Singh, A. K., Issac, J., & Narayanan, K. G. S. (2019). Measurement of environmental sustainability index and its association with socio-economic indicators in selected Asian economies: An empirical investigation. *International Journal of Environment and Sustainable Development, 18*(1), 57–100. DOI: 10.1504/IJESD.2019.098641

Singh, A. K., & Jyoti, B. (2024). A review for assessing the impact of climate change on the agricultural sector, food security, and pests in India. In Zia Ul Haq, M., & Ali, I. (Eds.), *Revolutionizing Pest Management for Sustainable Agriculture* (pp. 53–84). IGI Global., DOI: 10.4018/979-8-3693-3061-6.ch003

Singh, A. K., Jyoti, B., & Ashraf, S. N. (2025). A theoretical review for assessing the determinants and challenges of agri-startups with reference to appropriate technology in India. In Gupta, V., & Gupta, C. (Eds.), *Navigating Strategic Partnerships for Sustainable Startup Growth* (pp. 159–190). IGI Global Scientific Publishing., DOI: 10.4018/979-8-3693-4066-0.ch007

Singh, A. K., Kumar, S., & Gupta, M. (2024). Relationship between digitalization and environmental sustainability in selected countries. In Yıldırım, S., Yıldırım, D., Demirtaş, I., & Kandpal, V. (Eds.), *Green Transition Impacts on the Economy, Society, and Environment* (pp. 83–107). IGI Global., DOI: 10.4018/979-8-3693-3985-5.ch006

Singh, A. K., Sharma, A. K., & Jyoti, B. (2023). Does economic development have a causal relationship with environmental degradation? Experience from different income group countries. In Sart, G. (Ed.), *Considerations on Education for Economic, Social, and Environmental Sustainability* (pp. 300–333). IGI Global., DOI: 10.4018/978-1-6684-8356-5.ch015

Singh, A. K., & Singh, R. (2025). Digital infrastructure and sustainable development goals in G20 countries. In Popescu, C. (Ed.), *Building Inclusive Global Knowledge Societies for Sustainable Development* (pp. 363–408). IGI Global Scientific Publishing., DOI: 10.4018/979-8-3693-6298-3.ch014

Singh, A. K., & Singh, S. (2023). A case study from Kanpur Nagar, Uttar Pradesh for assessing the economic value of water and its determinants. In Karmaoui, A. (Ed.), *Water-Soil-Plant-Animal Nexus in the Era of Climate Change* (pp. 212–242). IGI Global., DOI: 10.4018/978-1-6684-9838-5.ch010

Singh, A. K., Srinivasan, S., & Jyoti, B. (2025a). Implications of digitalization on public health sector across countries. In Wickramasinghe, N. (Ed.), *Digitalization and the Transformation of the Healthcare Sector* (pp. 359–394). IGI Global Scientific Publishing., DOI: 10.4018/979-8-3693-9641-4.ch013

Singh, D., Jain, N., Jain, P., Kayal, P., Kumawat, S., & Batra, N. (2019). PlantDoc: A dataset for visual plant disease detection. arXiv. https://arxiv.org/abs/1911.10317

Singh, G., & Kaur, G. (2021). Digital technologies for smart agriculture. In Tomar, P., & Kaur, G. (Eds.), *Artificial Intelligence and IoT-Based Technologies for Sustainable Farming and Smart Agriculture* (pp. 40–53). IGI Global., DOI: 10.4018/978-1-7998-1722-2.ch004

Singh, P., & Kaur, A. (2022). A systematic review of artificial intelligence in agriculture. In Poonia, R. C., Singh, V., & Nayak, S. R. (Eds.), *Deep learning for sustainable agriculture* (pp. 57–80). Elsevier Inc., https://doi.org/https://doi.org/10.1016/C2020-0-02363-7 DOI: 10.1016/B978-0-323-85214-2.00011-2

Sinha, B. B., & Dhanalakshmi, R. (2022). Recent advancements and challenges of Internet of Things in smart agriculture: A survey. *Future Generation Computer Systems*, *126*, 169–184. DOI: 10.1016/j.future.2021.08.006

Siomau, M. (2019). Any quantum network is structurally controllable by a single driving signal. *Quantum Information Processing*, *18*(1), 1–26. DOI: 10.1007/s11128-018-2112-6

Sirisha, U., Praveen, S. P., Srinivasu, P. N., Barsocchi, P., & Bhoi, A. K. (2023). Statistical Analysis of Design Aspects of Various YOLO-Based Deep Learning Models for Object Detection. *International Journal of Computational Intelligence Systems*, *16*(1), 126. DOI: 10.1007/s44196-023-00302-w

Sishodia, R. P., Ray, R. L., & Singh, S. K. (2020). Applications of remote sensing in precision agriculture: A review. *Remote Sensing (Basel)*, *12*(19), 3136. DOI: 10.3390/rs12193136

Smidt, H. J., & Jokonya, O. (2021). Factors affecting digital technology adoption by small-scale farmers in agriculture value chains (AVCs) in South Africa. *Information Technology for Development, 28*(3), 558–584. DOI: 10.1080/02681102.2021.1975256

Smith, A., & Zhang, L. (2018). Smart irrigation: A comprehensive review. *Journal of Environmental Management, 220*, 1–10.

Smith, J., Brown, A., & Wilson, K. (2018). Grass species and root depths in urban landscapes: A comparative study. *Journal of Urban Ecology, 12*(3), 145–160. DOI: 10.1016/j.jue.2018.03.007

Smith, M. J. (2018). Getting value from artificial intelligence in agriculture. *Animal Production Science, 60*(1), 46–54. DOI: 10.1071/AN18522

Sohel, A., Rahman, M., Hasan, M. U., Islam, M. D. K., Rukhsara, L., & Rabeya, T. (1850). M., Hasan, M. U., Islam, M., Rukhsara, L., & Rabeya, T. (2025). Automated tomato leaf disease recognition using deep convolutional networks. *International Journal of Power Electronics and Drive Systems, 15*(2), 1850. Advance online publication. DOI: 10.11591/ijece.v15i2.pp1850-1860

Song, L., & Wang, X.; Merveille. (. (2021). N. Blockchain-Based Flex. Double-Chain Arch. and Perfo. Optimiz. for Better Sust. in Agri. *Computers, Materials & Continua, 68*, 1429–1446. DOI: 10.32604/cmc.2021.016954

Song, Y., Zhang, B., Wang, J., & Kwek, K. (2022). The impact of climate change on China's agricultural green total factor productivity. *Technological Forecasting and Social Change, 185*, 185. DOI: 10.1016/j.techfore.2022.122054

Sozzi, M., Pillan, G., Ciarelli, C., Marinello, F., Pirrone, F., Bordignon, F., Bordignon, A., Xiccato, G., & Trocino, A. (2023). Measuring Comfort Behaviours in Laying Hens Using Deep-Learning Tools. *Animals (Basel), 13*(1), 1. Advance online publication. DOI: 10.3390/ani13010033 PMID: 36611643

SP., P., & Balamurugan, P. (2022). Unmanned aerial vehicle in the smart farming systems: Types, applications and cyber-security threats. *2022 International Conference on Innovative Computing, Intelligent Communication and Smart Electrical Systems (ICSES)*

Spreer, W., & Müller, J. (2011). Estimating the mass of mango fruit (Mangifera indica, cv. Chok Anan) from its geometric dimensions by optical measurement. *Computers and Electronics in Agriculture, 75*(1), 125–131. https://doi.org/http://dx.doi.org/10.1016/j.compag.2010.10.007. DOI: 10.1016/j.compag.2010.10.007

Srinivasarao, C. (2021). Soil degra. Chall. for sust. Agri. in tropical India. *Current Science, 120*(3), 492–500. DOI: 10.18520/cs/v120/i3/492-500

Srinivasarao, C. (2022). Indian Agri. Tow. 2030. *India Stud.* In *Bus. and Eco.* Springer.

Srivastava, N., Hinton, G., Krizhevsky, A., Sutskever, I., & Salakhutdinov, R. (2014). Dropout: A simple way to prevent neural networks from overfitting. *Journal of Machine Learning Research, 15*(1), 1929–1958.

Steane, A. M. (1999). Efficient fault-tolerant quantum computing. *Nature, 399*(6732), 124–126. DOI: 10.1038/20127

Stein, M., Bargoti, S., & Underwood, J. (2016). Image based mango fruit detection, localisation and yield estimation using multiple view geometry. *Sensors (Basel), 16*(11), 1915. DOI: 10.3390/s16111915 PMID: 27854271

Strauss, E. D., Curley, J. P., Shizuka, D., & Hobson, E. A. (2022). The centennial of the pecking order: Current state and future prospects for the study of dominance hierarchies. *Philosophical Transactions of the Royal Society B: Biological Sciences, 377*(1845). *Philosophical Transactions of the Royal Society of London. Series B, Biological Sciences, 1845*(1845), 20200432. Advance online publication. DOI: 10.1098/rstb.2020.0432 PMID: 35000437

Subedi, P. P., & Walsh, K. B. (2020). Assessment of avocado fruit dry matter content using portable near infrared spectroscopy: Method and instrumentation optimisation. *Postharvest Biology and Technology, 161*, 111078. DOI: 10.1016/j.postharvbio.2019.111078

Subramanya, S., & Naveen, B. (2024). *Enhanced detection of tomato leaf diseases using ensemble deep learning: INCVX-NET model.* IAES International Journal of Artificial Intelligence., DOI: 10.11591/ijai.v13.i4.pp4757-4765

Suchithra, M. S., & Pai, M. L. (2020). Impro. the pred. accu. of soil nutrient class. by optimizing extreme learn. Mach. parameters. *Information Processing in Agriculture, 7*(1), 72–82. DOI: 10.1016/j.inpa.2019.05.003

Sun, B., . . . *A multi-channel representation for images on quantum computers using the RGBα color space.* in *2011 IEEE 7th International Symposium on Intelligent Signal Processing*. 2011. IEEE.

Sun, B., et al., An RGB multi-channel representation for images on quantum computers. J.

Sun, X., Du, Y., Nawaz, M. A., Abobatta, W. F., Lyu, Q., Liu, J., Chen, Z., & Feng, S. (2024). Apple SSC estimation using hand-held NIRS instrument for outdoor measurement with ambient light correction. *Postharvest Biology and Technology, 217*, 113101. DOI: 10.1016/j.postharvbio.2024.113101

Sun, X., Guo, F., Liu, J., Chen, Z., Abobatta, W. F., Nawaz, M. A., & Feng, S. (2024). From lab to orchard use for models of hand-held NIRS instrument: A case for navel orange quality assessment considering ambient light correction. *Computers and Electronics in Agriculture, 219*, 108797. DOI: 10.1016/j.compag.2024.108797

Sun, X., Subedi, P., Walker, R., & Walsh, K. B. (2020). NIRS prediction of dry matter content of single olive fruit with consideration of variable sorting for normalisation pre-treatment. *Postharvest Biology and Technology, 163*, 111140. DOI: 10.1016/j.postharvbio.2020.111140

Sun, X., Subedi, P., & Walsh, K. B. (2020). Achieving robustness to temperature change of a NIRS-PLSR model for intact mango fruit dry matter content. *Postharvest Biology and Technology, 162*, 111117. DOI: 10.1016/j.postharvbio.2019.111117

Sun, X., Wang, Z., Aydin, H., Liu, J., Chen, Z., & Feng, S. (2023). First step for hand-held NIRS instrument field use: Table grape quality assessment consideration of temperature and sunlight chemometrics correction. *Postharvest Biology and Technology, 201*, 112374. DOI: 10.1016/j.postharvbio.2023.112374

Su, S., Tang, Z., & Zhu, Y. (2024). High-density foreground object detection in optical remote sensing images via semantic fusion and box alignment. *The Visual Computer, 40*(6), 4355–4371. DOI: 10.1007/s00371-023-03086-4

Szegedy, C., Liu, W., Jia, Y., Sermanet, P., Reed, S., Anguelov, D., Erhan, D., Vanhoucke, V., & Rabinovich, A. (2015). Going deeper with convolutions. *Proceedings of the IEEE conference on computer vision and pattern recognition*

Tan, Q., He, R., Bing, L., & Ng, H. T. (2022). *Document-Level Relation Extraction with Adaptive Focal Loss and Knowledge Distillation* (No. arXiv:2203.10900). arXiv. /arXiv.2203.10900DOI: 10.18653/v1/2022.findings-acl.132

Tang, Y., Dananjayan, S., Hou, C., Guo, Q., Luo, S., & He, Y. (2021). A survey on the 5G network and its impact on agriculture: Challenges and opportunities. *Computers and Electronics in Agriculture, 180*, 105895. https://doi.org/https://doi.org/10.1016/j.compag.2020.105895. DOI: 10.1016/j.compag.2020.105895

Tanim, S. A., Shrestha, T. E., Tanvir, K., & Kabir, Md. S., Mridha, M. F., & Haq, M. K. (2024). Single-Level Fusion for Enhancing Meat Quality Classification with Explainable AI. *2024 IEEE International Conference on Computing, Applications and Systems (COMPAS)*, 1–6. DOI: 10.1109/COMPAS60761.2024.10796775

Tan, K., Lee, W. S., Gan, H., & Wang, S. (2018). Recognising blueberry fruit of different maturity using histogram oriented gradients and colour features in outdoor scenes. *Biosystems Engineering*, *176*, 59–72. https://doi.org/https://doi.org/10.1016/j.biosystemseng.2018.08.011

Tanvir, K., & Kabir, Md. S., Anik, Z. H., Hasan, Md. R., Tushar, MD. T., & Rahman, R. B. (2024). Forest Fire Detection Using Ensemble Deep Learning Model With XAi. *2024 IEEE International Conference on Computing, Applications and Systems (COMPAS)*, 1–6. DOI: 10.1109/COMPAS60761.2024.10796582

Taseer, A., & Han, X. (2024). Advancements in variable rate spraying for precise spray requirements in precision agriculture using Unmanned aerial spraying Systems: A review. *Computers and Electronics in Agriculture*, *219*, 108841.

Teterja, D., Garcia-Rodriguez, J., Azorin-Lopez, J., Sebastian-Gonzalez, E., Nedić, D., Leković, D., Knežević, P., Drajić, D., & Vukobratović, D. (2024). A Video Mosaicing-Based Sensing Method for Chicken Behavior Recognition on Edge Computing Devices. *Sensors (Basel)*, *24*(11), 11. Advance online publication. DOI: 10.3390/s24113409 PMID: 38894200

Tibbetts, E. A., Pardo-Sanchez, J., & Weise, C. (2022). The establishment and maintenance of dominance hierarchies. *Philosophical Transactions of the Royal Society B: Biological Sciences, 377*(1845). *Philosophical Transactions of the Royal Society of London. Series B, Biological Sciences*, *1845*(1845), 20200450. Advance online publication. DOI: 10.1098/rstb.2020.0450 PMID: 35000449

Tiurlikova, A., Stepanov, N., & Mikhaylov, K. (2019). Wireless power transfer from unmanned aerial vehicle to low-power wide area network nodes: Performance and business prospects for LoRaWAN. *International Journal of Distributed Sensor Networks*, *15*(11), 1550147719888165. Advance online publication. DOI: 10.1177/1550147719888165

Tohidi, F., Ul-Haq, A., Haque, E., Paul, M., & Debnath, T. "Enhancing image representation and compression: An innovative nz-nqer framework with block truncation quantum coding," in 2023 The International Conference on Digital Image Computing: Techniques and Applications (DICTA). IEEE Xplore, 2023.

Torres, I., Pérez-Marín, D., De la Haba, M.-J., & Sánchez, M.-T. (2017). Developing universal models for the prediction of physical quality in citrus fruits analysed on-tree using portable NIRS sensors. *Biosystems Engineering*, *153*, 140–148. DOI: 10.1016/j.biosystemseng.2016.11.007

Torres-Morales, E., Khatiwada, D., Xylia, M., & Johnson, F. X. (2023). Investigating biochar as a net-negative emissions strategy in Colombia: Potentials, costs, and barriers. *Current Research in Environmental Sustainability*, *6*(1), 1–12. DOI: 10.1016/j.crsust.2023.100229

Touvron, H., Cord, M., Douze, M., Massa, F., Sablayrolles, A., & Jégou, H. (2021). Training data-efficient image transformers & distillation through attention. *Proceedings of the International Conference on Machine Learning (ICML)*, 10347–10357. https://doi.org/DOI: 10.48550/arXiv.2012.12877

Trendov, N. M., Varas, S., & Zeng, M. (2019). *Digital technologies in agriculture and rural areas*. Food and Agricultural Organization of the United Nations., https://openknowledge.fao.org/server/api/core/bitstreams/885161de-dccf-4589-8376-07fe37b68799/content

Triyanto, W. A., Adi, K., & Suseno, J. E. (2023). Detection and Tracking of Broiler Flock Movements in The Chicken Coop using YOLO. *E3S Web of Conferences, 448*, 02064. DOI: 10.1051/e3sconf/202344802064

Tsvetanov, F., & Kuzmanov, K. (2024). Communication Challenges in Smart Irrigation Systems. 2024 9th International Conference on Energy Efficiency and Agricultural Engineering (EE&AE), Tychola, K. A., & Rantos, K. (2025). Cyberthreats and Security Measures in Drone-Assisted Agriculture. *Electronics (Basel)*, *14*(1), 149.

Tu, C., Shen, J., Dai, J., Zhang, L., & Wang, J. (2022). A lower size, weight acquisition and tracking system for airborne quantum communication. *IEEE Photonics Journal*, *14*(6), 1–8. DOI: 10.1109/JPHOT.2022.3212590

Tu, S., Pang, J., Liu, H., Zhuang, N., Chen, Y., Zheng, C., Wan, H., & Xue, Y. (2020). Passion fruit detection and counting based on multiple scale faster R-CNN using RGB-D images. *Precision Agriculture*, *21*(5), 1072–1091. DOI: 10.1007/s11119-020-09709-3

U.S. Bureau of Labor Statistics. (2023). Occupational employment and wage statistics: 2020 2023 [Data set]. U.S. Department of Labor. https://www.bls.gov/oes/current/oes453032.htm

Uijlings, J. R. R., van de Sande, K. E. A., Gevers, T., & Smeulders, A. W. M. (2013). Selective Search for Object Recognition [journal article]. *International Journal of Computer Vision*, *104*(2), 154–171. DOI: 10.1007/s11263-013-0620-5

Ullo, S. L., & Sinha, G. R. (2021). Advances in IoT and smart sensors for remote sensing and agriculture applications. *Remote Sensing (Basel)*, *13*(13), 2585. DOI: 10.3390/rs13132585

Ulsperger, J., & Ulsperger, K. (2017). Community, regional identity, and civic agriculture: A structural ritualization analysis of rural online farmers' market sellers. *Journal of Rural Social Sciences*, *32*(2), 98–124.

University of Washington. "Object and concept recognition for content-based image retrieval," Accessed on 12, Dec, 2022, http://imagedatabase.cs.washington.edu/

University Southern California School of Engineering. "The usc-sipi image database," Accessed on 12, Dec, 2022, https://sipi.usc.edu/database/

Utai, K., Nagle, M., Hämmerle, S., Spreer, W., Mahayothee, B., & Müller, J. (2019). Mass estimation of mango fruits (Mangifera indica L., cv. 'Nam Dokmai') by linking image processing and artificial neural network. *Engineering in Agriculture, Environment and Food*, *12*(1), 103–110. DOI: 10.1016/j.eaef.2018.10.003

Vallance, C. (2024). Google unveils 'mind-boggling' quantum computing chip. https://www.bbc.com/news/articles/c791ng0zvl3o

Van Evert, F. K.. (2012). Robot-assisted precision irrigation: A review. *Biosystems Engineering*, *113*(1), 76–90. DOI: 10.1016/j.biosystemseng.2012.10.007

Vangala, A., Das, A. K., Kumar, N., & Alazab, M. (2021). Smart Secu. Sens. for IoT-Based Agricu.: Blockcha. Perspective. *IEEE Sensors Journal*, *21*(16), 17591–17607. DOI: 10.1109/JSEN.2020.3012294

Varoquaux, G., & Colliot, O. (2023). Evaluating Machine Learning Models and Their Diagnostic Value. In Colliot, O. (Ed.), *Machine Learning for Brain Disorders* (pp. 601–630). Springer US., DOI: 10.1007/978-1-0716-3195-9_20

Varun, Y. C., Ray, B., Nanjappa, A., & Pramod, S. (2020). Application of Internet of Things (IoT) to develop a smart watering system for Cairns Parklands – A case study. *2020 IEEE Region 10 Symposium (TENSYMP)*, 1118–1122.

Venegas-Andraca, S. E., & Ball, J. L. (2010). Processing images in entangled quantum systems. *Quantum Information Processing*, *9*(1), 1–11. DOI: 10.1007/s11128-009-0123-z

Venegas-Andraca, S. E., & Bose, S. (2003). "Storing, processing, and retrieving an image using quantum mechanics," in Quantum information and computation. *SPIE*, *5105*, 137–147.

Verma, S., Gala, R., Madhavan, S., Burkule, S., Chauhan, S., & Prakash, C. (2018). An internet of things (IoT) architecture for smart agriculture. 2018 fourth international conference on computing communication control and automation (ICCUBEA)

Vilkov, A., & Tian, G. (2023). Blockchain's scope and purpose in carbon markets: A systematic literature review. *Sustainability (Basel), 15*(11), 1–14. DOI: 10.3390/su15118495

Virnodkar, S. S., Pachghare, V. K., Patil, V., & Jha, S. K. (2020). Remote sensing and machine learning for crop water stress determination in various crops: A critical review. *Precision Agriculture, 21*(5), 1121–1155.

Wadod, M., & Mohammed, F. (2023, July). Review on drone application methodologies in agriculture precision. *IOP Conference Series. Earth and Environmental Science, 1202*(1), 012001. Advance online publication. DOI: 10.1088/1755-1315/1202/1/012001

Walsh, J., Neupane, A., Koirala, A., Li, M., & Anderson, N. (2023). The evolution of chemometrics coupled with near infrared spectroscopy for fruit quality evaluation. II. The rise of convolutional neural networks. *Journal of Near Infrared Spectroscopy, 31*(3), 109–125. DOI: 10.1177/09670335231173140

Walsh, J., Neupane, A., & Li, M. (2024). Evaluation of 1D convolutional neural network in estimation of mango dry matter content. *Spectrochimica Acta. Part A: Molecular and Biomolecular Spectroscopy, 311*, 124003. DOI: 10.1016/j.saa.2024.124003 PMID: 38354673

Walsh, K. B. (2023). NIRS, fruit quality and Central Queensland University. *NIR News, 34*(7-8), 5–8. DOI: 10.1177/09603360231204541

Walsh, K. B., Blasco, J., Zude-Sasse, M., & Sun, X. (2020). Visible-NIR 'point' spectroscopy in postharvest fruit and vegetable assessment: The science behind three decades of commercial use. *Postharvest Biology and Technology, 168*, 111246. DOI: 10.1016/j.postharvbio.2020.111246

Walsh, K., McGlone, V., & Han, D. (2020). The uses of near infra-red spectroscopy in postharvest decision support: A review. *Postharvest Biology and Technology, 163*, 111139. DOI: 10.1016/j.postharvbio.2020.111139

Walsh, K., & Subedi, P. (2014). In-field monitoring of mango fruit dry matter for maturity estimation. *XXIX International Horticultural Congress on Horticulture: Sustaining Lives, Livelihoods and Landscapes (IHC2014)*: 1119

Walsh, K., & Wang, Z. (2018). Monitoring fruit quality and quantity in mangoes. In Galán Saúco, V., & Lu, P. (Eds.), *Achieving sustainable cultivation of mangoes* (pp. 313–338). Burleigh Dodds Science Publishing Limited., https://doi.org/, DOI: 10.19103/AS.2017.0026.14

Walter, A., Finger, R., Huber, R., & Buchmann, N. (2017). Smart farming is key to developing sustainable agriculture. *Proceedings of the National Academy of Sciences of the United States of America, 114*(24), 6148–6150. DOI: 10.1073/pnas.1707462114 PMID: 28611194

Wang, A., Chen, H., Liu, L., Chen, K., Lin, Z., Han, J., & Ding, G. (2024). Yolov10: Real-time end-to-end object detection. *arXiv preprint arXiv:2405.14458.*

Wang, B., et al., Quantum representation of indexed images and its applications.

Wang, L., et al., QRCI: A new quantum representation model of color digital images.

Wang, X., & Song, J. (2021). ICIoU: Improved Loss Based on Complete Intersection Over Union for Bounding Box Regression. *IEEE Access, 9*, 105686–105695. IEEE Access. DOI: 10.1109/ACCESS.2021.3100414

Wang, B., Hao, M., Li, P., & Liu, Z. (2020). Quantum Representation of Indexed Images and its Applications. *International Journal of Theoretical Physics, 59*(2), 374–402. DOI: 10.1007/s10773-019-04331-0

Wang, B., Xie, J., Li, S., Wan, Y., Fu, S., & Lu, K. (2018). Enabling high-performance onboard computing with virtualization for unmanned aerial systems. In *2018 International Conference on Unmanned Aircraft Systems (ICUAS)* (pp. 202-211). IEEE. DOI: 10.1109/ICUAS.2018.8453368

Wang, C., Han, J., Pu, Y., & Wang, X. (2022). Tea (Camellia sinensis): A Review of Nutritional Composition, Potential Applications, and Omics Research. *Applied Sciences (Basel, Switzerland), 12*(12), 12. Advance online publication. DOI: 10.3390/app13010012

Wang, C., & Rahman, A. (2022). Quantum-enabled 6G wireless networks: Opportunities and challenges. *IEEE Wireless Communications, 29*(1), 58–69. DOI: 10.1109/MWC.006.00340

Wang, F., Cui, J., Xiong, Y., & Lu, H. (2023). Application of deep learning methods in behavior recognition of laying hens. *Frontiers in Physics (Lausanne), 11*, 1139976. Advance online publication. DOI: 10.3389/fphy.2023.1139976

Wang, H.. (2020). IoT-based smart irrigation for sustainable agriculture. *Agricultural Water Management, 235*, 106162.

Wang, J., Li, H., & Zhao, T. (2023). Integrating machine learning paradigms and mixed-integer model predictive control for irrigation scheduling. ArXiv Preprint. Retrieved from https://arxiv.org/abs/2306.08715

Wang, L., Ran, Q., & Ma, J. (2020). Double quantum color images encryption scheme based on DQRCI. *Multimedia Tools and Applications*, *79*(9-10), 6661–6687. DOI: 10.1007/s11042-019-08514-z

Wang, L., Ran, Q., Ma, J., Yu, S., & Tan, L. (2019). QRCI: A new quantum representation model of color digital images. *Optics Communications*, *438*, 147–158. DOI: 10.1016/j.optcom.2019.01.015

Wang, M. (2021). Xu Z. and Zhang, Y., "Review of quantum image processing,". *Archives of Computational Methods in Engineering*, •••, 1–25.

Wang, Z., Walsh, K. B., & Verma, B. (2017). On-tree mango fruit size estimation using RGB-D images. *Sensors (Basel)*, *17*(12), 2738. DOI: 10.3390/s17122738 PMID: 29182534

Wang, Z., Walsh, K., & Koirala, A. (2019). Mango Fruit Load Estimation Using a Video Based MangoYOLO—Kalman Filter—Hungarian Algorithm Method. *Sensors (Basel)*, *19*(12), 2742. https://www.mdpi.com/1424-8220/19/12/2742. DOI: 10.3390/s19122742 PMID: 31216769

Warsame, Z. A. (2023). The significance of FDI inflow and renewable energy consumption in mitigating environmental degradation in Somalia. *International Journal of Energy Economics and Policy*, *13*(1), 443–453. DOI: 10.32479/ijeep.13943

Wedding, B. B., Wright, C., Grauf, S., White, R., Tilse, B., & Gadek, P. (2013). Effects of seasonal variability on FT-NIR prediction of dry matter content for whole Hass avocado fruit. *Postharvest Biology and Technology*, *75*, 9–16. DOI: 10.1016/j.postharvbio.2012.04.016

Wills, B., & Arundel, A. (2017). Inter.-enabled acc. to alter. food networks: A comp. of on-line and off-line food shop. and their differ. Interpret. of quality. *Agri Hum Vals.*, *34*(3), 701–712. DOI: 10.1007/s10460-017-9771-2

Wolfert, S., Ge, L., Verdouw, C., & Bogaardt, M.-J. (2017). Big data in smart farming–a review. *Agricultural Systems*, *153*, 69–80. DOI: 10.1016/j.agsy.2017.01.023

Wu, W., Dawson, D., Fleming-Muñoz, D., Schleiger, E., & Horton, J. (2019). *The future of Australia's Agricultural Workforce*. CSIRO Data61, Canberra, Australia

Xiao, Q., Bai, X., Zhang, C., & He, Y. (2022). Advanced high-throughput plant phenotyping techniques for genome-wide association studies: A review. *Journal of Advanced Research*, *35*, 215–230. DOI: 10.1016/j.jare.2021.05.002 PMID: 35003802

Xia, X., Chai, X., Li, Z., Zhang, N., & Sun, T. (2023). MTYOLOX: Multi-transformers-enabled YOLO for tree-level apple inflorescences detection and density mapping. *Computers and Electronics in Agriculture, 209*, 107803. https://doi.org/https://doi.org/10.1016/j.compag.2023.107803. DOI: 10.1016/j.compag.2023.107803

Xie, W., Li, J., & Zhang, H. (2024). SPARROW: Smart Precision Agriculture Robot for Ridding of Weeds. ArXiv. https://arxiv.org/abs/2405.20896

Xu, G., et al., Order-encoded quantum image model and parallel histogram specification.

Xu, M., Park, J. E., Lee, J. J., Yang, J., & Yoon, S. (2024). Plant disease recognition datasets in the age of deep learning: Challenges and opportunities. *Frontiers in Plant Science, 15*, 1452551. Advance online publication. DOI: 10.3389/fpls.2024.1452551 PMID: 39399537

Yaacoub, J. P., Noura, H., Salman, O., & Chehab, A. (2020). Security analysis of drones systems: Attacks, limitations, and recommendations. *Internet of Things : Engineering Cyber Physical Human Systems, 11*, 100218. DOI: 10.1016/j.iot.2020.100218 PMID: 38620271

Yadav, R., & Kumar Mishra, A. (2024). IoT and Mach. Learn.-Based Monit. of the Growth of Crops Using Blockchain Techn. In Verma, A., Verma, P., Pattanaik, K. K., Dhurandher, S. K., & Woungang, I. (Eds.), *Advanced Net. Tech. and Intell. Comput. ANTIC 2023. Commun. in Comp. and Info. Science* (Vol. 2090). Springer., DOI: 10.1007/978-3-031-64076-6_11

Yadav, R., Saroj, R., & Kumar Verma, A. (2023). *A Survey of IoT and Mach. Learn. Based Mon. of the Growth of Crops Using Blockchain Tech. 2023 Intern. Conf. on IoT, Comm. and Auto. Tech.* ICICAT., DOI: 10.1109/ICICAT57735.2023.10263755

Yan, F. (2013). F. Dong B. Sun, A. Iliyasu and K. Hirota, “An rgb multi-channel representation for images on quantum computers,”. *J. Adv. Comput. Intell. Intell. Inform, 17*(3).

Yao, X., Li, Y., Zhang, Q., & Wang, L. (2024). Automatic visual recognition for leaf disease based on enhanced attention mechanism. *Frontiers in Plant Science, 15*, 1382802. DOI: 10.3389/fpls.2024.1382802 PMID: 39650513

Yazdinejad, A., Zolfaghari, B., Azmoodeh, A., Dehghantanha, A., Karimipour, H., Fraser, E., Green, A. G., Russell, C., & Duncan, E. (2021). A review on security of smart farming and precision agriculture: Security aspects, attacks, threats and countermeasures. *Applied Sciences (Basel, Switzerland), 11*(16), 7518. DOI: 10.3390/app11167518

Yeung, M., Sala, E., Schönlieb, C.-B., & Rundo, L. (2022). Unified Focal loss: Generalising Dice and cross entropy-based losses to handle class imbalanced medical image segmentation. *Computerized Medical Imaging and Graphics*, *95*, 102026. DOI: 10.1016/j.compmedimag.2021.102026 PMID: 34953431

Yıldırım, Ş., & Ulu, B. (2023). Deep learning based apples counting for yield forecast using proposed flying robotic system. *Sensors (Basel)*, *23*(13), 6171. DOI: 10.3390/s23136171 PMID: 37448020

Yimyam, P., Chalidabhongse, T., Sirisomboon, P., & Boonmung, S. (2005). Physical properties analysis of mango using computer vision. *Proceeding of ICCAS*

Younes, A., N. Nasr and A. Elsayed, "Efficient representations of digital images on quantum computers," Multimedia Tools and Applications, vol. 80, no. 25, pp. 34019 34034, 2021.

Yuan, S., et al., SQR: a simple quantum representation of infrared images. Quantum

Yuan, H., Martin, P., Fullerton, C., Gould, N., Hall, A., & Burdon, J. (2020). Refining the estimation of kiwifruit size from linear fruit dimensions. *Scientia Horticulturae*, *262*, 108878.

Zaiser, S., Rendler, T., Jakobi, I., Wolf, T., Lee, S.-Y., Wagner, S., Bergholm, V., Schulte-Herbrüggen, T., Neumann, P., & Wrachtrup, J. (2016). Enhancing quantum sensing sensitivity by a quantum memory. *Nature Communications*, *7*(1), 12279. DOI: 10.1038/ncomms12279 PMID: 27506596

Zanella, A. R. A., Silva, E., & Albini, L. C. P. (2020). Security challenges to smart agriculture: Current state, key issues, and future directions. *Array (New York, N.Y.)*, *8*, 100048. DOI: 10.1016/j.array.2020.100048

Zeiler, M. D., & Fergus, R. (2014). Visualizing and understanding convolutional networks. European conference on computer vision

Zeng, N., Wu, P., Wang, Z., Li, H., Liu, W., & Liu, X. (2022). A Small-Sized Object Detection Oriented Multi-Scale Feature Fusion Approach With Application to Defect Detection. *IEEE Transactions on Instrumentation and Measurement, 71*, 1–14. *IEEE Transactions on Instrumentation and Measurement*. Advance online publication. DOI: 10.1109/TIM.2022.3153997

Zhang, X., Liu, Y., & Wang, Z. (2023). IoT Based Weed Detection and Removal in Precision Agriculture. IEEE Xplore. https://ieeexplore.ieee.org/document/1020075

Zhang, H., Liu, P., & Wang, X. (2022). Deep neural networks for precision irrigation: Applications and challenges. *Agrárinformatika Folyóirat*, *24*(3), 187–199.

Zhang, H., Wang, L., Jin, X., Bian, L., & Ge, Y. (2023). High-throughput phenotyping of plant leaf morphological, physiological, and biochemical traits on multiple scales using optical sensing. *The Crop Journal, 11*(5), 1303–1318. DOI: 10.1016/j.cj.2023.04.014

Zhang, J., Yu, F., Zhang, Q., Wang, M., Yu, J., & Tan, Y. (2024). Advancements of UAV and Deep Learning Technologies for Weed Management in Farmland. *Agronomy (Basel), 14*(3), 494. DOI: 10.3390/agronomy14030494

Zhang, S.. (2020). The role of IoT in precision agriculture: A review. *Computers and Electronics in Agriculture, 167*, 105113. DOI: 10.1016/j.compag.2019.105113

Zhang, X., & Yang, J. (2024). Advanced chemometrics toward robust spectral analysis for fruit quality evaluation. *Trends in Food Science & Technology, 104612*. Advance online publication. DOI: 10.1007/978-3-319-10590-1_53

Zhang, Y., Guo, Z., Wu, J., Tian, Y., Tang, H., & Guo, X. (2022). Real-Time Vehicle Detection Based on Improved YOLO v5. *Sustainability (Basel), 14*(19), 19. Advance online publication. DOI: 10.3390/su141912274

Zhang, Y., Yang, X., Cattani, C., Rao, R. V., Wang, S., & Phillips, P. (2016). Tea Category Identification Using a Novel Fractional Fourier Entropy and Jaya Algorithm. *Entropy (Basel, Switzerland), 18*(3), 3. Advance online publication. DOI: 10.3390/e18030077

Zhao, C.. (2017). Climate change impacts on crop yield and quality. *Science Advances, 3*(4), e1602117. DOI: 10.1126/sciadv.1602117 PMID: 28508038

Zhao, Y., Gong, L., Zhou, B., Huang, Y., & Liu, C. (2016). Detecting tomatoes in greenhouse scenes by combining AdaBoost classifier and colour analysis. *Biosystems Engineering, 148*, 127–137. https://doi.org/https://doi.org/10.1016/j.biosystemseng.2016.05.001. DOI: 10.1016/j.biosystemseng.2016.05.001

Zhao, Z., Chen, S., Ge, Y., Yang, P., Wang, Y., & Song, Y. (2024). RT-DETR-Tomato: Tomato Target Detection Algorithm Based on Improved RT-DETR for Agricultural Safety Production. *Applied Sciences (Basel, Switzerland), 14*(14), 6287. DOI: 10.3390/app14146287

Zheng, Z., Wang, P., Liu, W., Li, J., Ye, R., & Ren, D. (2020). Distance-IoU Loss: Faster and Better Learning for Bounding Box Regression. *Proceedings of the AAAI Conference on Artificial Intelligence, 34*(07), 07. Advance online publication. DOI: 10.1609/aaai.v34i07.6999

Zhou, B.. (2022). Incremental learning models for dynamic environmental data: Applications in smart irrigation. *Environmental Modelling & Software, 148*, 105226.

Zhou, N., Yan, X., Liang, H., Tao, X., & Li, G. (2018). Multi-image encryption scheme based on quantum 3D Arnold transform and scaled Zhongtang chaotic system. *Quantum Information Processing*, *17*(12), 1–36. DOI: 10.1007/s11128-018-2104-6

Zhou, R., Damerow, L., Sun, Y., & Blanke, M. M. (2012). Using colour features of cv. 'Gala' apple fruits in an orchard in image processing to predict yield [journal article]. *Precision Agriculture*, *13*(5), 568–580. DOI: 10.1007/s11119-012-9269-2

Zhou, X.. (2020). GPS-based autonomous systems in precision farming. *Computers and Electronics in Agriculture*, *170*, 105–113. DOI: 10.1016/j.compag.2019.05.005

Zinkevich, M., Weimer, M., Li, L., & Smola, A. (2010). Parallelized stochastic gradient descent. *Advances in Neural Information Processing Systems*, •••, 23.

Zolanvari, M., Jain, R., & Salman, T. (2020). Potential data link candidates for civilian unmanned aircraft systems: A survey. *IEEE Communications Surveys and Tutorials*, *22*(1), 292–319. DOI: 10.1109/COMST.2019.2960366

Zude, M., Herold, B., Roger, J.-M., Bellon-Maurel, V., & Landahl, S. (2006). Non-destructive tests on the prediction of apple fruit flesh firmness and soluble solids content on tree and in shelf life. *Journal of Food Engineering*, *77*(2), 254–260. DOI: 10.1016/j.jfoodeng.2005.06.027

Zude, M., Pflanz, M., Kaprielian, C., & Aivazian, B. L. (2008). NIRS as a tool for precision horticulture in the citrus industry. *Biosystems Engineering*, *99*(3), 455–459. DOI: 10.1016/j.biosystemseng.2007.10.016

Zude-Sasse, M., Fountas, S., Gemtos, T. A., & Abu-Khalaf, N. (2016). Applications of precision agriculture in horticultural crops.

About the Contributors

Biplob Ray will deliver an informative talk that delves into the possibilities and challenges involved in automating Unmanned Aerial Vehicles (UAVs) for intelligent and precise herbicide spraying, aiming to eliminate weeds. The talk will cover various topics, ranging from UAV image collection to processing using machine learning to detect weeds to prepare weed maps. The coordinates of the detected weeds can then be utilised by spraying UAVs for precision herbicide spraying. Attendees can expect to understand better how the entire process can be made more efficient and automated, resulting in reduced costs and increased crop yield. Bio: Biplob Ray (Senior Member, IEEE) is currently an Associate Professor and research cluster leader in CML-NET (Centre for Machine Learning - Networking and Education Technology), School of Engineering and Technology (SET), Central Queensland University (CQU), Melbourne, Australia. A/Prof. Ray is elected Vice-Chair (2024-25) of the IEEE Victorian Section. He is an outstanding researcher with a background mix of research, academic, and industry experience. A/Prof. Ray is highly interested in multidisciplinary research with core interests in smart farming and secure communication protocols of Cyber-Physical systems driven by Artificial Intelligence (AI), the Internet of Drones (IoD), and the Internet of Things (IoT). As a lead Chief Investigator (CI) or CI, Associate Professor Ray has received research funding of over three million from the Australian Federal Government and industry since 2016. He has received several awards, including the Dean's Awards for Outstanding Researchers in 2023 (Mid-Career Research Category) and 2019 (Early-Career Research Category), as well as a Vice-Chancellor's commendation Award for being outstanding researcher in both years. He has also served (or is serving) as a guest editor, editorial board member, and reviewer in reputable journals and as a keynote speaker, organising chair, PC member, and reviewer for several conferences since 2012.

***

Jyoti Agarwal, Founded by Prof. Kamal Ghanshala in 1997, Graphic Era (Deemed to be University) has grown immensely. As the Best University in Dehradun, we offer a high-quality education and a nurturing environment that encourages innovation, fosters critical thinking, and prepares you for the future. To offer a world-class education that focuses on cutting-edge technology, student professional development, critical thinking, and high-quality research. Graphic Era (Deemed to be University), India's premier university, has accomplished countless milestones in its illustrious history thanks to its academic rigor, continuously top-performing students and alumni, and a very strong and competent teaching faculty.

Nanjappa Ashwath is an academic and an active researcher. He has been researching Australian plants for over 40 years and is an acknowledged expert in Central Queensland botany. A/P Ashwath has screened 100's of Australian native species for stress tolerance, biodiesel production, revegetation and other uses. Additionally, he has developed an innovative technology called 'Phytocapping' for landfill remediation, and demonstrated its benefits via field trials conducted in QLD, NSW, VIC, SA and WA. As part of this project, he has established active collaboration with various scientists, environmental consultants, landfill owners and regulators. Currently, he is exploring Australian native species for biofuel (bioethanol; eg agave and biodiesel; eg beauty leaf tree) production from degraded landscapes. He has also helped establish a pyrolysis facility at CQUniversity to convert urban, industrial and agricultural wastes into biofuel. At present he is investigating the use of grape marc and beauty leaf tree in biofuel production. Associate Professor Ashwath teaches into both undergraduate and postgraduate courses. In the last 26 years, he has taught into almost all 'plant science' courses that were offered by CQUniversity. He now teaches into three courses, viz., "Australian Botany", "Life Science Laboratories" and "Mining, Industrial & Urban Lands Rehabilitation". Nanjappa is popular amongst undergraduate and postgraduate students, and has supervised more than 30 PhD, Masters' and Honours students. His passion for research, particularly on Australian native plants, has enabled him to win the highly prestigious "Vice Chancellor's Researcher of the Year" award in 2001. He has attracted considerable research funds from ARC, CRC, NHT, RIRDC, EPA, mining companies and federal/state governments. He has published over 250 research articles in journals, books and as technical reports. He has an h-index of 35 with 5260 citations. Associate Professor Ashwath collaborates with the researchers at Texas A&M University, QUT, RMIT and Khairun University, Indonesia. He is also an adjunct professor at Amrita University, India. As a Central Queensland expert on Australian plants, he also engages in community services via organising training sessions on plant identification, plant propagation and tissue culture. His contributions to science were recognised by the Rotary Club via the award of 'Rotary International University Fellowship' which helped him promote Australian plants in India.

Varun Yarehalli Chandrappa is a research PhD student at Central Queensland University. He has 19 years of industry experience in embedded, IoT, and database server development. He has recently been involved in a smart irrigation project with Cairns Regional Council for an efficient irrigation system. His expertise is in providing technology solutions that everyone can use. Varun is highly interested in Cyber-Physical Systems, Artificial Intelligence (AI), and providing project management solutions. He has five patents in his name and has published several journals and conference papers. He works at the Bureau of Meteorology as an IT Manager and is also an IEEE Victoria IoT event and publication manager.

Dipta Justin Gomes is a dedicated academic and researcher specializing in Machine Learning, Computer Vision, Algorithms, Image Processing, and Natural Language Processing. He is currently pursuing a Doctor of Philosophy (Ph.D.) in Computer Science and Engineering at Bangladesh University of Engineering and Technology (BUET), focusing on non-planar graphs and advanced graph algorithms. Mr. Gomes earned his Master of Science (M.Sc.) in Computer Science from the University of Ulster, Belfast, Northern Ireland, in January 2023. Prior to this, he completed another M.Sc. in Computer Science, specializing in Intelligent Systems, at American International University-Bangladesh (AIUB), where he graduated summa cum laude for achieving a CGPA above 3.95. He also holds a Bachelor of Science in Computer Information Systems from AIUB, completed in December 2017.His research encompasses deep learning algorithms, convolutional neural networks, scheduling algorithms, graceful labeling of graphs, machine learning for predictive analytics, image processing optimizations, data mining algorithms, underwater image processing, and natural language processing. He has presented his work at various international conferences, including the International Conference on Computing Advancements (ICCA) and the International Conference on Robotics, Electrical and Signal Processing Techniques (ICREST). In his professional career, Mr. Gomes has served as a Lecturer in the Department of Computer Science at AIUB since September 2019. Before this role, he was a Lecturer in the ICT Department at Notre Dame College, Dhaka, from February 2018 to August 2019. Throughout his academic journey, Mr. Gomes shown excellence, receiving Dean's List Honors during his undergraduate studies and the Daily Star Award for achieving more than six A's in his GCSE O Levels in 2013. He also secured the 2nd Runner-Up position at the National Hackathon 2016, organized by the ICT Ministry of Bangladesh. His fields of interest include Graph Theory, Computation Learning, Computer Vision, and Algorithms. Mr. Gomes continues to contribute to these areas through his research, teaching, and active participation in academic conferences and workshops.

Sk Muktadir Hossain is a passionate Data Scientist, Data Analyst, and Machine Learning enthusiast committed to leveraging data and computational technologies for impactful solutions. His research focuses on machine learning, data analytics, and computer vision, emphasizing data-driven frameworks to optimize decision-making and enhance visual intelligence. Adept at predictive modelling and advanced image analysis, Muktadir addresses critical challenges in automation, healthcare, and smart cities. A lifelong learner, he is driven by the transformative potential of data and computer vision to reshape industries and society.

lAhmed Imtiaz is a passionate researcher in Natural Language Processing (NLP), Computer Vision and Pattern Recognition (CVPR), and Data Analysis. He works fundamentally on creating new solutions by bringing advanced technologies from these fields together. Specializing in text manipulation, vision understanding and analytics, Ahmed hopes to make a significant impact solving the real-world problems spanning across a multitude of fields.

Rukshani Puvanendran is a dedicated researcher specializing in machine learning, deep learning, natural language processing (NLP), and medical image analysis. Currently, her research focuses on bioacoustic detection for elephant conservation using low-resourced hardware. She explores efficient lightweight deep learning models that balance accuracy and computational cost, making them suitable for real-time deployment on edge devices. With a strong academic background and a passion for innovation, Rukshani has contributed to various research areas, including environmental sound analysis, and energy harvesting through piezoelectric sensors. Her work aims to develop AI-driven solutions for environmental and wildlife conservation, particularly in resource-constrained settings.

Rahat Rihan is a Machine Learning and NLP enthusiast dedicated to harnessing data and computational technologies to drive impactful solutions. His research centers on machine learning, data analytics, and computer vision, with a strong focus on developing data-driven frameworks to improve decision-making and advance visual intelligence.

Tej Bahadur Shahi is a dedicated researcher and academic with over ten years of experience in teaching and research within Computer Science and Information Technology across Nepal and Australia. His doctoral research focused on the innovative integration of Remote Sensing, Image Processing, and Machine Learning to optimize agricultural systems through artificial intelligence and drone technology, aiming for enhanced efficiency and sustainability. His scholarly impact is evidenced by his significant presence in the academic community, including an H-index of 19 and an i10-index of 26 on Google Scholar. His research findings have been disseminated through high-quality publications in esteemed journals such as AI Review, Remote Sensing, and IEEE Transactions.

Pramod Shrestha, Ph.D. in microbial biochemistry has over 10 years of experience on several different projects within the natural resource management field. Although his area of expertise is Microbial Biochemistry, Pramod has worked on multiple disciplines of sciences, such as biochemistry, microbiology, microbial ecology, genetic engineering, protein, mine rehabilitation, biofuel, composting, and soil 'health'. He has recently been involved smart irrigation project with Cairns Reginal Council for the efficient irrigation system. He was involved in research in a mine site rehabilitation project (establishing sustainable native plant communities on overburdened mine dump and assessing the soil health of revegetated coal mine sites. As a researcher, he was also involved in several soils and plant-related projects such as oxygation and biofuel production from native plant species including Beauty Leaf Tree. He has also worked on the project comparing the oil content and fatty acid composition of Calophyllum inophyllum (Beauty leaf tree) from different provinces. Currently, He is actively involved in guiding RHD students studying "Provenance variation and application of NIR/FTIR technologies to select salt-tolerant genotypes of Beauty Leaf Tree (Calophyllum inophyllum)". He has also worked as a Lecturer at Academic Learning Services Unit/Industry CQUniversity, Australia. To date, Pramod has published 15 international journal articles, 4 of which were published in high quality refereed international journals (ERA rating A)

Ajay K. Singh is working as an Associate Professor (Research) in the Department of Humanities and Social Science, Graphic Era (Deemed to be University) Dehradun. He has worked as Assistant Professor (Economics) in the School of Liberal Arts & Management, DIT University Dehradun for 6 years. He did Post-Doctorate Research with EDI of India, Ahmedabad, Gujarat (India). He received MPhil (Economics) from DAVV Indore (India), and PhD (Economics) from IIT Indore (India. He has published several research papers in the diversified area such as climate change, agricultural productivity, assessment of food security; estimation of GFSI, development of environmental sustainability index (ESI) and its association with socio-economic indicators; measurement and determinants of entrepreneurship ecosystem, and dimension of sustainable development and its interlinkages with economic development.

Sameer Sitoula, A computer engineering graduate with a master's degree in software engineering (Gold Medalist), experienced in both industry and academia. Worked as a Graduate Technical Intern at Intel Corporation, Bangalore, for 1.5 years. Served as a lecturer for undergraduate computer engineering students at Advanced College of Engineering and Management, affiliated with IOE, Tribhuvan University, for 1.5 years. Currently teaching IT and computer application undergraduate courses at Mechi Multiple Campus, Tribhuvan University, for the past 2.5 years. Supervised multiple undergraduate projects throughout the academic career.

Xudong Sun received bachelor and master degrees from China Agriculture University in 2002 and 2007, respectively. He received doctor degree from Donghua University in 2017. He visited Future Farming group in Center Queensland University as a visiting fellow during 2019 to 2020. Now he is a professor in East China Jiaotong University. He focuses on precision orchard management, especially on fruit load estimation with machine vision and fruit quality prediction with near infrared spectroscopy.

Kazi Tanvir is currently doing an MSc in Data Science at Vellore Institute of Technology (VIT), Vellore, Tamil Nadu, India. He has completed his BSc in Computer Science and Engineering majoring in Computational Theory from American International University-Bangladesh (AIUB). He has done an internship at CommJuncture; an IT solutions provider company in their software department. He has passed his Secondary and Higher Secondary from Birshrestha Noor Mohammad Public College, Dhaka. He aims to become a successful researcher in his domain. His fields of interest are Data Science, Image Processing, and Biostatistics.

Kazi Tanvir is currently doing an MSc in Data Science at Vellore Institute of Technology (VIT), Vellore, Tamil Nadu, India. He has completed his BSc in Computer Science and Engineering majoring in Computational Theory from American International University-Bangladesh (AIUB). He has done an internship at CommJuncture; an IT solutions provider company in their software department. He has passed his Secondary and Higher Secondary from Birshrestha Noor Mohammad Public College, Dhaka. He aims to become a successful researcher in his domain. His fields of interest are Data Science, Image Processing, and Biostatistics.

Faranak Tohidi, I have a PhD degree in Computer Science from Yazd University and I am also a PhD scholar at Charles Sturt University, Australia. My areas of research interest are Image and Video Processing and 3D Point Cloud Compression. Information Security and Image Authentication.

Anwaar Ulhaq is a Senior Lecturer in Artificial Intelligence at Central Queensland University, Sydney Campus, Australia. Holding the position of Vice President of the Australian Pattern Recognition Society, he is also a member of the Australian Academy of Sciences, the Australian Computer Society, and IEEE Signal Processing. With a diverse academic background, Dr. Ulhaq has played significant roles at Charles Sturt University and Victoria University, Melbourne, before joining CQU. In 2023, he served as the General Chair of DICTA, a major computer vision conference, showcasing his dedication to advancing academic discussions. Dr. Ulhaq has contributed to various prominent computer vision conferences, serving as a reviewer for ICCV, CVPR, WACV, and ECCV. His expertise is further recognized as an Associate Editor for Springer Nature Computer Science (SNCS) journal. Acknowledged for his contributions, Dr. Ulhaq has received Teaching Excellence Awards, Research Excellence Awards, and Best Paper Awards. His scholarly work includes over 80 peer-reviewed journal and conference papers, reflecting the depth of his research. Securing approximately 3 million dollars in research projects as a Chief and Co-Investigator, Dr. Ulhaq's impact extends beyond academia. He has been featured in national and international media over 30 times, highlighting the relevance and significance of his research. Dr. Anwaar Ulhaq's research interests span computer vision, image and signal processing, responsible AI, and remote sensing. With a focus on advancing knowledge and technology, he brings a wealth of experience and expertise to his professional roles, contributing significantly to the computer vision community.

Srinivasan Vasan has been working in the Department of Humanities and Social Sciences at Graphic Era (Deemed to be University) as an Assistant Professor. With a wealth of experience, Mr. Srinivasan has accumulated 14 years of extensive teaching and field-level expertise. Prior to his current role, he served in the Ministry of Health and Family Welfare, Government of India, and gained valuable experience. He also contributed as a Guest Faculty at the university level and served for nearly 2 years in the State Government Public Works Department as a Mobilization Training Specialist. In the past year, Mr. S. Srinivasan has enriched his experience by actively participating in a tuberculosis project funded by GFATM and USAID, providing a platform for further exploration of his research and practical field-based knowledge. His scholarly contributions include publishing 65 papers, with 18 more under peer review in various UGC-CARE, SCOPUS, and peer-reviewed journals.

Zhenglin Wang is an accomplished academic and industry professional who has made significant contributions to the fields of computer science and engineering. He obtained his Master of Computer Science by Research and PhD degrees from the University of South Australia in 2012 and 2016, respectively. During his career,

Zhenglin has worked as a software engineer at TCL, UTStarcom, and Hitachi Construction Machinery Australia for several years, where he gained extensive experience in developing software solutions for various applications. Following this, he served as a Postdoctoral Research Fellow in the Institute for Future Farming Systems at CQU for over five years. During this time, he conducted pioneering research in the areas of agriculture automation, and his team developed the world's first mango auto-harvester. In recognition of his research excellence, Zhenglin was awarded the "Advanced Queensland Industry Research Fellowship" in 2019. Zhenglin currently holds the position of lecturer-ICT with the School of Engineering and Technology at CQUniversity Sydney, where he is actively engaged in teaching and research. Zhenglin is a member of the Australian Computer Society (ACS). His research interests include computer vision, machine learning, unmanned ground vehicle, and precision agriculture.

Index

A

B

C

D

E

G

I

M

N

O

P

Q

R

S

T

V

Y

www.ingramcontent.com/pod-product-compliance
Lightning Source LLC
LaVergne TN
LVHW080611300926
849162LV00005B/1117